2 - 95

29/05

Parasitic nematodes —
antigens, membranes and genes

Parasitic nematodes —
antigens, membranes and genes

Edited by
M. W. Kennedy
Wellcome Laboratories for Experimental Parasitology
University of Glasgow

Taylor & Francis
London • New York • Philadelphia
1991

UK   Taylor & Francis Ltd, 4 John St, London WC1N 2ET

USA   Taylor & Francis Inc., 1900 Frost Road, Suite 101,
            Bristol, PA 19007

British Library Cataloguing in Publication Data
Parasitic nematodes.
    1. Nematodes
    I. Kennedy, Malcolm W.
    595.182

    ISBN 0-85066-772-0

Library of Congress Cataloging-in-Publication Data
is available

Phototypesetting by
Chapterhouse, The Cloisters, Formby, England.

Printed in Great Britain by Burgess Science Press, Basingstoke
on paper which has a specified pH value on final paper
manufacture of not less than 7.5 and therefore 'acid free'.

Contents

radio-iodination
Cuticular collagens
Temporal regulation of protein synthesis and turnover
Antigenicity and cross-reactivity of cuticular proteins
Gp29 homologues are present on all species of lymphatic
filariae

A model for surface damage and repair
Surface antigens can be replaced after removal by detergents
Inhibition of antigen replacement
Site of synthesis and replacement of surface antigens
Concluding remarks

The design of strategies to interrupt the life cycle
Potential diagnostic antigens
Potential protective antigens
Recent studies on cuticular antigens
The molecular biology of hookworm antigens
Recent progress in the immunoepidemiology of hookworm
infection
Concluding remarks

D. P. Knox and D. G. Jones
Introduction
Ruminant nematodiasis — the problems
Enzymes in the diagnosis and control of nematodiasis
Is differential serodiagnosis important?
Serodiagnosis — the present
Serodiagnosis — the future
Acetylcholinesterase
Proteinases
Superoxide dismutase (SOD)
Concluding remarks

W. Harnett
Introduction
Detecting *Onchocerca volvulus*
Diagnosis of onchocerciasis by detection of antibody
Defining specific antigens
Cloning of antigens for use in antibody diagnosis
Improved specificity of immunoglobulin isotype-restricted
assays
Serological differentiation of *O. volvulus*
Serological differentiation of disease status
Detection of antigen
Attempts to measure circulating antigen
Improvement of assay sensitivity
Improvement of assay specificity
Detection of infective larvae in blackflies
DNA probes for differentiation of infective larvae
Conclusions and future prospects

Preface

The purpose of this book is to provide a cross-section of the immunological, biochemical and molecular biological research currently being undertaken on parasitic nematodes. Each author was asked to review their subject area and to point to new approaches which are likely to be widely applicable. The book is not, therefore, meant to be a comprehensive review of nematode research but deals with a wide variety of topics and is selective for the species on which research is most advanced. Most of the major nematodiases of humans and domestic animals are covered, and research on those which are not is currently advancing rapidly using many of the approaches dealt with in the book. For instance, should this book ever go into a second edition in a few years time, much will then need to be said of *Trichinella, Trichuris, Strongyloides, Loa, Haemonchus* and *Ostertagia*.

The economic and medical importance of parasitic nematodes has prompted the increasing application of advanced techniques in immunology, biochemistry and molecular genetics to their study and this has led to radical changes in formerly accepted ideas about the biology of this group of organisms. An example of this would be the belief that nematodes present an inert, non-provocative surface to their hosts and can hence survive in an otherwise hostile environment. It is now clear, however, that the very opposite is the case. Not only is the nematode's exposed surface highly antigenic but it is also a dynamic and complex structure which is capable of rapid repair following external assault. Moreover, it is also becoming clear that the surface of these organisms probably represents a unique type of biological interface. As a site for immunological attack and as a barrier to chemotherapeutic agents, the surface has received a great deal of attention and this is reflected in the number of chapters devoted to it. These encompass studies on the surface antigens and associated subsurface components, and the synthesis and replacement of surface materials after detergent-mediated damage. Much of this kind of work has concentrated for technical reasons on cuticular proteins or glycoproteins, but one of the chapters deals with lipid and considers the evidence for the existence of polymorphic lipid and domains in a putative surface lipid layer.

The later chapters deal with a variety of topics including the nature and function of glycoconjugates and enzymes secreted by nematode parasites, taking in both those of medical and those of veterinary importance. Secreted molecules, while important in themselves, have also been the tools used to reveal a hitherto poorly understood aspect of nematode immunology, the genetic control of the immunological repertoire. This will have immediate implications for the selection and application of simple vaccines containing recombinant polypeptides as well as contributing to some of the unanswered aspects of nematode infections. For instance, how is it that immunity to nematodes can take so long to develop and how do genetic factors in resistance or susceptibility to infection operate?

Malcolm Kennedy

Contributors

Eileen Devaney Department of Parasitology, Liverpool School of Tropical Medicine, Pembroke Place, Liverpool L3 5QA, UK.

William Harnett Division of Parasitology, National Institute for Medical Research, The Ridgeway, Mill Hill, London NW7 1AA, UK.

David G. Jones Department of Biochemistry, Moredun Research Institute, 408 Gilmerton Road, Edinburgh EH17 7JH, UK.

Malcom W. Kennedy Wellcome Laboratories for Experimental Parasitology, University of Glasgow, Bearsden, Glasgow G61 1QH, UK.

I. Barry Kingston Department of Pathology, University of Cambridge, Tennis Court Road, Cambridge CB2 1QP, UK.

David P. Knox Department of Biochemistry, Moredun Research Institute, 408 Gilmerton Road, Edinburgh EH17 7JH, UK.

John Kusel Department of Biochemistry, University of Glasgow, Glasgow G12 8QQ, UK.

Paul G. McKean Department of Zoology, The University, University Park, Nottingham NG7 2RD, UK.

Rick M. Maizels Department of Biology, Imperial College of Science and Technology and Medicine, Prince Consort Road, London SW7 2B, UK.

Christine M. Preston-Meek School of Pure and Applied Biology, University of Wales College of Cardiff, Newport Road, Cardiff CF2 1TA, UK.

David I. Pritchard Department of Zoology, The University, University Park, Nottingham NG7 2RD, UK.

Lorna Proudfoot Department of Biochemistry, University of Dundee, Dundee DD1 4HN, UK.

Rupert J. Quinnell Department of Zoology, University of Oxford, South Parks Road, Oxford OX1 2RD, UK.

Brian D. Robertson Department of Biology, Imperial College of Science, Technology and Medicine, Prince Consort Road, London SW7 2BB, UK.

Murray E. Selkirk Department of Biochemistry, Imperial College of Science, Technology and Medicine, Prince Consort Road, London SW7 2AZ, UK.

Huw V. Smith Scottish Parasite Diagnostic Laboratory, Department of Bacteriology, Stobhill General Hospital, Glasgow G21 3UW, UK.

Patrick J. Tighe Department of Zoology, The University, University Park, Nottingham NG7 2RD, UK.

Abbreviations

2D	two-dimensional (electrophoresis)
2ME	2-mercaptoethanol
AChE	acetylcholinesterase
AF18	5-(N-octadecanoyl)aminofluorescein
AP	alkaline phosphatase
BALT	bronchial associated lymphoid tissue
bp	base pair (in DNA or RNA)
cAMP	adenosine $3'$, $5'$-cyclic monophosphate
CBB	coomassie brilliant blue
cDNA	complementary DNA
CMI	cell-mediated immunity
CNS	central nervous system
Con A	concanavalin A
CPC	cetylpyridinium chloride
c.p.m	counts per minute
CRD	cross-reacting determinant
CT	covert toxocariasis
CTAB	cetyltrimethylammonium bromide
DiI18	dioctadecyltetramethylindocarbocyanine
D_L	lateral diffusion coefficient
DOC	(sodium) deoxycholate
d.p.i.	days post-infection
DTAF	dichlorotriazinylaminofluorescein
EDTA	ethylenediaminetetraacetic acid
EGTA	Ethylene glycol-bis(β-aminoethylether) N,N,N',N'-tetraacetic acid
ELISA	enzyme-linked immunosorbent assay
e.p.g.	eggs per gram (faeces)
ES	excretory–secretory material; strictly, *in vitro* culture supernatant
Fc	Fc fragment of immunoglobulin
FITC-WGA	fluorescein isothiocyanate-conjugated wheatgerm agglutinin
FM	fluorescein maleimide
FRAP	fluorescence recovery after photobleaching
GALT	gut associated lymphoid tissue

GI	gastrointestinal
H-2	MHC complex of the mouse
HLA	MHC complex of the human
HRP	horseradish peroxidase
IEF	isoelectric focusing
IFAT	indirect fluorescent antibody test
Ig	immunoglobulin
IMP	intramembranous particle
Ir	immune response (genes)
IRS	immune rabbit serum
IVR	*in vitro* released parasite products (equivalent to ES)
kb	kilobase (in DNA or RNA)
kDa	kilodalton (units of molecular mass)
L1, L2, L3, L4	larval stages (instars) of nematodes
M	molar concentration
mAb	monoclonal antibody
mf	microfilaria
MHC	major histocompatibility complex
M_r	relative molecular mass, usually as relative mobility, expressed in daltons
MSL	muscle stage larva (of *Trichinella spiralis*)
NaDOC	sodium deoxycholate
NBD	nitrobenzoxadiazolamine
NBD-chol	25-(NBD-methylamino)-27-cholesterol
NBD-PC	NBD-phosphatidylcholine
NBD-PE	NBD-phosphatidylethanolamine
NBL	newborn larvae (of *Trichinella spiralis*)
OD	optical density
OLM	ocular larva migrans
OT	ocular toxocariasis
PAGE	polyacrylamide gel electrophoresis
PCR	polymerase chain reaction
PEG	polyethylene glycol
pI	isoelectric point
p.i.	post-infection
PNA	peanut agglutinin
RH18	octadecylrhodamine B
RAST	radioallergosorbent assay
RIA	radioimmunoassay
%R	percentage recovery (of fluorescence in FRAP)
SDS-PAGE	sodium dodecyl sulphate polyacrylamide gel electrophoresis
RIPEGA	radioimmunoprecipitation–PEG assay
s.e.	standard error (of the mean)
SHAM	salicylhydroxamic acid
SOD	superoxide dismutase

TBS	tris buffered saline
TCA	trichloroacetic acid
TES	*Toxocara canis* excretory–secretory products
Tx-100/114	Triton X-100/114
VLM	visceral larva migrans
VSG	variable surface glycoprotein (of trypanosomes)
WGA	wheatgerm agglutinin

1. Biophysical properties of the nematode surface

L. Proudfoot, J. R. Kusel, H. V. Smith and M. W. Kennedy

INTRODUCTION

The nematode cuticle is a complex multilayered extracellular structure that completely encloses the organism except for small openings into the pharynx, anus, sensory, secretory and reproductive pores. There is considerable variation in cuticle structure between different nematode species and between different stages of the same species and differences can also occur between the sexes (reviewed by Bird, 1980). Despite this heterogeneity, a basic pattern exists in which the outermost layer, the epicuticle, is always present.

As the major interface with the environment, the epicuticle is potentially important in the nutrition of the parasite and its protection against immunological assault. The ability of parasitic nematodes to survive for prolonged periods in immunocompetent hosts testifies to the success with which they have adapted to a potentially hostile environment, and the inherent resistant properties of the nematode surface probably contribute to this success. In the face of evidence for dynamic exchange of low molecular weight substances across the body wall of some nematodes, the patent antigenicity of the surface, and the shedding of bound antibody, it would be misleading to argue that the epicuticle is a totally inert layer. Increasing interest is being drawn to this dynamism, its biological significance and its mechanisms. In this chapter we review aspects of this before concentrating on new biophysical information on the organization of the nematode surface and its transformation in response to conditions which mimic the infection process.

WHAT IS THE EPICUTICLE?

In the past, the epicuticle has been compared to a highly modified plasma membrane, largely because of its transmission electron microscopic resemblance to this structure (Lumsden, 1975; Bird, 1980, 1984). However, it does have properties which distinguish it from such a membrane and it has been compared to an extra-cellular envelope (Locke, 1982) and to the cuticulin membrane of insects (Howells and Blainey, 1983).

In recent years, the comparison between the epicuticle and inert biological structures has not been held unanimously. Jain (1988) stated that 'In the hierarchy of biological organization, the structure and function of membranes lies somewhere between macromolecules and cells. As non-covalently aggregated macroscopic structures arising from amphipathic molecules, membranes in general have their unique characteristics that are not shared by other molecules and their aggregates'. This quotation could be seen to defend the inclusion of the epicuticle in the class of structures to which plasma membranes belong, given that 'membranes' vary widely in their properties. For example, the plasma membrane of fertilized *Xenopus* eggs is peculiar in several respects including a very high specific resistance (De Laat *et al.*, 1973), lack of selectivity to cations (De Laat *et al.*, 1975), insensitivity to cytochalasins (De Laat and Bluemink, 1974), possession of large intramembranous particles (IMPs) (Bluemink *et al.*, 1976) and a substantial fraction of immobilized lipid (Dictus *et al.*, 1984).

Current concepts of membrane structure and function are based largely on information derived from a limited number of micro-organisms and mammalian cell types. Consideration of other organisms, such as parasites, might alter the consensus as to what constitutes a membrane. It might then be that the epicuticle will be considered a structure apart from other biological surfaces, or that it represents an extreme in a limited but continuous range of specializations of a basic membrane structure.

THE EPICUTICLE AS A MEMBRANE

Membranes are lipid bilayers containing proteins and glycoproteins, forming a structure capable of controlling the internal cellular environment by selective inter-actions with the external environment. The question is whether the epicuticle is capable of such selective interactions. What is certain, however, it that proteins and glycoproteins are associated with the surface (reviewed by Maizels and Selkirk, 1988) and that they could play a role in internal homeostasis or nutrient acquisition. The uncertainty lies in the precise location and function of these proteins and whether they are integral to the epicuticle. Devaney (1988) and Scott *et al.* (1988) noted that the major surface-associated peptides of adult *Brugia pahangi* and *Dirofilaria immitis* were extracted in buffer without the aid of detergents. This would indicate that these components are not anchored on the surface of the worm in the manner of a conventional integral membrane component, although recently it has been discovered that there is a new class of membrane protein which is anchored by a glycolipid moiety (Ferguson and Williams, 1988) and can be rapidly released from the surface by the action of specific phospholipases (Noda *et al.*, 1987).

The presence of integral proteins is believed to be one of the factors associated with the appearance of IMPs on freeze-fracture electron microscopy (Verkleij and Ververgaert, 1978). For nematodes, contrasting results for different species and larval stages have left us with no consensus as to whether a characteristic of the epicuticle is a lack of IMPs or whether they occur variably. This could in part be due to technical difficulties in obtaining large fracture faces in the epicuticle and that the plane of fracture commonly occurs between the epicuticle and the rest of the cuticle, rather than within the epicuticle itself. IMPs have been reported in the epicuticle of *Nippostrongylus* (Lee and Bonner, 1982) and *Meloidogyne* (Bird, 1984), but are reported to be absent in microfilariae of *Onchocerca volvulus* (Martinez-Palomo, 1978) and in various stages of *Trichinella spiralis* (Lee *et al.*, 1984, 1986). However, the absence of IMPs does not preclude integral membrane proteins because some atypical membranes, such as the myelin sheath, tend to fracture smoothly without particles (Branton and Deamer, 1972), although integral proteins are known to be present (Boggs and Moscarello, 1978). This is also true for the outer membrane of Gram-negative bacteria (Lugtenberg and van Alphen, 1983; Nikaido, 1989) which only produces small fracture faces due to its unusual bilayer organization.

The transepicuticular penetration of many polar compounds (Howells *et al.*, 1983) is thought to be evidence that the epicuticle is not an uninterrupted lipid bilayer, but one punctuated by components involved in nutrient translocation. However, the precise site of this activity is not known. Rutherford *et al.* (1977) showed that sites for amino acid transport were probably located in the epicuticle of *M. nigrescens*, a larval parasite which has a much reduced cuticle. In the majority of nematode species, however, it is probable that nutrients are transported at the hypodermal membrane (Howells, 1987) and that the epicuticle and cuticle are selectively permeable to such materials.

By way of comparison, the bacterial outer membrane is able to transport nutrients to the cytoplasmic membrane despite the apparent impermeability (Nikaido, 1989). An analogy could, therefore, be drawn between the outer membrane or envelope of the *Enterobacteriaceae* and the nematode epicuticle in that both form the physical and functional barrier between internal membranes and the environment. The bacterial cytoplasmic membrane (*cf.* nematode hypodermal membrane) is responsible for the metabolic functions, including bio-synthetic activities, while the major known functions of the outer membrane (*cf.* nematode epicuticle) are primarily physical. In *Escherichia coli* the low perme-ability of the outer membrane is modulated by porins, a class of 30–40 kDa proteins that produce water-filled, non-specific, transmembrane diffusion channels (Nikaido and Vaara, 1985; Benz, 1985). This is in the same molecular mass range as the predominant surface-associated (glyco)protein of the surface of filarial nematodes (Selkirk *et al.*, 1986; Devaney, 1988; Scott *et al.*, 1988). As mentioned above, this group of parasites is known to absorb nutrients transcuticularly, but there is no evidence as yet for porin-like molecules.

The protective capacity of the epicuticle is impressively demonstrated by dauer larvae of *Caenorhabditis elegans* which survive for prolonged periods in 2 per cent sodium dodecyl sulphate (SDS) (Cassada and Russell, 1974). Likewise, *E. coli* can grow in the presence of as much as 5 per cent SDS (Lugtenberg and van Alphen, 1983). This could be taken to indicate that the bacterial outer membrane and the nematode epicuticle do not have a typical membrane structure, or that a conventional membrane structure is in some way shielded from the environment, perhaps by glycoconjugates.

THE EPICUTICLE AS AN ENVELOPE

Many workers now believe that there is no homology between the epicuticle and plasma membranes and that it may be an 'envelope' structure. Locke (1982) stated that 'some cells in almost all groups from bacteria to vertebrates have the ability to secrete an envelope directly through or above their plasma membranes, and that envelopes are not plasma membranes'. In the present context, an envelope would form a barrier to the environment that allows the hypodermis to control the intervening space, that is, the cuticle. However, it is true to say that, in parasites, hypothetical envelope structures must vary greatly in function. For instance, Podesta (1982) described the surface layer of the adult schistosome as an envelope overlying a membrane, although many workers have described this as a double membrane (Kusel, 1972; Hockley and McLaren, 1973; Foley *et al.*, 1986). At the other extreme, there is the cystacanth envelope of the acanthocephalan parasite *Moniliformis* (Lackie and Rotheram, 1972; O'Brien, 1988) which is considered to be a stable, almost inert, structure.

There is, therefore, still considerable confusion in the literature as to whether the outermost surface of the nematode cuticle should be considered as a membrane, an envelope, or some type of intermediate structure. This has led to double-think in the literature, as exemplified by reference to the 'surface membrane' of the infective larva but to the 'epicuticle' in adults, sometimes in the same paper (Storey *et al.*, 1989). Such confusion is forgivable in the light of current knowledge and an attempt has recently been made to reach a consensus in which it was proposed that the epicuticle be regarded as the 'trilaminate differentiation containing lipids' (Wright, 1987). We will, however, retain the use of the term epicuticle here, and it is the purpose of this chapter to review its known structure, then concentrate on the biophysical properties of its lipid components.

BIOCHEMICAL NATURE OF THE NEMATODE EPICUTICLE

Very little is known about the biochemical composition of the epicuticle, mainly due to difficulties in its isolation or solubilization by methods conventionally employed for membranes (Bird, 1957; Cox *et al.*, 1981; Murrell and Graham, 1982; Betschart and Jenkins, 1987). In spite of this difficulty, surface labelling with extrinsic agents, for example with radio-iodine, has provided important details on surface-associated proteins and glycoproteins, many of which are antigenic (reviewed by Philipp and Rumjaneck, 1984; Maizels and Selkirk, 1988). Recently, attention has turned towards the non-proteinaceous components of the epicuticle (Kennedy *et al.*, 1987a; Scott *et al.*, 1988), and future investigations will, we are sure, establish the importance of lipids.

LIPIDS

Evidence for epicuticular lipids is as follows. Firstly, radio-iodination of whole parasites catalysed by 1, 3, 4, 6-tetrachloro-3α-6α-diphenylglycoluril (IODO–GEN), which is assumed to be relatively surface restricted in action, leads to iodination of lipid double bonds in unsaturated lipids. A similar situation has been noted previously for *Schistosoma mansoni* (Hayunga *et al.*, 1979). Radio-iodinated phospholipids and heavily labelled non-polar lipids are found on the surface of adult *Acanthocheilonema viteae* and adult *Litomosoides carinii* and a similar profile of labelled lipids can be obtained by labelling the external face of pieces of cuticle obtained from the large intestinal nematode, *Ascaris suum* (our unpublished results). In 1957, Bird noted that Sudan-black-stainable lipid was extracted from *Ascaris*. Secondly, Scott *et al.* (1988) isolated glycolipid from the surface of adults of the filarial nematode *Dirofilaria immitis*. This lipid was observed in the low molecular weight region by means of SDS polyacrylamide gel electrophoresis (PAGE), after radio-iodination of the intact surface and subsequent extraction. The glycolipid was thought to be potentially important in the provision of a non-immunogenic, protease-resistant barrier between the cuticular surface and the host environment. The authors considered that epicuticular lipid might then be vital for the maintenance of an effective defence in an immunocompetent host.

Indirect evidence for the presence of lipid in the epicuticle comes from the observation that certain fluorescent lipid probes can be inserted into the epicuticle (Kennedy *et al.*, 1987a). This established its lipophilic nature, although the authors noted that the insertion of lipid probes was highly selective and that this could be related to the length of the acyl chain and to the general physico-chemical characteristics of the probe. Further indirect evidence comes from the observation that the permeability of the surface of the anhydrobiotic nematode *Ditylenchus dipsaci* increases sharply between 40°C and 50°C (Wharton *et al.*, 1988), suggesting that this is caused by some phase change or transition in epicuticular components, probably lipid. A similar transition has been observed for muscle-stage larvae of *T. spiralis* (our unpublished observations) using the biophysical technique of differential scanning microcalorimetry, which can detect temperature-induced lipid phase transitions (Sturtevant, 1987).

SURFACE-ASSOCIATED PROTEINS

The epicuticle and cortical layers of the cuticle appear to contain proteins which are extensively cross-linked by non-reducible bonds (Cox *et al.*, 1981; Betschart and Jenkins, 1987). This insoluble matrix may be common among nematode surfaces and could have similarities with the non-collagenous cuticlin of *Ascaris* cuticle, described by Fujimoto and Kanaya (1973). A cuticlin-like matrix could account for the remarkable detergent resistance properties of the intact epicuticle in many species of nematode (Pritchard *et al.*, 1985; Mok *et al.*, 1988).

The extent of detergent solubilization of surface antigenic material depends on the type of detergent employed and the nematode species and stage. The cationic detergent cetyltrimethylammonium bromide (CTAB) is effective in the removal of antigens from both *Necator* (Pritchard *et al.*, 1985, 1988) and *Trichinella* (Grencis *et al.*, 1986), but precisely which part of the cuticle from which these proteins derive is controversial. A strong possibility is, however, that their immediate source is the surface coat.

SURFACE COAT AND GLYCOCONJUGATES

Under transmission electron microscopy, an amorphous, glycocalyx-like layer has been observed immediately overlying the epicuticle (Himmelhoch and Zuckerman, 1978). This layer is particularly prominent in the infective larvae of *Toxocara canis* (Badley *et al.*, 1987), and is thought to be composed of antigenic material which is continuously released from the surface (Smith *et al.*, 1981; Maizels *et al.*, 1984). In this parasite, the coat can be observed under scanning electron microscopy to break away from the epicuticle, potentially effecting the removal of immunoglobulin, adherent cells and their toxic products (Badley *et al.*, 1987). A surface coat is apparent in the infective stage of most parasitic species (Grove *et al.*, 1984, Lee *et al.*, 1986; Abraham *et al.*, 1988) and in free-living nematodes (Zuckerman *et al.*, 1979), but adult parasitic nematodes do not appear to have a surface coat. Lee *et al.* (1986) noted that adult *T. spiralis* lacked one, and the same has been noted by Kieffer *et al.* (1989) for adult *A. viteae* using gold labelled lectins.

A remarkable feature of some nematode parasites is that sugar residues are apparently not exposed on the intact surface for lectin binding. For instance, the 47 kDa antigen of *T. spiralis* (Parkhouse *et al.*, 1981; Ortega-Pierres *et al.*, 1984), the 30 kDa antigen of *B. pahangi* (Devaney, 1988), and the 49 kDa antigen of *Dirofilaria immitis* (A. Scott, personal communication) have carbohydrate determinants which are not available for lectin binding on the intact surface. We have also noted that adult *Ostertagia ostertagi* does not bind any of the three lectins that we tested (concanavalin A (Con A), peanut agglutinin (PNA) and wheatgerm agglutinin (WGA)) except around the region of the copulatory bursa. Several possibilities exist to explain this: first, glycoproteins are near to, but not exposed on, the surface, yet are iodinatable. Secondly, they are positioned such that lectin binding is sterically hindered. Lastly, that glycoproteins are inserted with their sugar side-chains in inverted lipid micelles (Kennedy *et al.*, 1987a; Wright and Hong, 1988). The last explanation would require an unusual lipid composition and/or phase, which is discussed later in the light of new biophysical evidence.

Sugars are apparently exposed on the intact surface of the infective larva of the dog ascarid *Toxocara canis*, to which anti-carbohydrate monoclonal antibodies can bind (Maizels *et al.*, 1987). Such monoclonal antibodies also bind to *Toxocara cati* but which of these are exposed, hidden or present at all differs between the two species (Kennedy *et al.*, 1987b).

The cuticular surface is sometimes enclosed within a loosely attached sheath derived from the moulted cuticle of the previous developmental stage. Ensheathment involving a cuticle occurs most commonly in the infective third-stage larvae of nematode parasites, and in the developmentally arrested dauer larvae of free-living nematodes. We find that this sheath has exposed sugars which can be bound by fluorescently labelled lectins, whereas the underlying epicuticle of the third-stage cuticle of *O. ostertagi* and the dauer larvae of *C. elegans* does not (our unpublished observations).

Certain microfilariae, such as those of *Wuchereria*, *Brugia* and *Litomosoides*, possess a sheath of a different origin from that of the infective stage, having ultra-structural characteristics of the egg-shell membrane (Zaman, 1987). The exposed sugars of this sheath are of particular interest as they could be the primary targets of the host's immune response. The microfilarial sheath prevents entry of immuno-globulins which could potentially damage the epicuticular surface, but will allow the entry of smaller proteins, such as the 36 kDa wheat-germ lectin (Devaney, 1985). Wheatgerm agglutinin (WGA) is specific for the sugar *N*-acetylglucosamine which is present in the egg-shell chitin. Sheathed microfilariae fluoresce very strongly after labelling with this lectin (Kaushal *et al.*, 1984; Paulson *et al.*, 1988), more strongly than with other lectins, suggesting that WGA could be binding to egg-shell-derived sugars on the surface of the larva. Paulson *et al.* (1988) studied the lectin-binding of six species of microfilariae and found that only the sheath, and not the larval surface, would bind lectins. In addition, all lectin-binding *in utero* to derived *Onchocerca volvulus* larvae was associated with the sheath because hatched microfilariae showed no ability to bind any of the lectins tested.

It is particularly notable that sugars have not been found exposed on many of the stages which migrate through mammalian tissues (*Toxocara* being a major exception), since we know that other tissue-dwelling parasites such as trypanosomes (Vickerman, 1974) and schistosomes (Lumsden, 1975; Podesta, 1982) have polyanionic carbohydrate in their surface coat. It is possible that parasitic nematodes mask their surface sugars in order to reduce antigenicity, or that this is a surface specialization related to protection of structural proteins from parasite or host-derived protease. How this masking could be achieved by non-antigenic material is unknown, but lipids, especially those derived from the host, are a possibility.

DYNAMICS OF THE NEMATODE SURFACE

It is now clear that the nematode surface is not merely an inert extracellular covering; surface-associated antigens can be shed rapidly and can alter radically in composition without a moult taking place (Philipp *et al.*, 1980; Smith *et al.*, 1981;

Maizels *et al.*, 1983, 1984; Marshall and Howells, 1986; Carlow *et al.*, 1987). The process of antigen shedding perhaps best illustrates the dynamic properties of the nematode surface. The release of surface antigens of the second larval stage of *Toxocara canis* is rapid, and Maizels *et al.* (1984) showed that larvae release about 25 per cent of the extrinsically radiolabelled surface components in less than 1 h *in vitro*. Smith *et al.* (1981) demonstrated that this shedding is an energy-dependent process which is inhibited in the presence of antimetabolites or if the larvae are maintained at a low temperature. Whether this energy dependency is associated with the aggregation and shedding process itself or is related to worm motility remains to be established.

In *B. pahangi*, it appears that the surface of the third-stage larva (L3) is more dynamic than that of the fourth-stage larva or that of the adult, in that the turnover of radiolabelled surface proteins in the L3 is considerably faster (Marshall and Howells, 1986). In addition, in *in vitro* experiments with the L3 of this parasite, a monoclonal immunoglobulin M (IgM) against surface component(s) was lost from the surface of these parasites in minutes (Carlow *et al.*, 1987). It has also become clear that nematodes can replace shed antigen, or antigen artificially stripped off by detergent (see Chapter 5), but this might not generally apply (Ibrahim *et al.*, 1989).

These properties of the nematode surface might bear some analogy to the mobility and modulation of proteins in the plasma membrane of certain eukaryotic cells, for example in the release of egg proteins at fertilization (Johnson and Epel, 1975; Mazia *et al.*, 1975). If so, then it could be predicted that they are associated in a layer of lipid with properties which could be studied using membrane probes. This will be dealt with in the following sections, but it must first be stressed that there is no firm evidence that surface antigens are associated with a lipid layer. For example, the major exposed antigens could be confined to the amorphous glycocalyx/surface coat.

BIOPHYSICAL STUDIES OF THE EPICUTICLE

Many of the techniques which are currently employed to study the cuticle involve the use of strong detergents or potent chemicals for the purpose of extraction. The resultant preparation is usually purified further and subjected to sophisticated methods of analysis, and then conclusions are made about the possible function of these molecules and how they might be organized on the surface. The problem is that extraction procedures are rarely surface restricted and will commonly encompass subsurface components and yield little information on the organization of the surface's native state. An alternative approach is to use non-invasive biophysical techniques which can deal with the live, intact parasite. Such techniques involve either the insertion of a probe whose physico-chemical properties are influenced by its environment, or the direct measurement of some physical property of the organism itself.

SELECTIVE INSERTION OF FLUORESCENT LIPID PROBES

This work is based on the use of fluorescent lipid probes which can be obtained with different fluorophores and lipidic moities which are assumed to insert into the epicuticle (Figure 1.1). The nematode surface exhibits a high selectivity for the insertion of fluorescent lipid probes (Kennedy *et al.*, 1987a) and we have since found this to apply for a wide range of nematode species and stages. Application of these techniques has revealed particularly interesting differences within the infective stage in that the surface of the infective larva has properties which differ from those of later stages (Tables 1.1 and 1.3). This may have important implications for the adaptation of the infective-stage parasite to the mammalian tissue environment. This will be discussed later in relation to a possible process of transformation in the nature of surface lipid.

The observed selectivity of insertion of fluorescent lipid probes in adult surfaces is thought to be primarily due to the physico-chemical properties of the probe. Anionic lipid probes such as 5–(*N*-octadecanoyl) aminofluorescein (AF18) can be easily inserted in the nematode surface, whereas cationic lipid probes such as octadecylrhodamine B (RH18) and dioctadecyltetramethylindocarbocyanine (Dil18) cannot. Since AF18 and RH18 have the same hydrophobicity, selectivity must be largely head-group dependent, although within the aminofluorescein analogues there is also a strict dependence on acyl chain length (Kennedy *et al.*, 1987a). Molecules that contain both charged and hydrophobic residues such as the

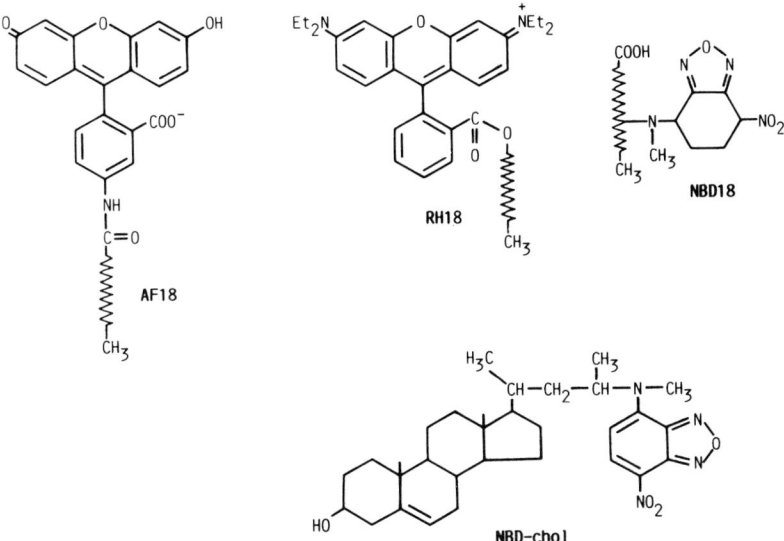

Figure 1.1. A range of fluorescent lipid analogues can be obtained with different fluorophores and lipidic moieties, including 5-(*N*-octadecanoyl)aminofluorescein (AF18), 25-(nitrobenzoxadiazolamine methylamino)-27-norcholesterol (NBD-chol), NBD-octadecanoate (NBD18), and octadecyl rhodamine B (RH18).

Table 1.1. The binding and surface specificity of fluorescent probes varies between nematode species and developmental stage.[a]

Species	Stage	Fluorescent lipid probe[b]		
		AF18	RH18	NBD-chol
T. canis	Infective stage	+ + +	–	+ +
O. ostertagi	Exsheathed L3	—[c]	–	–
	L4	+ + +	–	+ +
	Adult	+ + +	–	+ +
T. spiralis	Newborn larva	+ + +	–	+
	Infective larva	—[c]	–	–
	Adult	+ + +	–	+ +
A. viteae	Microfilaria (unsheathed)	+ + +	–	+ + +
	L3	—[c]	–	–
	Adult	+ + +	–	+ + +
B. pahangi	Microfilaria (sheathed)	internal	–	–
	L3	—[c]	–	–
	Adult	+ +	–	+ +
L. carinii	Microfilaria (sheathed)	internal	–	–
	Adult	+ +	–	+
C. elegans	Dauer (L3)	—[d]	–	–
	Adult	+ + +	–	+ +

[a] The intensity of fluorescence is represented by an arbitrary scale of ' – ' to ' + + + '.

[b] AF18, octadecylaminofluorescein, RH18, octadecylrhodamine; NBD-chol, nitrobenzoxadiazolamine cholesterol.

[c] AF18 insertion only occurs after exposure to appropriate environmental stimuli, e.g. RPMI 1640 at 37°C for *A. viteae* (see Table 1.3).

[d] AF18 insertion only occurs after exposure to a fresh bacterial food source.

fatty acid analogues AF18 and RH18, would normally occupy the polar–non-polar interface region of a membrane (Radda, 1975), so that differently charged probes would have slightly different vertical partitioning in the epicuticle. It should be emphasized at this point that the epicuticle will generally not allow the insertion of cationic lipid probes, a property which is shared by the unusual plasma membrane of fertilized *Xenopus* eggs (Dictus *et al.*, 1984). In both cases this could be due to interfacial interactions with a polyanionic surface coat.

Nitrobenzoxadiazolamine (NBD) is an uncharged fluorophore and can be used to sense lipid environments. It also has useful spectral properties in being essentially non-fluorescent unless incorporated into a lipid layer, in which case its emission is similar to fluorescein. Non-polar lipid probes such as 25-(NBD-methyl-amino)-27-cholesterol (NBD-chol) will likely be distributed in the lipophilic interior of the epicuticle, with little or no interaction with a charged surface coat.

Some lipid probes interact very differently with the nematode epicuticle, regardless of fluorophore structure. For example, the fluorescent phospholipid NBD-phosphatidylethanolamine (NBD-PE) is rapidly internalized by the adult filarial nematodes *A. viteae* and *L. carinii*, whereas NBD-phosphatidylcholine (NBD-PC) is surface restricted. It can be concluded from these observations that the physical properties of choline and ethanolamine phospholipid head-groups affect the uptake of the whole molecule into the worm, since the process of NBD-PE internalization is not slowed by low temperature or metabolic inhibitors (our unpublished results).

The depth of insertion of fluorescent probes can be estimated by measuring the extent to which a non-permeant molecule included in the medium (Trypan Blue) can quench the fluorescence of the probe by Forster resonance energy transfer (Foley *et al.*, 1986). Fluorescent lipid probes, which were used in fluorescence recovery after photobleaching (FRAP) experiments, were found to be surface restricted by this method (Kennedy *et al.*, 1987a).

FLUORESCENCE RECOVERY AFTER PHOTOBLEACHING

The freedom of a fluorescent lipid probe to diffuse within the plane of a surface can provide important information on the physical state of the lipid in its immediate environment. The biophysical technique of FRAP is used extensively in the study of lateral diffusion of plasma membrane components (for reviews see Peters, 1981; Wolf, 1988). In all versions of FRAP, the lateral mobility of molecules in the membrane surface is determined by measuring the rate of local changes in the con-

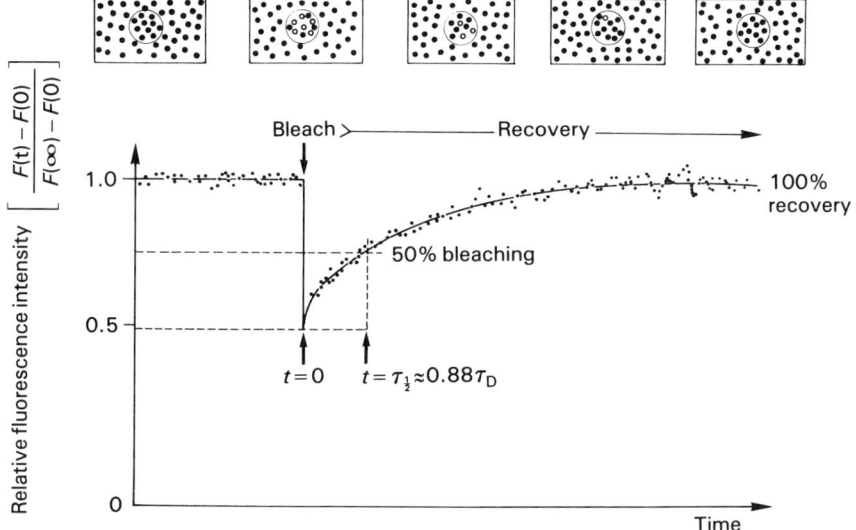

Figure 1.2. Typical FRAP curve. In all versions of FRAP, the lateral mobility of molecules on the surface is estimated by measuring the rate of local changes in the concentration of fluorescent markers of surface molecules. A small region of the surface, which is labelled with fluorescent molecules, is illuminated with a focused laser beam of diameter $< 2\mu m$. A brief (about 125 ms) and intense enhanced pulse of laser light causes an irreversible photobleaching of a proportion (ideally 50 per cent) of the fluorophores in the illuminated region. The laser is then returned to low-level illumination to monitor the subsequent fluorescence intensity within the spot without causing further bleaching. As fresh fluorophores diffuse into the bleached region, the fluorescence intensity increases and the diffusion coefficients of the labelled molecules can be calculated from the rate of this process. The ordinate gives the normalized, $F(t)$-$F(0)$, of the fluorescence intensities measured at time t and time $t = 0$.

centration of fluorescent markers (Figure 1.2). FRAP provides us with two measures of diffusion: first, the fraction of the component that is free to diffuse (per cent recovery, %R) and second, the lateral diffusion coefficient (D_L) of that fraction (Axelrod *et al.*, 1976).

FRAP studies of lipid probe diffusion in typical eukaryotic plasma membranes show rapid lateral diffusion ($D_L = 10^{-6}$ to 10^{-7} cm^2 s^{-1}) with 90–100 per cent of the probe free to diffuse. However, it appears that a significant fraction of plasma membrane lipid is not free to diffuse in certain specialized membranes such as those of *Xenopus* eggs (Dictus *et al.*, 1984), sea-urchin eggs (Peters and Richter, 1981), spermatozoa (Wolf and Voglmayr, 1984; Wolf *et al.*, 1988), and in certain other differentiating or highly polarized cells.

One potential problem in the application of these techniques to nematodes is that their surface is not planar or even cylindrical. However, diffusion coefficients are thought not to be significantly affected for curved or annulated surfaces such as those of concern here (Aizenbud and Gershon, 1982). For instance, diffusion rates of a lipid analogue in a microvillous surface is similar to that in featureless cell surfaces (Wolf *et al.*, 1982).

Most membrane proteins diffuse about 50 times slower than lipid molecules ($D_L \simeq 4 \times 10^{-10}$ cm^2 s^{-1}), and a substantial fraction of membrane proteins are immobile (Schlessinger, 1983). However, in the absence of a restraining network, such as the cytoskeleton, it is possible for membrane proteins to diffuse with $D_L = 5 \times 10^{-9}$ cm^2 s^{-1} (Peters, 1981; Saffman and Delbruck, 1975).

LATERAL DIFFUSION OF LIPID IN THE PLANE OF THE EPICUTICLE

FRAP studies on the nematode surface immediately showed it to have a highly unusual biological surface (Kennedy *et al.*, 1987a). A large non-diffusing fraction of lipid is detected on the surface of many adult parasitic nematodes, including *A. viteae*, *B. pahangi*, *L. carinii*, *O. ostertagi*, and *T. spiralis*, using the anionic lipid probe AF18 (Kennedy *et al.*, 1987a; and Table 1.2). But it was also found that adults of *L. carinii* have small patches of mobile AF18 distributed in no discernible pattern over their surface (Kennedy *et al.*, 1987c).

The general immobility of AF18 in mammalian-parasitic forms was in stark contrast to the surface of the adult free-living nematode *C. elegans*, in which there is the unrestricted lateral diffusion of AF18 (Table 1.2). This fundamental difference in lipid diffusibility prompts the suggestion that, in the parasitic forms, it is an adaptation to physico-chemical and biological environments encountered in the host.

Immobility of the AF18 lipid probe in the surface of adult parasites implies a totally rigid lipid surface, but the non-polar lipid probe NBD-chol detects mobile lipid (Table 1.2). This observation, and the horizontal heterogeneity in the surface of adult *L. carinii* (Kennedy *et al.*, 1987c) suggested that in the adult nematode surface there is differential horizontal partitioning of the lipid probes.

Table 1.2. Lateral diffusion of fluorescent markers[a] in the nematode surface.

Species of nematode	Stage	Diffusion rate of mobile fraction[b](%R)		
		AF18	NBD-chol	DTAF (or FM)
A. viteae	Adult	Immobile	Mobile	Mobile
		15	60	80
L. carinii	Adult and L4	Immobile / mobile patches	Mobile	Mobile (FM)
		< 10–80	60	40
B. pahangi	Adult	Immobile	Mobile	Mobile (FM)
		15	80	75
T. spiralis	Adult	Immobile	Mobile	Mobile
		5	80	75
O. ostertagi	Adult and L4	Intermediate	Not done	Mobile
		20–30		80
C. elegans	Adult	Mobile	Mobile	Mobile
		65	90	80
	Dauer larva	Immobile	Mobile	Immobile
		20	80	5

[a] AF18 octadecanoylaminofluorescein; NBD-chol, nitrobenzoxadiazolamine cholesterol; DTAF, dichlorotriazinylaminofluorescein (an amine reactive probe); FM, fluorescein maleimide (a sulphydryl group reactive probe).

[b] The freedom of lipid to diffuse within the plane of a surface is expressed as the lateral diffusion coefficient D_L (in cm^2 s^{-1}). In this table, immobile is taken as a D_L value between 10^{-11} and 10^{-10} cm^2 s^{-1}, and mobile as D_L of between 10^{-9} and 10^{-7} cm^2 s^{-1}. The percent recovery of fluorescence (%R) after photobleaching is also given.

HETEROGENEITY OF SURFACE LIPID — DOMAINS?

For many years the predominant concept of membrane organization has been embodied in the fluid mosaic model of Singer and Nicolson (1972). This model was proposed to stress the dynamic aspects of membrane structure, although it belies the complexities now recognized as features of many membranes. Rather than being a homogeneous sea of lipid, membrane lipids exist in discrete domains which may differ greatly in their composition and properties. Radda (1975) stated that, 'even a given membrane cannot be described by a single structure but at any one instance must be considered as an ensemble average of several rapidly interconvertible forms'.

Limited recovery upon photobleaching of AF18 in adult parasites indicates the existence of gel-phase lipid microdomains (Karnovsky *et al.*, 1982) which may be stabilized by interactions with a subsurface structural element (e.g. the insoluble cuticlin matrix), or by lipid–lipid interactions, or due to an unusual lipid composition. Diffusion studies in model systems (Derzko and Jacobson, 1980; Klausner and Wolf, 1980) indicate that non-diffusing fractions of lipid may be caused by co-existent gel and fluid lipid phases. If interaction with a resistant insoluble protein matrix were to be the explanation, then it would still be necessary to understand how it is that free-living nematodes, which also possess this insoluble protein matrix, do not show restricted mobility of surface lipids. There may of course be some fundamental difference in epicuticle protein structure between free-living and parasitic nematodes which has not yet become apparent.

If lipid factors are the cause of non-diffusing lipid fractions, and not structural proteins, then the next question is: what distinguishes nematode surface lipids from those of host plasma membranes? Nematodes are peculiar in their lipid composition in that some have been found to contain large amounts of plasmalogens and other ether-linked phospholipids (Subrahmanyam, 1967; Chitwood and Krusberg, 1981). Plasmalogens appear to play a role in membrane fluidity (Russell, 1984) and they also appear to enhance transitions to non-bilayer structures (see Figure 1.3) (Goldfine and Langworthy, 1988). It is also noteworthy that plasma membranes of both *Leishmania* (Wassef *et al.*, 1985), and spermatozoa (Evans *et al.*, 1980) possess unusually high levels of plasmalogens. Ether phospholipids (which include plasmalogens) are highly resistant to enzymatic hydrolysis and clearly this lipid stability could be advantageous to parasites. Another function of these ether lipids may be that they are used in a membrane anchor for surface proteins or glycoconjugates, in a manner similar to that exhibited by trypanosomes (Ferguson and Homans, 1988), *Leishmania* (Turco, 1988) and schistosomes (Sauma and Strand, 1990). See Chapter 3 for further discussion of lipid anchors.

Archaebacteria have membranes which are adapted to cope with environmental stress and contain high levels of ether lipids (Goldfine and Langworthy, 1988), but whether the nematode epicuticle comprises these lipids for similar reasons would be speculative at present. Another possibility is that there is a lipid component of the epicuticle which causes it to be partially solidified while maintaining the capability of rapid response to environmental changes. The primary candidates for such unusual lipid composition would be waxes and ascaroside-like lipids. The melting point of ascaroside esters is about 40°C and they are fluid at the body temperature of the host, whereas free ascarosides have melting points in the range 70–80°C and so produce a solid membrane (Barrett, 1981). Although ascarosides are mainly thought of as being present in the vitelline membrane of ascarid eggs, it has been suggested that they could also be present in the cuticle since small amounts were found in adult males (Fairbairn and Passey, 1955). After all, there have been recent

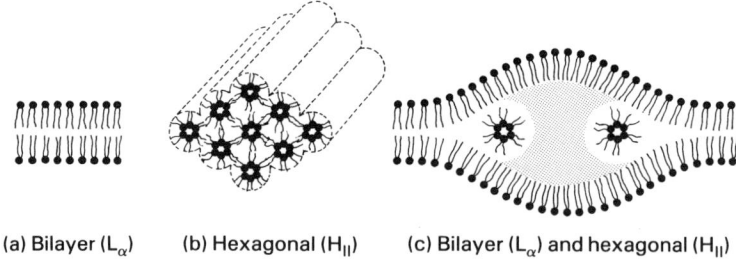

(a) Bilayer (L$_\alpha$)   (b) Hexagonal (H$_{II}$)   (c) Bilayer (L$_\alpha$) and hexagonal (H$_{II}$)

Figure 1.3. A modification to the fluid mosaic model of Singer and Nicolson (1972) to include lipid polymorphism was proposed by Cullis *et al.* (1985). Such membrane structures include inverted micelles, cylinder and hexagonal phase lipids. When lipids with a small polar head-group and an extended hydrophobic part are dispersed in water, they adopt an inverted hexagonal (H$_{II}$) phase in which the lipid molecules are arranged in hexagonally packed cylinders. This type of non-bilayer phase lipid can affect the permeability barrier and increase ion permeability, transbilayer (flip-flop) movement of the lipid molecules, and fusion events (De Gier, 1988).

reports which might suggest some homology between the epicuticle and the vitelline membrane of eggs (see Perry, 1989). Waxes and ascarosides can exist in different physical states, i.e. solid or fluid, depending on the temperature or action of modifying enzymes, and could allow the epicuticle some degree of flexibility in interactions with the environment while retaining resistance to chemical and physical agents.

The immobile fraction of adult epicuticular lipid presents a paradox when one considers the dynamic functions of the cuticle which include antigen shedding. However, when surface proteins are directly labelled with sulphydryl group or amine reactive fluorescent probes, their lateral diffusion is rapid ($D_L \approx 1 \times 10^{-7}$cm^2 s^{-1}) with a large mobile fraction (%R between 70 and 80 per cent). Assuming that the fluorescent labelling does not disrupt any anchoring mechanism, then there is the possibility that the proteins are loosely associated by ionic interactions with polar lipids. Alternatively, the proteins could be partitioned into microdomains which contain mainly neutral lipids and are free to diffuse.

The concept of horizontal partitioning of lipid and lipophilic molecules between domains may have profound implications for the host–parasite interface. A variety of lipophilic molecules and proteins may interact with the epicuticle in such a way that there is the involvement of a unique pattern of partitioning into domains, which is illustrated in Figure 1.4. Recently, Mountford and Wright (1988) proposed a similar structural model whereby neutral lipid domains are intercalated with the bilayer lipid in the plasma membrane of malignant cells. These authors were particularly interested in the possibility of particles being expelled or 'jettisoned' from the intact surface by mechanisms using the domain organization.

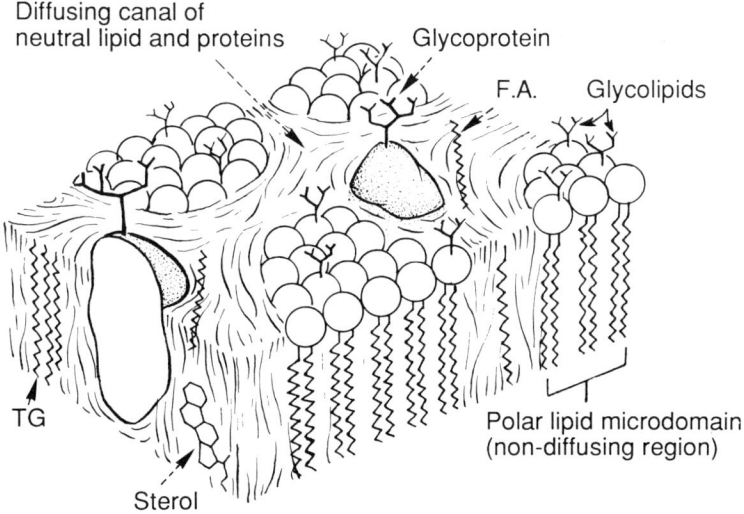

Figure 1.4. A topological view of the epicuticle, where regions of differing composition and organization may co-exist. Only the outer leaflet of the epicuticle is shown, but it is presumed to have a modified bilayer structure. FA, fatty acids; TG, triglycerides.

WHAT IS THE ADVANTAGE OF THE UNUSUAL PROPERTIES OF THE SURFACE LIPID?

Immobility of surface lipid and the capacity to exclude certain lipophilic molecules may be fundamental features of parasitic nematodes which protect them against the chemical and physical onslaughts of tissue environments. Surface-damaging molecules such as phospholipases, fatty acids, and complement components may be unable to perturb the predominantly stable lipid organization of the adult epicuticle (Kato and Bito, 1978; Taylor, 1983). Too little is known at present, but the composition of the lipid might also have profound effects on interaction with host defence mechanisms (Mold, 1989). Chemotherapy might then be improved greatly by drugs which are designed to circumvent the restricted entry of certain molecules, perhaps by means of a modification to the charge and/or hydrophobicity of the drug, or means found by which selectivity and immobility could be abrogated in order to permit effective immune attack. Such effects might explain the action of certain drugs which are harmless to filarial nematodes in the absence of host components, but known to alter their surface to permit cell binding and parasite attrition (Piessens and Beldekas, 1979).

CHANGES IN THE LIPOPHILICITY OF THE THIRD-STAGE EPICUTICLE UPON EXPOSURE TO THE ENVIRONMENT OF THE MAMMALIAN HOST

A critical stage for both the physiological processes of the parasite and the defence mechanisms of the host is the process of infection. Possession of a surface which is capable of responding rapidly to environmental shifts may be essential when a poikilotherm-infecting parasite is transmitted to a homeotherm, or when a free-living larva penetrates the skin of a mammal. Since any modifications which might be undergone by the parasite during the process of infection could provide targets for chemo- or immuno-therapy, it is important that the basic biology and biochemistry of the transition from the pre- to the post-parasitic state should be understood.

In our recent experiments, larvae which enter the mammalian host by different routes and from different environments were examined. First, those that enter via an arthropod vector, e.g. the L3 of *A. viteae* from the argasid tick vector, and the L3 of *B. pahangi* from the mosquito vector. Secondly, those which have a free-living infective larva which actively penetrates tissue, e.g. *Nippostrongylus brasiliensis* and *Strongyloides ratti*. Thirdly, those which are taken directly into the gastrointestinal tract, e.g. *O. ostertagi* and *T. spiralis*. In all cases, the surfaces of infective larvae before exposure to mammalian tissue conditions were found to be remarkable in having no affinity for any of the fluorescent lipid probes described above. Even more remarkable, however, was the fact that on exposure to the environment of the mammalian host, the biophysical properties of the surface changed, sometimes within a matter of a few minutes, and the AF18 probe could then insert. This is described in more detail for the infective L3 of *A. viteae* in Table 1.3. In addition, the infective L3 of *S. ratti* which had been surface labelled with fluorescent cationized ferritin (King and Preston, 1977) began to lose this anionic material from the surface at the same time as they became lipophilic for the AF18 probe (our unpublished results).

It has often been considered that the dauer stage of free-living nematodes, such as *Caenorhabditis*, is similar to the developmentally arrested L3 stage of parasitic nematodes. This parallel also seems to apply to the affinity for AF18 in that the dauer larvae of *C. elegans* excluded this lipid probe. It is also possible that they respond to environmental changes by similar mechanisms to infective larvae, as they allow the insertion of AF18 after only 30 min exposure to fresh *E. coli*, which will act as a stimulus for this stage to begin feeding and restart development. This change is considerably more rapid than for any other recorded sign of recovery from the dauer state (Cassada and Russell, 1975).

In infective larvae, when the vector stage L3 of *A. viteae*, for example, is transferred from insect culture medium (Grace's medium) at 27 °C to mammalian tissue culture medium (RPMI 1640) at 37 °C, its surface begins to show affinity for AF18 within a matter of minutes (Proudfoot *et al.*, 1990). This effect appears to be due to synergism between increased temperature and exposure to the new medium, since temperature elevation to 37 °C or medium change alone have only slight effects on lipophilicity. There is some evidence that the stimuli involved in this process might include an increased sodium ion (Na^+) concentration, an increased bicarbonate ion (HCO_3^-) concentration, and a more alkaline pH, since manipulation of these parameters can affect transformation. That is to say, by increasing the pH of Grace's insect medium from pH 6.0 to that of RPMI (pH 7.4) one can achieve

Table 1.3. Temperature, pH and ionic balance elicit the surface change which permits uptake of the AF18 fluorescent lipid probe into the surface of infective-stage larvae of *A. viteae*.[a]

Medium	Temperature (°C)	Level of AF18 binding
RPMI 1640	4	+
RPMI 1640	27	+ +
RPMI 1640	37	+ + + + +
RPMI 1640, pH 6.0	27	+
Grace's insect medium	27	−
Grace's insect medium	37	+ + +
Grace's insect medium, pH 7.4	27	+ + + +
Na^+ (300 mOsm)	27	+ +
Mannitol (300 mOsm)	27	−
HCO_3^- (20 mM)	27	+ + + +

[a] Infective (L3) larvae of the parasite were freshly removed from ticks and subjected to the indicated conditions before being tested for affinity for the AF18 surface probe. The results show that mammalian tissue environment and temperature are required for induction of maximal uptake of the probe, but that exposure to the appropriate temperature, pH and ionic conditions can elicit the change to some degree. The pH of RPMI 1640 is normally 7.4 and that of Grace's insect medium 6.0. The components of the Na^+ solution were 140 mM NaCl and 10 mM NaH_2PO_4 which approximates the mammalian osmolarity of 300 mOsm. The same osmolarity of mannitol (280 mM mannitol + 10 mM NaH_2PO_4) was used as a control.

transformation; and by reducing the pH of RPMI from pH 7.4 to that of the insect medium (pH 6.0) the normal process of transformation is inhibited.

Transformation of cercariae to schistosomula in *S. mansoni* has been found to be dependent on an increasing saline concentration (Samuelson and Stein, 1989), and dauer larvae of *C. elegans* are stimulated to emerge from the dauer state when exposed to NaCl (Cassada and Russell, 1975). As for HCO_3^-, Petronijevic and Rogers (1987) have shown that the undissociated acid H_2CO_3 induces the Ca^{2+} dependent development of the infective-stage larvae of *Haemonchus contortus* and that this leads to a more alkaline intracellular pH, perhaps by Ca^{2+}/H^+ exchange. A similar developmental change occurs in *Plasmodium berghei* where Na^+/H^+ exchange results in a more alkaline intracellular pH and induces gametogenesis (Kawamoto *et al.*, 1990).

The time taken for the post-infection increase in lipophilicity in the L3 nematode appears to correspond to the time taken for a given species of parasite to enter host tissue during the natural infection process. For instance, the L3 of *A. viteae* which fall directly onto the wound caused by their arthropod vector, allow the insertion of AF18 within less than 10 min. Skin-penetrating L3 larvae of *N. brasiliensis*, however, take up to 30 min to become lipophilic for AF18, possibly reflecting a requirement for a slightly longer transition period during skin penetration. Similarly, the sheathed L3 of *O. ostertagi* does not become lipophilic for AF18 until approximately 30 min after it has exsheathed in RPMI or in an artificial stomach solution.

The longest recorded time for lipophilicity to appear is in the infective larva of *T. spiralis*, in which AF18 insertion does not occur until 4–6 h. The surface layer of the cuticle of infective muscle-stage larvae of *T. spiralis* is unusually complicated and has been referred to as the accessory layer (Lee *et al.*, 1984). It is a heterogeneous layer distinct from the epicuticle with sensitivity to trypsin and bile at 37°C (Stewart *et al.*, 1987). When *T. spiralis* infected muscle is ingested by another animal, the infective larvae undergo a very rapid series of moults that are completed within 24 h. During this period it is thought that components of the host environment act on the accessory layer to change its organization such that it becomes more permeable to nutrients (Stewart *et al.*, 1987). Those workers also pointed out that an important indicator of the surface change was a simultaneous change in behaviour, switching from coiling/uncoiling movement to sinusoidal movement. From our own experiments, it appears that the AF18 insertion takes place simultaneously with the behavioural changes.

Alteration of the surface biophysical properties in *T. spiralis* is unlikely to be due to the loss of the accessory layer since Capo *et al.* (1984) have shown, by electron microscopy, that it remains on the larva until the entire cuticle is lost at the first moult. Therefore, the effect is more likely to be due to the physical alteration of the surface, as we have observed that rapid behavioural changes and AF18 insertion can be induced using the cationic detergent cetyltrimethylammonium bromide (CTAB), but it is likely that this is accompanied by the loss of some antigenic material (Grencis *et al.*, 1986).

Peculiarities in lipid composition could result in lateral phase separations that could enable the parasite to cope with a rapid temperature change without loss of the integrity of a lipid layer. Such lipid polymorphisms are thought to operate in the outer membrane of Gram-negative bacteria for precisely this purpose (Burnell *et al.*, 1980; Borovjagin *et al.*, 1987). Wright and Hong (1988) postulated that the inner layer of the accessory layer on *T. spiralis* is composed of filaments of non-bilayer phase lipid (see also Figure 1.3) in an attempt to explain the 'hidden' glycosylated sites on surface proteins. Such lipid phases might also exist in order to accommodate rapid changes in the physical environment.

An interesting but enigmatic feature which was revealed in our experiments is that, if lipid-rich serum from the definitive host is included in the culture medium and then removed before labelling, there is a dramatic decrease in the affinity of the surface of the nematode for the fluorescent lipid probe AF18. Sera from an inappropriate host species (e.g. lipid-rich human serum for *N. brasiliensis* which normally infects rats) can have similar inhibitory effects but delipidated human serum has no such effect. The inference is, therefore, that host lipid is taken up from serum and saturates the surface to the exclusion of the fluorescent lipid probe. Saturation of the surface with normal levels of host lipid would probably not occur, although our experiments do indicate that there will be a close interaction between the nematode surface and blood lipids upon infection. This prompts the question as to whether an interaction between host lipids and surface antigens could promote their release from the surface, or, alternatively, cause them to be masked.

The increase in lipophilicity for AF18 and possibly serum lipids, and the loss of anionic material in *S. ratti*, suggests that there is a general change in the biophysical properties of the surface. The following possible explanations are suggested: firstly, there may be a rapid reorganization of the surface which changes its affinity for lipids. This could be accompanied by the release of a surface coat which may have been attributed to the exclusion of lipids while it was present on the surface. Secondly, the fact that increased temperature alone had rapid surface effects suggests that there is either a temperature-induced lipid-phase change, or that certain enzymes are activated at this temperature to cleave specific molecules. Thirdly, changes in ionic composition or pH could affect the operation of cellular signalling pathways in the hypodermis which could rapidly produce signals to be relayed to the surface. The sensory apparatus of the nematode is highly specialized, so it is also possible that the signal for epicuticle modification comes via nervous stimulation. This explanation seems unlikely, however, since attempts to block surface lipophilicity for AF18 with the acetylcholinesterase inhibitors eserine sulphate and carbamyl choline were unsuccessful.

In recent years, attention has been focused on signalling mechanisms in parasites, with particular emphasis on those mechanisms which might govern developmental changes. In mammalian cells, external signals such as hormones and growth factors are mediated by signal transduction mechanisms of cyclic adenosine 3′,5′-monophosphate (cAMP) and/or calcium dependent pathways (Nishizuka, 1984, 1986). In this regard, there is little information on the roles of these systems in nematode development. However, the knowledge that certain drugs which can activate or interfere with signalling mechanisms can actually alter or modify development in schistosomes (Kawamoto *et al.*, 1989) and *Plasmodium* (Kawamoto *et al.*, 1990) encourages us to believe that these mechanisms exist in nematodes and that developmental changes may be mediated initially by alterations in the biophysical properties of the epicuticle. Our recent work would indicate that these signalling pathways are indeed involved in the lipid change in the infective larvae of nematodes in ways that previously would have been thought unlikely or impossible.

CONCLUDING REMARKS

Fluorescence-based biophysical methods provide a powerful tool in understanding the organization of the lipid in the nematode surface. There remains, however, a need for new information on the lipid composition of the epicuticle and the interaction of lipid with surface proteins. Nuclear magnetic resonance and electron spin resonance would be more definitive biophysical techniques, and freeze–fracture electron microscopy ought to be pursued, although previous findings did not provide clear indications of epicuticle structural organization. It has become clear, however, that parasitic nematodes present a highly unusual surface to their hosts, and that this extends to the characteristics of its lipid. This is most notable for the transition to the environment of the mammalian host, indicating that the nematode epicuticle can adjust dynamically to its environment.

ACKNOWLEDGEMENTS

Our work is supported by the Wellcome Trust through a grant to M.W.K., J.R.K., and H.V.S. We must also thank the following people who have provided considerable help: Dr William Harnett, Dr Michael Worms, Dr Eileen Devaney, Dr Taff Jenkins, Mrs Anne McIntosh, Mr David McLaughlin and Dr David Hayes. Thanks must also go to Dr David Wolf for his invaluable technical advice.

REFERENCES

Abraham, D., Grieve, R. B. and Mika-Grieve, M., 1988, *Dirofilaria immitis*: surface properties of third- and fourth-stage larvae, *Experimental Parasitology*, **65**, 157–67.

Aizenbud, B. M. and Gershon, N. D., 1982, Diffusion of molecules on biological membranes of non-planar form. A theoretical study, *Biophysical Journal*, **38**, 287–93.

Axelrod, D., Koppel, D. E., Schlessinger, J., Elson, E. and Webb, W. W., 1976, Mobility measurement by analysis of fluoresence photobleaching recovery kinetic, *Biophysical Journal*, **16**, 1055–69.

Badley, J. E., Grieve, R. B., Rockey, J. H. and Glickman, L. T., 1987, Immune-mediated adherence of eosinophils to *Toxocara canis* infective larvae: the role of excretory–secretory antigens, *Parasite Immunology*, **9**, 133–43.

Barrett, J., 1981, Nutrition and biosynthesis: Ascaroside synthesis, in *Biochemistry of Parasitic Helminths*, pp. 204–5, London: Macmillan.

Benz, R., 1985, Porin from bacterial and mitochondrial outer membranes, in *Critical Reviews in Biochemistry*, vol. 19, pp. 145–90, Boca-Raton, Florida: CRC Press.

Betschart, B. and Jenkins, J. M., 1987, Distribution of iodinated proteins in *Dipetalonema viteae* after surface labelling, *Molecular and Biochemical Parasitology*, **22**, 1–8.

Bird, A. F., 1957, Chemical composition of the nematode cuticle. Observations on individual layers and extracts from these layers in *Ascaris lumbricoides* cuticle, *Experimental Parasitology*, **6**, 383–403.

Bird, A. F., 1980, The nematode cuticle and its surface, in Zuckerman, B. M. (Ed.) *Nematodes as Biological Models*, pp. 213–36, New York: Academic Press.

Bird, A. F., 1984, Nematoda, in Bereiter-Hahn, J., Matoltsy, A. G. and Richards, K. S. (Eds) *Biology of the Integument*, pp. 212–31, Berlin: Springer.

Bluemink, J. G., Tertoolen, L. G. J., Ververgaert, P. H. J. T. and Verkleij, A. J., 1976, Freeze fracture electron microscopy of preexisting a nascent cell membrane in cleaving eggs of *Xenopus laevis*, *Biochimica et Biophysica Acta*, **443**, 143–55.

Boggs, J. M. and Moscarello, M. A., 1978, Structural organisation of the human myelin membrane, *Biochimica et Biophysica Acta*, **515**, 1–21.

Borovjagin, V. L., Sabelnikov, A. G., Tarahovsky, Y. S. and Vasilenko, I. A., 1987, Polymorphic behaviour of Gram-negative bacteria membranes, *Journal of Membrane Biology*, **100**, 229–42.

Branton, D. and Deamer, D. W., 1972, *Membrane Structure*, Berlin: Springer.

Burnell, E., van Alphen, L., Verkleij, A., de Kruijff, B. and Lugtenberg, B., 1980, Non-bilayer phase lipid in ^{31}P – n.m.r. spectra of *Escherichia coli* outer membrane, *Biochimica et Biophysica Acta*, **597**, 518–32.

Capo, V. A., Despommier, D. D. and Silberstein, D. S., 1984, The site of ecdysis of the L1 larva of *Trichinella spiralis*, *Journal of Parasitology*, **70**, 992–4.

Carlow, C. K. S., Perronne, J., Spielman, A. and Philipp, M., 1987, A developmentally regulated surface epitope expressed by the infective larva of *Brugia malayi* which is rapidly lost after infection, *UCLA Symposium on Molecular and Cellular Biology*, **60**, 301–10.

Cassada, R. C. and Russell, R. L., 1974, A positive selection for behavioural and developmental mutants of a nematode (*Caenorhabditis elegans*), *Federation Proceedings*, **33**, 1476.

Cassada, R. C. and Russell, R. L., 1975, The dauer larva, a post-embryonic developmental variant of the nematode *Caenorhabditis elegans*, *Developmental Biology*, **46**, 326–42.

Chitwood, D. J. and Krusberg, L. R., 1981, Diacyl, alkylacyl and alkenylacyl phospholipids of the nematode *Turbatrix aceti*, *Comparative Biochemistry and Physiology*, **69B**, 115–20.

Cox, G. N., Kusch, M. and Edgar, R. S., 1981, Cuticle of *Caenorhabditis elegans*: its isolation and partial characterization, *Journal of Cell Biology*, **90**, 7–17.

Cullis, P. R., Hope, M. J., de Kruijff, B., Verkleij, A. J. and Tilcock, C. P. S., 1985, Structural properties and functional roles of phospholipids in biological membranes, in Kuo, J. F. (Ed.) *Phospholipids and Cellular Regulation*, Vol. 1, pp. 3–49, Boca Raton, Florida: CRC Press.

De Gier, J., 1988, The use of liposomes in a search for an understanding of the significance of membrane lipid diversity, *Biochemical Society Transactions*, **16**, 912–14.

De Laat, S. W. and Bluemink, J. G., 1974, New membrane formation during cytokinesis in normal and cytochalasin B treated eggs of *Xenopus laevis*. 2. Electrophysiological observations, *Journal of Cell Biology*, **60**, 529–40.

De Laat, S. W., Luchtel, D. and Bluemink, J. G., 1973, The action of cytochalasin B during egg cleavage in *Xenopus laevis*: dependance on cell membrane permeability, *Developmental Biology*, **31**, 163–77.

De Laat, S. W., Wouters, W., Marques Da Silva Pimenta Guarda, M. M. and Da Silva Guarda, M. A., 1975, Intracellular ionic compartmentation, electrical membrane properties and cell membrane permeability before and during first cleavage in the *Ambystoma* egg, *Experimental Cell Research*, **91**, 15–30.

Derzko, Z. and Jacobson, K., 1980, Comparative lateral diffusion of fluorescent lipid analogues in phospholipid multibilayers, *Biochemistry*, **19**, 6050–7.

Devaney, E., 1985, Lectin-binding characteristics of *Brugia pahangi* microfilariae, *Tropenmedizin und Parasitologie*, **36**, 25–8.

Devaney, E., 1988, The biochemical and immunological characterisation of the 30 kilodalton surface antigen of *Brugia pahangi*, *Molecular and Biochemical Parasitology*, **27**, 83–92.

Dictus, W.J.A.G., van Zoelen, E.J.J., Tetteroo, P.A.T., Tertoolen, L.G.J., de Laat, S.W. and Bluemink, J.G., 1984, Lateral mobility of plasma membrane lipids in *Xenopus* eggs: regional differences related to animal/vegetal polarity become extreme upon fertilization, *Developmental Biology*, **101**, 201–11.

Evans, R.W., Weaver, D.E. and Clegg, E.D., 1980, Diacyl, alkenyl and alkyl ether phospholipids in ejaculated, *in utero*-, and *in vitro*-incubated porcine spermatozoa, *Journal of Lipid Research*, **21**, 223–8.

Fairbairn, D. and Passey, B.I., 1955, The lipid components in the vitelline membrane of *Ascaris lumbricoides* eggs, *Canadian Journal of Biochemistry and Physiology*, **33**, 130–4.

Ferguson, M.A.J. and Homans, S.W., 1988, Parasite glycoconjugates: towards the exploitation of their structure, *Parasite Immunology*, **10**, 465–79.

Ferguson, M.A.J. and Williams, A.F., 1988, Cell-surface anchoring of proteins via glycosylphosphatidylinositol structures, *Annual Reviews of Biochemistry*, **57**, 285–320.

Foley, M., MacGregor, A.N., Kusel, J.R., Garland, P.B., Downie, T. and Moore, I., 1986, The lateral diffusion of lipid probes in the surface membrane of *Schistosoma mansoni*, *Journal of Cell Biology*, **103**, 807–18.

Fujimoto, D. and Kanaya, S., 1973, Cuticlin: a noncollagen structural protein from *Ascaris* cuticle, *Archives of Biochemistry and Biophysics*, **157**, 1–6.

Goldfine, H. and Langworthy, T.A., 1988, A growing interest in bacterial ether lipids, *Trends in Biochemical Sciences*, **13**, 217–21.

Grencis, R.K., Crawford, C., Pritchard, D.I., Behnke, J.M. and Wakelin, D., 1986, Immunization of mice with surface antigens from the muscle larvae of *Trichinella spiralis*, *Parasite Immunology*, **8**, 587–96.

Grove, D.I., Northern, C., Warwick, C. and Lovegrove, F.T., 1984, Loss of surface coat by *Strongyloides ratti* infective larvae during skin penetration: evidence using larvae radiolabelled with 67-gallium, *Journal of Parasitology*, **70**, 689–93.

Hayunga, E.G., Murrell, K.D., Taylor, D.W. and Vannier, W.E., 1979, Isolation and characterization of surface antigens from *Schistosoma mansoni*. 1. Evaluation of techniques for radioisotope labeling of surface proteins from adult worms, *Journal of Parasitology*, **65**, 488–96.

Himmelhoch, S. and Zuckerman, B.M., 1978, *Caenorhabditis briggsae*: ageing and the structural turnover of the outer cuticle surface and the intestine, *Experimental Parasitology*, **45**, 208–14.

Hockley, D.J. and McLaren, D.J., 1973, *Schistosoma mansoni*: changes in the outer membrane of the tegument during development from cercariae to adult worm, *International Journal of Parasitology*, **3**, 13–25.

Howells, R.E. and Blainey, L.J., 1983, The moulting process and the phenomenon of intermoult growth in the filarial nematode *Brugia pahangi*, *Parasitology*, **87**, 493–505.

Howells, R.E., Mendis, A.M. and Bray, P.G., 1983, The mechanisms of amino acid uptake by *Brugia pahangi in vitro*, *Zeitschrift für Parasitenkunde*, **69**, 247–53.

Howells, R.E., 1987, Dynamics of the filarial surface, in Evered, D. and Clark, S. (Eds) Filariasis, Ciba Foundation Symposium 127, pp. 94–102, Chichester: Wiley.

Ibrahim, M.S., Tamashiro, W.K., Moraga, D.A. and Scott, A.L., 1989, Antigen shedding from the surface of the infective stage larvae of *Dirofilaria immitis*, *Parasitology*, **99**, 89–97.

Jain, M. K., 1988, *Introduction to Biological Membranes*, 2nd Edn, New York: Wiley-Interscience.

Johnson, J. D. and Epel, D., 1975, Relationship between release of surface proteins and metabolic activation of sea urchin eggs at fertilization, *Proceedings of the National Academy of Sciences (USA)*, 72, 4474–8.

Karnovksy, M. J., Kleinfeld, A. M., Hoover, R. L. and Klausner, R. D., 1982, The concept of lipid domains in membranes, *Journal of Cell Biology*, 94, 1–6.

Kato, K. and Bito, Y., 1978, Relationship between bactericidal action of complement and fluidity of cellular membranes, *Infection and Immunity*, 19, 12–17.

Kaushal, N. A., Simpson, A. J. G., Hussain, K. and Ottesen, E. A., 1984, *Brugia malayi*: stage specific expression of carbohydrates containing N-acetyl-D-glucosamine as the sheathed surfaces of microfilariae, *Experimental Parasitology*, 58, 182–7.

Kawamoto, F.,Shozawa, A., Kumada, N. and Kojima, K., 1989, Possible roles of cAMP and Ca^{2+} in the regulation of miracidial transformation in *Schistosoma mansoni*, *Parasitology Research*, 75, 368–74.

Kawamoto, F., Alejo-Blanco, R., Fleck, S. L., Kawamoto, Y. and Sinden, R. E., 1990, Possible roles of Ca^{2+} and cGMP as mediators of the exflagellation of *Plasmodium berghei* and *P. falciparum*, in press.

Kennedy, M. W., Foley, M., Kuo, Y.-M., Kusel, J. R. and Garland, P. B., 1987a, Biophysical properties of the surface lipid of parasitic nematodes, *Molecular and Biochemical Parasitology*, 22, 233–40.

Kennedy, M. W., Maizels, R. M., Meghji, M., Young, L., Qureshi, F. and Smith, H. V., 1987b, Species-specific and common antigens of *Toxocara cati* and *Toxocara canis* infective larvae, *Parasite Immunology*, 9, 407–20.

Kennedy, M. W., Foley, M., Knox, K., Harnett, W., Worms, M. J., Kusel, J. R., Birmingham, J. and Garland, P. B., 1987c, Are the biophysical properties of the surface lipid of filariae different from other parasitic nematodes? In MacInnis, A. J. (Ed.) *Molecular Paradigms for Eradicating Helminthic Parasites*, pp. 289–300, New York: Alan R. Liss.

Kieffer, E., Rudin, W. and Hecker, H., 1989, Cytochemical demonstration of lectin binding-sites in the cuticle and tissues of *Acanthocheilonema viteae* (Filarioidea), *Acta Tropica*, 46, 3–15.

King, C. A. and Preston, T. M., 1977, Fluoresceinated cationised ferritin as a membrane probe for anionic sites at the cell surface, *FEBS Letters*, 73, 59–63.

Klausner, R. D. and Wolf, D. E., 1980, Selectivity of fluorescent lipid analogues for lipid domains, *Biochemistry*, 19, 6119–203.

Kusel, J. R., 1972, Protein composition and protein synthesis in the surface membranes of *Schistosoma mansoni*, *Parasitology*, 65, 55–62.

Lackie, J. M. and Rotheram, S., 1972, Observations on the envelope surrounding *Moniliformis dubius* (Acanthocephala) in the intermediate host, *Periplaneta americana*, *Parasitology*, 65, 303–8.

Lee, D. L. and Bonner, T. P., 1982, Freeze etch studies on nematode body wall, Proceedings of the British Society of Parasitology, 12–16 April 1981, The Netherlands, *Parasitology*, 84, xliv.

Lee, D. L., Wright, K. A. and Shivers, R. R., 1984, A freeze–fracture study of the surface of the infective-stage larva of the nematode *Trichinella*, *Tissue and Cell*, 16, 819–28.

Lee, D. L., Wright, K. A. and Shivers, R. R., 1986, A freeze–fracture study of the body wall of adult *in utero* larvae and infective-stage larvae of *Trichinella* (Nematoda), *Tissue and Cell*, 18, 219–30.

Locke, M., 1982, Envelopes at cell surfaces — a confused area of research of general importance, and Desser, S. S. (Eds) in Mettrick, D. F. *Parasites–Their World and Ours*, pp. 73–8, Amsterdam: Elsevier Biomedical Press.

Lugtenberg, B. and van Alphen, L., 1983, Molecular architecture and functioning of the outer membrane of *Escherichia coli* and other Gram-negative bacteria, *Biochimica et Biophysica Acta*, **737**, 51–115.

Lumsden, R. D., 1975, Surface ultrastructure and cytochemistry of parasitic helminths, *Experimental Parasitology*, **37**, 267–339.

Maizels, R. M. and Selkirk, M. E., 1988, Immunobiology of nematode antigens, in Englund, P. T. and Sher, A. (Eds) *The Biology of Parasitism*, pp. 285–308, New York: Alan Liss.

Maizels, R. M., Meghji, M. and Ogilvie, B. M., 1983, Restricted sets of parasite antigens from the surface of different stages and sexes of the nematode *Nippostrongylus brasiliensis, Immunology*, **48**, 107–21.

Maizels, R. M., de Savigny, D. and Ogilvie, B. M., 1984, Characterization of surface and excretory–secretory antigens of *Toxocara canis* infective larvae, *Parasite Immunology*, **6**, 23–7.

Maizels, R. M., Kennedy, M. W., Meghji, M., Robertson, B. D. and Smith, H. V., 1987, Shared carbohydrate epitopes on distinct surface and secreted antigens of the parasitic nematode *Toxocara canis, Journal of Immunology*, **139**, 207–14.

Marshall, E. and Howells, R. E., 1986, Turnover of surface proteins of third and fourth larval stages of *Brugia pahangi, Molecular and Biochemical Parasitology*, **18**, 17–24.

Martinez-Palomo, A., 1978, Ultrastructural characterization of the cuticle of *Onchocerca volvulus* microfilariae, *Journal of Parasitology*, **64**, 127–36.

Mazia, D., Schatten, G. and Steinhardt, R., 1975, Turning on of activities in unfertilized sea-urchin eggs: correlation with changes of the surface, *Proceedings of the National Academy of Sciences (USA)*, **72**, 4469–73.

Mok, M., Grieve, R. B., Abraham, D. and Rudin, W., 1988, Solubilization of epicuticular antigen from *Dirofilaria immitis* third-stage larvae, *Molecular and Biochemical Parasitology*, **31**, 173–82.

Mold, C., 1989, Effect of membrane phospholipids on activation of the alternative pathway of complement, *Journal of Immunology*, **143**, 1663–8.

Mountford, C. E. and Wright, L. C., 1988, Organizations of lipids in the plasma membranes of malignant and stimulated cells: a new model, *Trends in Biochemical Sciences*, **13**, 172–7.

Murrell, K. D. and Graham, C. E., 1982, Solubilization studies on the epicuticular antigens of *Strongyloides ratti, Veterinary Parasitology*, **10**, 191–203.

Nikaido, H., 1989, Role of the outer membrane of Gram-negative bacteria in antimicrobial resistance, in Bryan, L. E. (Ed.) *Microbial Resistance to Drugs. Handbook of Experimental Pharmacology*, Vol. 91, Berlin: Springer.

Nikaido, H. and Vaara, M., 1985, Molecular basis of bacterial outer membrane permeability, *Microbiological Reviews*, **45**, 1–32.

Nishizuka, Y., 1984, Turnover of inositol phospholipids and signal transduction, *Science*, **225**, 1365–70.

Nishizuka, Y., 1986, Studies and perspectives of protein kinase C, *Science*, **233**, 305–12.

Noda, M., Yoon, K., Gideon, A. R. and Koppel, D. E., 1987, High lateral mobility of endogenous and transfected alkaline phosphatase: a phosphatidylinositol-anchored membrane protein, *Journal of Cell Biology*, **105**, 1671–7.

O'Brien, V., 1988, The role of the envelope of *Moniliformis moniliformis* in immune evasion, unpublished PhD Thesis, University of Glasgow.

Ortega-Pierres, G., Chayen, A., Clark, N. W. T. and Parkhouse, R. M. E., 1984, The occurrence of antibodies to hidden and exposed determinants of surface antigens of *Trichinella spiralis, Parasitology*, **88**, 359–69.

Parkhouse, R. M. E., Philipp, M. and Ogilvie, B. M., 1981, Characterization of the surface antigens of *Trichinella spiralis* infective larvae, *Parasite Immunology*, **3**, 339–52.

Paulson, C. W., Jacobson, R. H. and Cupp, E. W., 1988, Microfilarial surface carbohydrates as a function of developmental stage and ensheathment status in six filariids, *Journal of Parasitology*, **74**, 743–7.

Perry, R. N., 1989, Dormancy and hatching of nematode eggs, *Parasitology Today*, **5**, 377–83.

Peters, R., 1981, Translational diffusion in the plasma membrane of single cells as studied by fluorescence microphotolysis, *Cell Biology International Reports*, **5**, 733–60.

Peters, R. and Richter, H. P., 1981, Translational diffusion in the plasma membrane of sea-urchin eggs, *Developmental Biology*, **86**, 285–93.

Petronijevic, T. and Rogers, W. P., 1987, The physiology of infection with nematodes: the role of intracellular pH in the development of the early parasitic stage, *Comparative Biochemistry and Physiology*, **88A**, 207–12.

Philipp, M. and Rumjaneck, F. D., 1984, Antigenic and dynamic properties of helminth surface structures, *Molecular and Biochemical Parasitology*, **10**, 245–68.

Philipp, M., Parkhouse, R. M. E. and Ogilvie, B. M., 1980, Changing proteins on the surface of a parasitic nematode, *Nature*, **287**, 538–40.

Piessens, W. F. and Beldekas, M., 1979, Diethylcarbamazine enhances antibody-mediated cellular adherence to *Brugia malayi* microfilariae, *Nature (London)*, **282**, 845–7.

Podesta, R. B., 1982, Membrane biology of helminths, in Podesta, R. B. (Ed.) *Membrane Physiology of Invertebrates*, p. 121, New York: Dekker.

Pritchard, D. I., Crawford, C. I. and Behnke, J. M., 1985, Antigen stripping from the nematode epicuticle using the cationic detergent cetyltrimethylammonium bromide (CTAB), *Parasite Immunology*, **7**, 575–85.

Pritchard, D. I., McKean, P. G. and Rogan, M. T., 1988, Cuticle preparations from *Necator americanus* and their immunogenicity in the infected host, *Molecular and Biochemical Parasitology*, **28**, 275–84.

Proudfoot, L., Kusel, J. R., Smith, H. V., Harnett, W., Worms, M. J. and Kennedy, M. W., 1990, The surface lipid of parasitic nematodes: organisation, and modifications during transition to the mammalian host environment, *Acta Tropica*, **47**, 323–31.

Radda, G. K., 1975, Dynamic aspects of membrane structure, in Parsons, D. S. (Ed.) *Biological Membranes*, pp. 81–95, Oxford: Clarendon Press.

Russell, N. J., 1984, Mechanisms of thermal adaptation in bacteria: blueprints for survival, *Trends in Biochemical Sciences*, **9**, 108–12.

Rutherford, T. A., Webster, J. M. and Barlow, J. S., 1977, Physiology of nutrient uptake by the entomophilic nematode *Mermis nigrescens* (Mermithidae), *Canadian Journal of Zoology*, **55**, 1773–81.

Saffman, P. G. and Delbruck, M., 1975, Brownian motion in biological membranes, *Proceedings of the National Academy of Sciences (USA)*, **72**, 3111–13.

Samuelson, J. C. and Stein, L. D., 1989, *Schistosoma mansoni*: increasing saline concentration signals cecariae to transform to schistosomula, *Experimental Parasitology*, **69**, 23–9.

Sauma, S. Y. and Strand M., 1990, Identification and characterization of glycosylphosphatidylinositol-linked *Schistosoma mansoni* adult worm immunogens, *Molecular and Biochemical Parasitology*, **38**, 199–210.

Schlessinger, J., 1983, Mobilities of cell membrane proteins: how are they modified by the cytoskeleton? *Trends in Neurological Sciences*, **8**, 360–3.

Scott, A. L., Chamberlain, D., Moraga, D. A., Ibrahim, M. S., Redding, L. and Tamashiro, W. K., 1988, *Dirofilaria immitis*: biochemical and immunological characterization of the surface antigens from adult parasites, *Experimental Parasitology*, **67**, 301–7.

Selkirk, M. E., Denham, D. A., Partono, F., Sutanto, I. and Maizels, R. M., 1986, Molecular characterization of antigens of lymphatic filariae, *Parasitology*, **91**, S15–38.

Singer, S. J. and Nicolson, G. L., 1972, The fluid mosaic model of the structure of cell membranes, *Science*, **175**, 720–31.

Smith, H. V., Quinn, R., Kusel, J. R. and Girdwood, R. W. A., 1981, The effect of temperature and antimetabolites on antibody-binding to the outer surface of second-stage *Toxocara canis* larvae, *Molecular and Biochemical Parasitology*, **4**, 183–93.

Stewart, G. L., Despommier, D. D., Burnham, J. and Raines, K. M., 1987, *Trichinella spiralis*: behaviour, structure and biochemistry of larvae following exposure to components of the host enteric environment, *Experimental Parasitology*, **63**, 195–204.

Storey, D. M., Ogbogu, V. C. and Kershaw, W. E., 1989, The fine structure of the cuticle of third-stage larvae and adults of *Litomosoides carinii*, *Tropical Medicine and Parasitology*, **39**, 83.

Sturtevant, J. M., 1987, Biochemical applications of differential scanning calorimetry, *Annual Review of Physical Chemistry*, **38**, 463–88.

Subrahmanyam, D., 1967, Glyceryl ether phospholipids and plasmalogens of the filarial parasite *Litomosoides carinii*, *Canadian Journal of Biochemistry*, **45**, 1195–7.

Taylor, P. W., 1983, Bactericidal and bacteriolytic activity of serum against Gram-negative bacteria, *Microbiological Reviews*, **47**, 46–83.

Turco, S. J., 1988, The lipophosphoglycan of *Leishmania*, *Parasitology Today*, **4**, 255–7.

Verkleij, A. J. and Ververgaert, P. H. J. Th., 1978, Freeze–fracture morphology of biological membranes, *Biochimica et Biophysica Acta*, **515**, 303–27.

Vickerman, K., 1974, Antigenic variation in African trypanosomes, in *Parasites in the Immunized Host: Mechanisms of Survival*, Ciba Foundation Symposium No. 25, pp. 53–70, Amsterdam: Elsevier.

Wassef, M. K., Fioretti, T. B. and Dwyer, D. M., 1985, Lipid analyses of isolated surface membranes of *Leishmania donovani* promastigotes, *Lipids*, **20**, 108–15.

Wharton, D. A., Preston, C. M., Barrett, J. and Perry, R. N., 1988, Changes in cuticular permeability associated with recovery from anhydrobiosis in the plant parasitic nematode *Ditylenchus dipsaci*, *Parasitology*, **97**, 317–30.

Wolf, D. E., 1988, Probing the lateral organization and dynamics of membranes, in Loew, L. (Ed.) *Spectroscopic Membrane Probes*, CRC Critical Reviews, pp. 194–216, Boca Raton, Florida: CRC Press.

Wolf, D. E. and Voglmayr, J. K., 1984, Diffusion and regionalization in membranes of maturing ram spermatozoa, *Journal of Cell Biology*, **98**, 1678–84.

Wolf, D .E., Handyside, A. H. and Edidin, M., 1982, Effect of microvilli on lateral diffusion measurements made by the fluoresence photobleaching recovery technique, *Biophysical Journal*, **38**, 295–7.

Wolf, D. E., Lipscomb, A. C. and Maynard, V. M., 1988, Causes of non-diffusing lipid in the plasma membrane of mammalian spermatozoa, *Biochemistry*, **27**, 860–5.

Wright, K. A., 1987, The nematode's cuticle — its surface and the epidermis: function, homology, analogy, *Journal of Parasitology*, **73**, 1077–83.

Wright, K. A. and Hong, H., 1988, Characterization of the accessory layer of the cuticle of muscle larvae of *Trichinella spiralis*, *Journal of Parasitology*, **74**, 440–51.

Zaman, V., 1987, Scanning electron microscopy of *Brugia malayi*, in Evered, D. and Clark, S. (Eds) *Filariasis*, Ciba Foundation Symposium, No. 127, pp. 77–90, Chichester: Wiley.

Zuckerman, B. M., Kahane, I. and Himmelhoch, S., 1979, *Caenorhabditis briggsae* and *C. elegans*: partial characterization of cuticle surface carbohydrates, *Experimental Parasitology*, **47**, 419–24.

2. Structure and biosynthesis of cuticular proteins of lymphatic filarial parasites

M. E. Selkirk

INTRODUCTION

There has been a recent revival of interest in the cuticle of parasitic nematodes, as (i) it represents the main body of tissue which comes into direct contact with the immune system of the mammalian host, and (ii) in some cases it appears to be of physiological importance as a site of nutrient acquisition. The cuticle can be regarded as a tough, elastic extracellular matrix synthesized and secreted by an underlying hypodermis which, in the case of the *Filarioidea*, is syncytial (Bird, 1971). Although great diversity in morphology is exhibited by cuticles from different species or between stages of the same organism, the basic biochemical composition appears to be well conserved. Thus the major structural components of this complex matrix have been defined as collagens, cross-linked by disulphide bonds and localized primarily to the basal and inner cortical layers which frequently contain readily discernible fibrils (McBride and Harrington, 1967; Cox *et al.*, 1981a; Leushner *et al.*, 1979; Pritchard *et al.*, 1988; Selkirk *et al.*, 1989). The external cortex, containing the epicuticle, is composed primarily of covalently cross-linked, high insoluble protein termed 'cuticlin' (Fujimoto and Kanaya, 1973).

Although it has long been the subject of debate, the consensus view is now that the cuticle is truly extracellular (Wright, 1987). Nevertheless, it is a dynamic organ in which a number of enzyme activities (e.g. esterase, ATPase, acid and alkaline phosphatase) have been localized (Lee, 1962; Anya, 1966; Howells and Chen, 1981; Sayers *et al.*, 1984). Enzymatic activity in extracellular matrices is not unusual, as both collagen and elastin are cross-linked by mechanisms based on lysyl oxidase mediated aldehyde formation from lysine or hydroxylysine side-chains following secretion (Eyre *et al.*, 1984). Nevertheless, the cuticle is highly ordered, with regular arrangements of fibrous blocks and struts (Bird, 1971), and thus the assembly of the cuticle must be tightly regulated by the hypodermal cell layer.

We are still relatively ignorant of the effector mechanisms of immunity operative against parasitic nematodes. Views on this subject were heavily influenced by a series of experiments done in the late 1970s and early 1980s which promoted the concept of antibody dependent cellular cytotoxicity (ADCC). In this scheme, antibodies of various isotypes can direct the adherence of granulocytes and macrophages which can kill parasites *in vitro*. Complement appears to be generally ineffective in mediating lysis, but serves to amplify cell adherence via receptors for C3 (reviewed by Maizels *et al.*, 1982). The specificity of killing is regulated by antibodies to the surface of the parasite, i.e. the cuticle, hence the interest in the constituent molecules as target antigens. It is worth noting that these experiments were of necessity carried out *in vitro*, and there is still some doubt as to the *in vivo* significance of the cuticle as a target, and ADCC as a general mechanism, of immunity against nematodes.

The adult stage of the filarial nematodes *Brugia pahangi* and *Dirofilaria immitis* take up glucose, amino acids and adenosine via a transcuticular route (Chen and Howells, 1981; Howells and Chen, 1981). These results, when first introduced, were heretical to the long-established concept of the cuticle as an impermeable barrier to large molecules or low molecular weight polar solutes (Pappas and Read, 1975), although transcuticular uptake of glucose and amino acids had been previously described in larval stages of *Mermis nigrescens* (Rutherford and Webster, 1974) and *Trichinella spiralis* (Stoner and Hankes, 1958). In the Mermithoidea, the cuticle is rudimentary and the oesophagus atrophied. Adult *Brugia*, however, have a cuticle which is relatively thick (approximately 2.5 μm), but is endowed with a high degree of permeability as judged by the rapid penetration of fixatives (Vincent *et al.*, 1975) and Bolton–Hunter iodination reagent (Marshall and Howells, 1985).

The gut lumen of *B. pahangi* is occluded at the oesophagopharyngeal junction in third stage infective larvae (Collin, 1971), but is functional by 4 days post-infection (Howells and Chen, 1981), after which time uptake of glucose and other solutes may presumably occur via both an oral and a transcuticular route. Competition experiments indicate that at least two separate loci exist for transcuticular uptake of amino acids (Howells *et al.*, 1983b), and that the stereospecific uptake of D-glucose involves both diffusion and a saturable (carrier transport) component (Howells *et al.*, 1983a). As the specificity (i.e. inhibition profile) of this glucose transport site in *B. pahangi* is markedly divergent from that of mammalian tissues, it has been proposed that selective inhibition of glucose uptake in filarial nematodes could be explored as a potential target for chemotherapy (Howells *et al.*, 1983a). Although the carrier transport systems involved are likely to be situated in the hypodermal membrane, one can see that the structure and function of the components of the nematode cuticle are of fundamental interest in terms of parasite physiology, as well as applied interest for potential targets of immunity and chemotherapy.

GENERAL CUTICULAR ARCHITECTURE

A schematic representation of the cuticle of adult *Brugia malayi* is shown in Figure 2.1. This figure, and the measurements of the thickness of components which appear as structurally distinct layers under the electron microscope, is adapted from an ultrastructural study of the adult worm by Vincent *et al.* (1975). This figure is necessarily generalized in order to provide a source of reference. As pointed out by Vincent *et al.*, variations in the thickness of the cuticle, the component layers, and the form of the annulations in the external cortex occur in different parts of the nematode body.

The cuticle is synthesized by an underlying syncytial hypodermis, and separated from it by a unit membrane. Wright (1987) has argued that the term '*epidermis*' is preferable for this syncytium, based on vertebrate terminology, given that the cuticle is an extracellular matrix, although the term '*hypodermis*' accurately describes a layer inferior to the tissue which provides the main structural support (i.e. the cuticle is interpreted here to act as a functional dermis). Although the former term is thus strictly more correct, the latter term (hypodermis) is more commonly used.

There has been frequent debate as to whether the nematode epicuticle is bounded by a cell membrane. This was initially thought to be the case due to the resolution by electron microscopy of a trilaminate pattern somewhat reminiscent of a cell membrane, but more recent data do not support the contention that this structure is a membrane (see Chapter 1). Freeze–fracture of microfilariae of *Onchocerca volvulus* (Martinez-Palomo, 1978), several stages of *T. spiralis* (Lee *et al.*, 1984, 1986), *Caenorhabditis elegans* and *Anguina agrostis* (Wright, 1987) failed to reveal intramembranous particles, and, most frequently, fracture planes lay along the exterior face of the epicuticle and an outer surface coat or 'glycocalyx'.

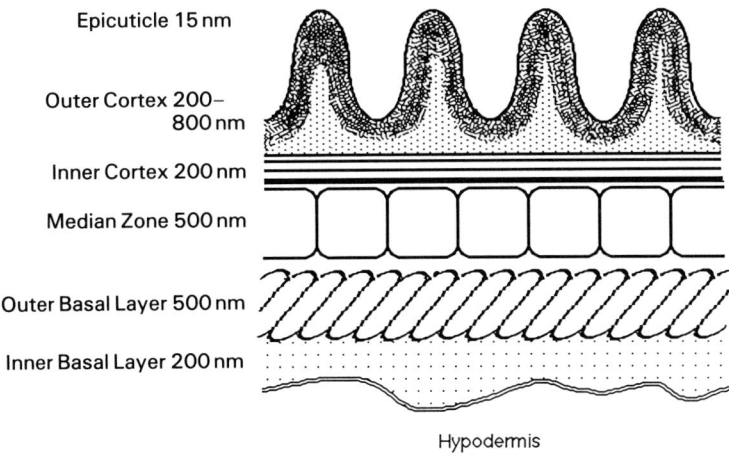

Epicuticle 15 nm

Outer Cortex 200– 800 nm

Inner Cortex 200 nm

Median Zone 500 nm

Outer Basal Layer 500 nm

Inner Basal Layer 200 nm

Hypodermis

Figure 2.1. Diagrammatic representation of the cuticle of adult *B. malayi* (after Vincent *et al.*, 1975).

Biophysical measurements of the fluidity of lipid in the epicuticle of nematodes, including that of adult *B. pahangi*, have been made by the application of fluorescence recovery after photobleaching (FRAP; Kennedy *et al.*, 1987; Proudfoot *et al.*, 1990). In the case of adult *B. pahangi*, only one fluorescently tagged lipid of those examined (5-*N*-octadecanoylaminofluoroscein) would insert into the epicuticle, and almost no repopulation of a defined area was observed following bleaching with laser light. This apparent immobility of lipid in the epicuticle is atypical of a cell membrane (see Chapter 1 for a detailed discussion). Thus one must conclude that the true limiting cell membrane occurs at the interface between the hypodermis and the cuticle, and that, as suggested by Wright (1987), the epicuticle represents a specialized form of envelope which allows the regulated extracellular organization of the cuticular components, but it is not a true cell membrane.

CHARACTERIZATION OF CUTICULAR COMPONENTS BY EXTRINSIC RADIO-IODINATION

Iodogen-mediated extrinsic iodination of adult *Brugia* has been used to define a limited set of proteins which, by autoradiographic evidence, appeared to be restricted to the cuticle (Sutanto *et al.*, 1985; Marshall and Howells, 1985; Selkirk *et al.*, 1986). Alternative methods of solubilization allows the separation of these proteins into two sets, and these results are summarized in Figure 2.2. Solubilization of labelled proteins via homogenization of worms in detergent reveals two major proteins; a glycoprotein with an estimated molecular weight of 29–30 kDa, and a non-glycosylated protein of 15 kDa (Figure 2.2b). The biochemical characteristics of this 29 kDa glycoprotein are currently under investigation by a number of groups, some of whom have sized the molecule at 30 kDa (Devaney, 1988; Maizels *et al.*, 1989; see also Chapter 3). If this extraction step is performed under harsh conditions (in boiling sodium dodecyl sulphate (SDS)), it is possible to prepare cuticles which are intact and free of cellular or membranous contamination and the 29 kDa molecule. Thus neither the 29 kDa nor the 15 kDa protein appear to perform any major structural role in the cuticle architecture.

Treatment of cuticles prepared in this manner with a reducing agent such as 2-mercaptoethanol (2ME) results in the dissolution of the basal and inner cortical layers, leaving the outer cortex and epicuticle intact (Figure 2.2d), and the solubilization of a series of proteins ranging in molecular weight from approximately 50 kDa to 160 kDa (Figure 2.2e). These proteins are all extensively hydrolysed by collagenase from *Clostridium histolyticum* (Figure 2.2f), in contrast to the 29 kDa and 15 kDa proteins (Figure 2.2c). Thus, as in other nematodes, the main structural components of the cuticle are collagenous proteins, cross-linked by disulphide bonds and localized in the basal and inner cortical layers.

The outer cortex and epicuticle (Figure 2.2d) is highly insoluble and appears to be composed of protein(s) cross-linked by non-reducible covalent bonds (Selkirk *et*

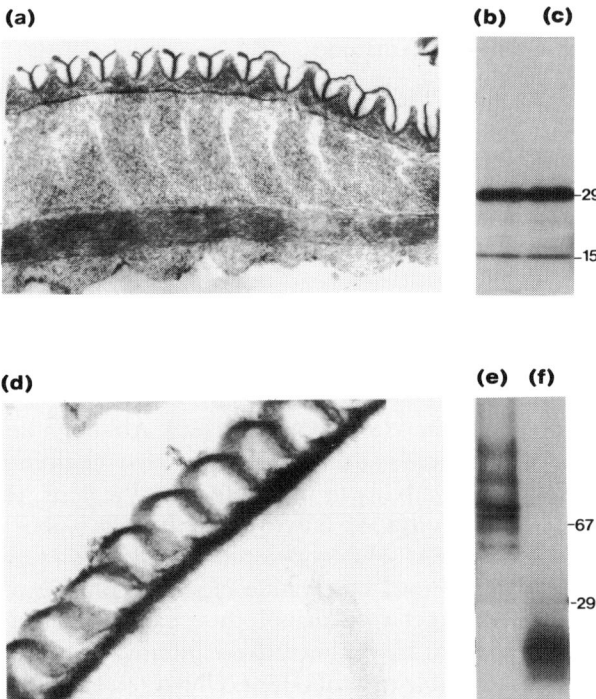

Figure 2.2. Cuticular proteins of adult *B. malayi* (from Selkirk *et al.*, 1989). Worms labelled with Iodogen, and solubilized by means of sonication in Tris buffered saline (TBS), pH 6.8 / 1 per cent SDS, show two major proteins on SDS polyacrylamide gel electrophoresis (PAGE) (b), which are insensitive to digestion with clostridial collagenase (c). The insoluble pellet contains intact cuticles (a). When treated with 2ME, the basal and inner cortical layers dissolve, yielding a set of higher molecular weight proteins (e), which are susceptible to digestion with collagenase (f). The outer cortex and epicuticle are insoluble and can be pelleted in long sheets (d).

al., 1989). Similar insoluble preparations of outer cortex and epicuticle have been isolated from *Ascaris lumbricoides* (McBride and Harrington, 1967; Fujimoto and Kanaya, 1973); *C. elegans* (Cox *et al.*, 1981a); *Acanthocheilonema viteae* (Betschart *et al.*, 1985; Betschart and Jenkins, 1987) and *Necator americanus* (Prichard *et al.*, 1988). Fujimoto and Kanaya (1973) termed this insoluble proteinaceous component 'cuticlin'. The protein/proteins which made up this structure were insoluble in all the solvents tested, including 8 M urea, 1 per cent SDS, 88 per cent formic acid and 0.1 per cent sodium hydroxide. The amino acid content, though broadly similar, differed from *Ascaris* cuticular collagens in having a relatively low content of glycine (15 per cent compared with 27 per cent), and the absence of hydroxyproline. In addition, cuticlin was insensitive to clostridial collagenase and showed an X-ray diffraction pattern distinct from that of the cuticular collagen. The 'cuticlin' from adult *B. malayi* is also relatively insensitive to clostridial collagenase, but can be digested in part by a number of proteolytic enzymes, notably elastase (unpublished data). The same result has been reported for *C. elegans* (Cox *et al.*,

1981a), perhaps suggesting a similar chemical composition for the epicuticle and external cortex in a range of nematodes.

The clear identification of the epicuticle/cuticlin as an insoluble, peripheral proteinaceous structure in the majority of nematodes examined means that the predominant 'surface proteins' of most nematode parasites remain to be defined, and the importance of this structure as a potential target of immunity is reflected by the fact that Betschart *et al.* (1987) have reported that antibodies raised to such an insoluble fraction from *Ascaris* bind the surface of intact parasites.

CUTICULAR COLLAGENS

In situ treatment of live, iodinated adult *B. malayi* with clostridial collagenase fails to release any cleavage products (Selkirk *et al.*, 1989). Although this result implies that the cuticular collagens are not accessible to the enzyme in intact worms, in a strict sense this defines accessibility to the triple-helical regions. However, tryptic treatment of live, labelled worms also fails to digest the collagenous proteins, whilst cleaving the 29 kDa detergent-soluble glycoprotein (Maizels *et al.*, 1989). Taken together with the ultrastructural observation of the dissolution of the basal and inner cortical layers with 2-mercaptoethanol, these data convincingly demonstrate that the collagenous proteins are confined to the internal matrix of the cuticle. As such, they must be inaccessible to antibody or cellular recognition, and unlikely to represent targets of a protective immune response.

Resolution of the 2ME soluble fraction of iodinated proteins from adult

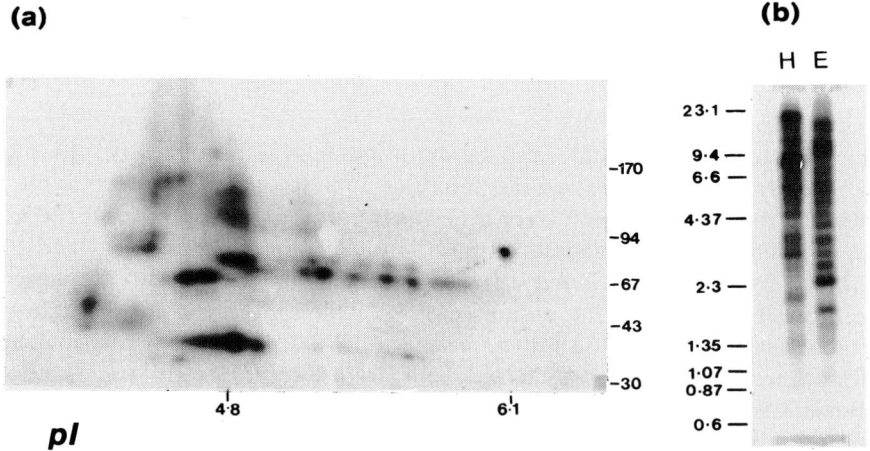

Figure 2.3. The filarial cuticle contains numerous collagenous proteins which are encoded by a multigene family (from Selkirk *et al.*, 1989). (a) Cuticular collagens of adult *B. malayi* labelled with Iodogen and resolved by two-dimensional PAGE (first dimension isoelectric focusing, second dimension SDS-PAGE). Molecular weights are indicated in kilodaltons. (b) Southern blot of *B. pahangi* DNA restricted with Hind111 (H) or EcoR1 (E), and hybridized with a cDNA clone coding for the α2 chain of chicken type 1 collagen. Size markers are indicated in kilobases.

B. malayi in two dimensions identifies at least 16 collagenase sensitive species (Figure 2.3a). An additional protein of approximately 35 kDa can be visualized in this figure, due to the use of Bolton–Hunter reagent, which penetrates throughout the soma (Marshall and Howells, 1985). This protein is structurally distinct from the cuticular collagens visualized by Iodogen labelling as defined by limited peptide mapping, and cross-reacts with antibodies raised to mammalian type IV (basement membrane) collagen (Selkirk *et al.*, 1989). It may thus represent a collagenous protein from the muscle layers or the intestinal basement membrane, as has been described in *Ascaris* (Fujimoto, 1968; Peczon *et al.*, 1975).

Comparison of the cuticular collagens of *B. malayi* by two-dimensional gel electrophoresis indicates that each stage in the mammalian host has a characteristic set of between 12 (L3), through 16 (adult) to 25 (L4) acidic structural proteins, although some would appear to be shared between stages (unpublished data). This result is again analogous to the situation seen in *C. elegans*, where each stage displays a characteristic but overlapping set of polypeptides (Cox *et al.*, 1981b; Politz and Edgar, 1984).

It appears that in *C. elegans* the mature cuticular collagens, which range in molecular weight between 60 and 210 kDa, are formed by covalent cross-linkage of numerous small precursors. This was deduced initially via collagenase digestion of *in vitro* translation products (Politz and Edgar, 1984), and subsequently by western blotting experiments with antisera directed to the 2-mercaptoethanol soluble components of adult cuticle. This antibody bound to small proteins of molecular weight 38–52 kDa in worms at the start of lethargus, and to a size class analogous to mature collagens (60–210 kDa) at the end of the moult (Politz *et al.*, 1986).

Different results have been obtained with another free-living nematode, *Panagrellus silusiae*. In this case, four collagenase sensitive products of *in vitro* translation corresponded broadly in molecular weight to those of the mature cuticular proteins (Leushner *et al.*, 1979). It is difficult to duplicate these experiments with *Brugia*, due to the lack of strict synchronicity of moulting in mammalian hosts. Nevertheless, procollagens identified via collagenase digestion of *in vitro* translation products correlate broadly in molecular weight with those seen by metabolic labelling of worms *in vitro* (Selkirk *et al.*, 1989). Thus the data are more compatible with a similar mode of collagen synthesis to that of *P. silusiae* than *C. elegans*.

A cDNA clone (pCg45) which codes for the α2 chain of type 1 collagen from chicken (Lehrach *et al.*, 1978) has been used to clone collagen genes from *C. elegans* (Kramer *et al.*, 1982). This heterologous DNA probe hybridizes at low stringency to multiple restriction fragments (17 or more) of genomic DNA from *B. pahangi* (Figure 2.3b). This result could be interpreted as indicating that *Brugia* either contains a large number of collagen genes, or a few genes with multiple introns. Given the multiplicity of collagenous proteins of distinct molecular weight and pI in *Brugia*, the former case would seem to be most likely. Multiple introns are common in vertebrate collagen genes, but sequence analysis of two collagen genes from *C. elegans* revealed only one and two introns, respectively (Kramer *et al.*, 1982). The total number of collagen genes from *C. elegans* has been estimated at between 50 and 200 (Cox *et al.*, 1984), and hybridization with specific probes indicates that their expression is developmentally regulated (Kramer *et al.*, 1985).

Nematode cuticular collagens show broad similarity to each other, but differ from vertebrate collagens in their relatively small molecular size, and cross-linkage by disulphide bridges, a phenomenon which is displayed only by type VI collagen in vertebrates (Hessle and Engvall, 1984). Vertebrate fibrillar collagens are predominantly cross-linked by a mechanism based on aldehyde formation from lysine or hydroxylysine side chains (Eyre *et al.*, 1984). Although in some cases the nature of the mature cross-linking residues is unknown, it is thought that hydroxylysines in the telopeptides align with sites in the helical regions of adjacent molecules, thus forming a lattice of molecules staggered by 67 nm which makes up the collagen fibril. Cuticular collagens from *C. elegans* and *Ascaris* have essentially no hydroxylysine, and workers have failed to demonstrate the presence of lysyl derived cross-links (Bailey, 1971; Fujimoto *et al.*, 1981; Ouzana *et al.*, 1984). Nevertheless, precise conservation of both lysine and cysteine residues in identical positions deduced from sequence analysis of the col-1 and col-2 genes of *C. elegans* (Kramer *et al.*, 1982) suggest that lysyl residues may play a role in cross-linking in addition to disulphide bridges.

Runnegar (1985) has suggested that the utilization of disulphide rather than lysyl derived cross-links in *Ascaris* cuticular collagens may be due to the dysaerobic environment of this parasite (molecular oxygen is not required for the formation of disulphide bonds). This would not be a factor for filarial parasites in the mammalian host, where the environment is relatively oxygen rich, and whose cuticular collagens are nevertheless also disulphide cross-linked, but may be important for stages in the insect host, and may also have been critical for ancestral nematodes.

A significant difference between the derived amino acid sequence of col-1 and col-2 from *C. elegans* (Kramer *et al.*, 1982) and vertebrate interstitial collagens is the occurrence of short regions (24–66 amino acids) of Gly–X–Y repeats (potential triple-helical regions), interrupted by stretches where glycine is not present in every third position, and cysteine residues provide the potential for intermolecular disulphide bridges. Vertebrate collagens contain long, uninterrupted regions of 300 or more Gly–X–Y repeats, resulting in a long, rod-like molecule. Thus it is possible that the interruption of the Gly–X–Y repeats in the nematode collagens may confer greater flexibility on the molecules, and that this represents an adaptation which confers the exoskeleton with considerable tensile strength, but allows longitudinal flexibility for locomotion.

TEMPORAL REGULATION OF PROTEIN SYNTHESIS AND TURNOVER

The synthesis of cuticular collagens in *B. malayi* has recently been examined by removal of parasites from infected hosts (jirds) at different time points in the life cycle and metabolic labelling *in vitro* with [³H]proline (Selkirk *et al.*, 1989). It was only possible to take individual time points representative of each stage, but the results parallel those obtained for *C. elegans* (Politz and Edgar, 1984). Thus there is minimal synthesis of collagens by mature adult male worms and the fourth larval stage (L4) in the intermoult period, but substantial synthesis at the L4–adult moult, and an extremely high level of collagen production by fertile adult females, presumably destined for microfilarial cuticles and indicative of the fecundity of these organisms. A simplified representation of these results is shown in Figure 2.4. It is assumed that the same pattern of temporal regulation of collagen synthesis occurs in the third larval stage (L3), although we have no direct data to confirm this. Given the fact that each stage in the parasite life cycle has a distinct set of cuticular collagens, one can see that collagen synthesis is regulated both temporally, with respect to moulting, and developmentally, with respect to the particular stage in the life cycle.

Temporal regulation of collagen synthesis was to be expected, as a new cuticle is laid down beneath the existing one in a brief period of intense synthetic activity prior to the actual physical act of sloughing off the old coat (ecdysis). Howells and Blainey (1983) described the growth of the L4 of *B. pahangi*, which was accompanied by an increase in the thickness of the median and basal layers of the cuticle. Thus it is clear that there is some intermoult synthesis of structural cuticular matrix proteins, but this occurs at a low level relative to that observed during moulting. Multivesicular bodies have been observed in the hypodermis, and their number is greatly increased at the time of moult, leading to the assumption that they perform a secretory function, and contain precursors of matrix proteins. Lee (1970) identified small vesicles fusing with the hypodermal membrane in

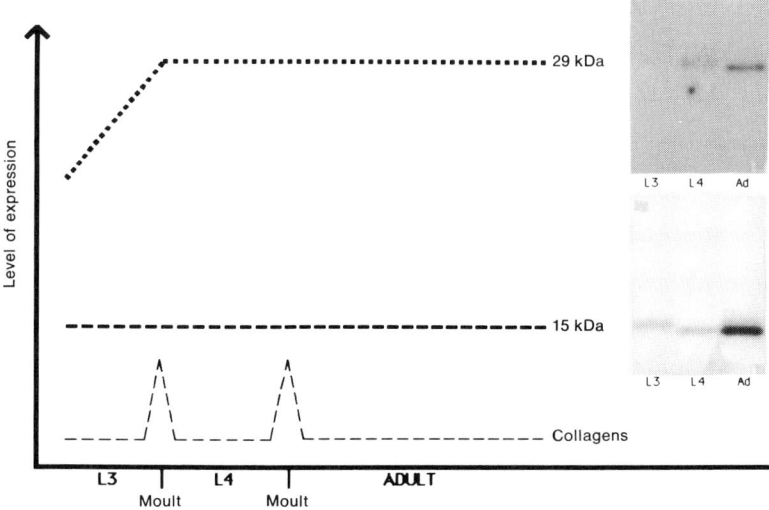

Figure 2.4.   Diagrammatic representation of the time course of synthesis of surface-labelled proteins of *B. malayi* in the mammalian host. Panels indicate immunoprecipitation analyses of surface labelled preparations of each stage with polyclonal sera directed to the 29 kDa and the 15 kDa proteins.

Nippostrongylus brasiliensis, and correlated an increase in numbers of these vesicles with cuticular growth.

An elegant morphometric study of development of *B. pahangi* in the mammalian host (Howells and Blainey, 1983) estimated that there was no increase in cuticular surface area of adult worms following the final moult, but that the epicuticular invaginations expanded like a concertina. This is consistent with the observation that the epicuticle is composed of a highly cross-linked proteinaceous structure (Figure 2.2d) rather than a fluid cell membrane. The authors pointed out that, in contrast, there is a 325 per cent increase in the area of the surface epithelium of the trematode *Schistosoma mansoni* (which is bounded by a true plasma membrane) during the first 3 days of development in the vertebrate host (Samuelson *et al.*, 1980). Thus the major increases in surface area of *Brugia* larvae occur at ecdysis (an eight-fold increase in surface area is achieved at the third moult) when the new epicuticle and associated cortical layers are laid down prior to the matrix which forms the median and basal layers of the cuticle (Howells and Blainey, 1983).

The non-collagenous proteins (Figure 2.2b) amenable to extrinsic iodination on adult *Brugia* are currently the subject of great interest, as they may represent reasonable targets for attempts at immunoprophylaxis (Maizels and Selkirk, 1988). The 29 kDa protein, which we have termed gp29 for convenience, is a glycoprotein with at least 2 *N*-linked oligosaccharide side-chains of 1.5–2 kDa molecular weight. It is susceptible to cleavage with trypsin in intact parasites (Figure 2.5) (see Devaney, 1988; Maizels *et al.*, 1989), and this observation, combined with antibody binding and absorption studies (Devaney, 1987) suggests that the protein

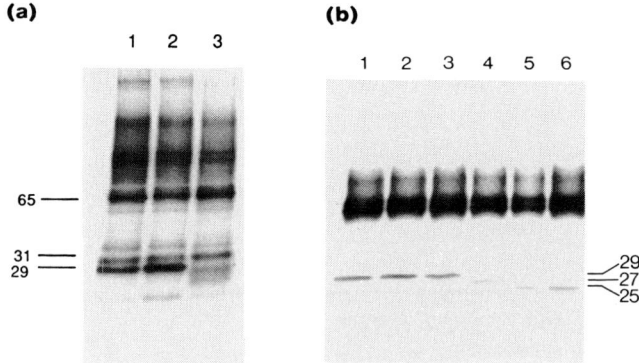

Figure 2.5. The 29 kDa protein is glycosylated and susceptible to tryptic cleavage *in situ* (from Maizels *et al.*, 1989). (a) Adult *B. malayi* were labelled with Bolton–Hunter reagent and incubated for 60 min at 37°C with increasing concentrations of enzyme. At the end of the incubation, worms were washed, incubated briefly in 2ME and extracted in SDS-PAGE sample buffer for analysis: (1) incubation in PBS; (2) incubation in 10 μg ml⁻¹ trypsin; (3) incubation in 100 μg ml⁻¹ trypsin. (b) Cleavage of oligosaccharide side-chains from Iodogen-labelled adult *B. malayi* surface proteins by peptide: *N*-glycosidase F (*N*-glycanase): (1) no enzyme; (2) 1 mU ml⁻¹; (3) 10 mU ml⁻¹; (4) 100 mU ml⁻¹; (5) 1 U ml⁻¹.

is peripherally located. In fact, this is the only defined protein for which this data is available (see also Chapter 3).

A summary of the biochemical characteristics and dynamics of expression of gp29 in *B. pahangi* is given in Chapter 3. The protein has been estimated to resolve with a molecular weight of 30 kDa. Some differences have been observed in *B. malayi*, however, and are worth mentioning. The protein was not detected in L3 removed from the peritoneum of jirds 2 days following infection, but was detected in both L4 and adult worms (Figure 2.4) (Selkirk *et al.*, 1990). A similar result has been reported by Philipp *et al.* (1988) in *B. malayi*, and this may thus represent a difference between the two species in the timing of expression of this protein. In addition, a slight difference in molecular weight was noted between stages; it thus resolved at 29 kDa in adults and 30 kDa in L4 (Philipp *et al.*, 1988; Selkirk *et al.*, 1990).

The 15 kDa protein was detected via surface labelling in all stages in the jird (Figure 2.4). Again, a slight difference in molecular weight was noted between stages. Both the 15 kDa protein and gp29 labelled in different life cycle stages focus at the same pH in two-dimensional electrophoresis, suggesting molecular identity (unpublished data). Both proteins, once expressed, are synthesized continuously, as judged by metabolic labelling of worms removed to *in vitro* culture. Thus, in contrast to the cuticular collagens, there is no apparent temporal regulation of synthesis with respect to moulting.

It has been observed that gp29 is released into culture medium following labelling (Devaney, 1988; Kwan-Lim *et al.*, 1989; Selkirk *et al.*, 1990), the dynamics of which is illustrated in Figure 2.6. The protein can be detected in

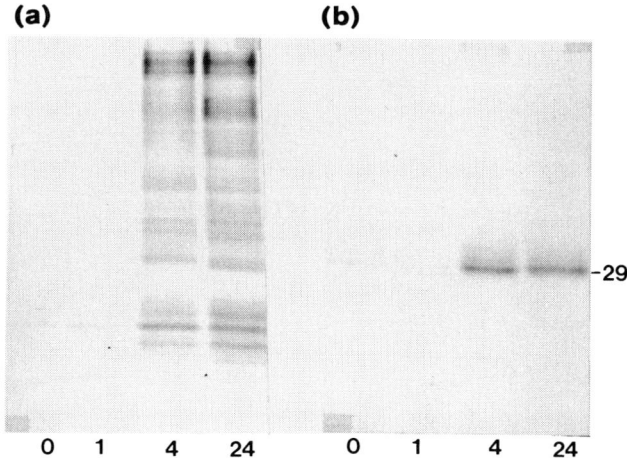

Figure 2.6. Time course of release of proteins into culture medium. Adult *B. malayi* were labelled for 1 h *in vitro* with [35S]methionine, washed, and equivalent amounts of culture media collected and concentrated after increasing periods of time. Figures indicate the duration (in hours) of the cold chase. Molecular weights are given in kilodaltons. (a) Total profile of secreted proteins. (b) Immunoprecipitation with an antiserum to the 29 kDa surface glycoprotein.

medium 4 h after the onset of metabolic labelling, indicative of a relatively high rate of turnover. This result runs counter to that obtained by Marshall and Howells (1985) in an *in vivo* study, so it is possible that the turnover *in vitro* is artefactual. Another possibility is that under natural conditions a proportion of gp29 is turned over, but that the majority remains in the cuticle or hypodermis.

Polyclonal antisera raised to both gp29 and the 15 kDa protein have been used to localize these molecules *in situ* by immunoelectron microscopy. Figure 2.7 shows that gp29 is distributed throughout the cuticle of adult *B. malayi*, and that the hypodermal cell layer is also heavily stained. This is to be expected, as the hypodermis is the presumptive site of synthesis of all cuticular components. In some sections, the gold particles appear to be concentrated on the infoldings of the hypodermal membrane (Figure 2.7b). These infoldings have been described previously (Rogers *et al.*, 1974; Vincent *et al.*, 1975; Howells and Blainey, 1983), and may function to increase the absorptive area of the animal, given that transcuticular uptake of nutrients is an established feature of the filariae. A similar concentration of gold particles on the hypodermal membrane has been observed with an antibody to cuticular collagens (Selkirk and Jenkins, unpublished observations), and one might expect to visualize vesicles which deliver extracellular proteins to the cuticle fusing with this membrane.

These results suggest that the turnover of gp29 observed *in vitro* most probably occurs through the cuticle. Although the protein was not observed to be localized at the epicuticular surface, this could be due to the mode of fixation and the surface

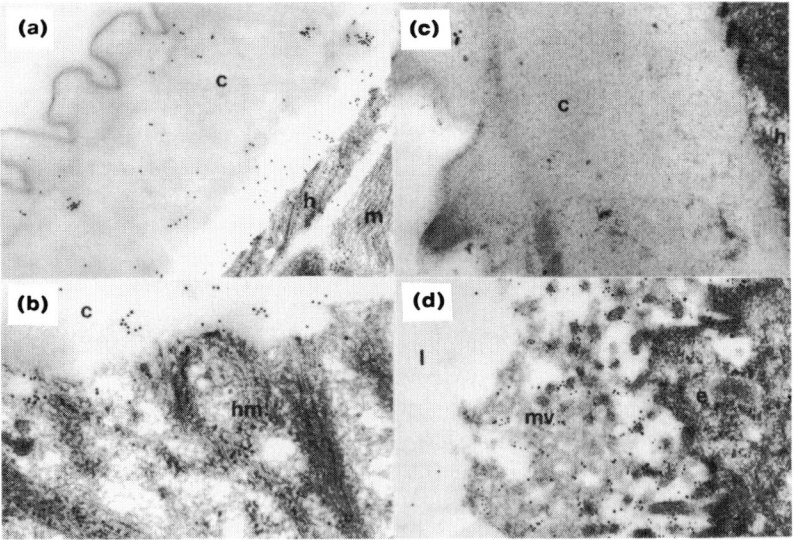

Figure 2.7. Immunogold localization of 29 kDa and the 15 kDa surface-labelled proteins in adult *B. malayi*. Sections from parasites embedded in London Resin (LR) were reacted either with a polyclonal rabbit anti-29 kDa antiserum (a and b), or a polyclonal rabbit anti-15 kDa antiserum (c and d). c, Cuticle; h, hypodermis; m, muscle; hm, hypodermal membrane; e, gut epithelial cell; mv, microvilli; l, gut lumen.

properties of the epicuticle, given that other workers have failed to demonstrate surface staining of *T. spiralis* larvae by the immunogold technique with sera from infected animals which would be expected to be reactive with the epicuticle (Takahashi *et al.*, 1988). Devaney has obtained similar results with an independent-ly produced antiserum to the *B. pahangi* homologue of gp29, which she has sized at 30 kDa (see Chapter 3).

Turnover of proteins from the cuticle has been described in other nematodes, most notably in larvae of *Toxocara canis* (Smith *et al.*, 1981; Maizels *et al.*, 1984). In these parasites, the turnover is so rapid that fluorescent staining with antibody is only possible by the inclusion of antimetabolites, or carrying out incubations at 4 °C (Smith *et al.*, 1981). Thus it is difficult to conceive of immune responses directed to these proteins being effective in damaging the parasite surface. *Toxocara* larvae are, for instance, resistant to ADCC mediated by eosinophils *in vitro*; although the cells degranulate, they are sloughed off together with extracuticular material (Fattah *et al.*, 1986; Bradley *et al.*, 1987). A basic question which is still unclear is whether immune-mediated damage to the epicuticle is sufficient to kill parasitic nematodes, or whether the peroxidases, pore-forming proteins and other toxic products from granulocytes must traverse the cuticle to act on the hypodermal membrane in order to affect worm viability. The latter scenario represents a tall order for immune effector cells, and the cuticle could thus be envisaged as an extracellular 'buffer' which, in addition to its structural properties, serves to protect the limiting membrane of the parasite.

A surprising result was obtained with the antiserum generated to the 15 kDa protein (Figure 2.7). The antiserum failed to bind to the cuticle, but reacted strongly with the microvilli of epithelial cells of the gut in adult males and females (Figures 2.7c and 2.7d). Subsequent work has shown that this particular antiserum reacts with a number of proteins of 15 kDa molecular mass (Blaxter and Selkirk, unpublished data), and thus this result is far from definitive. The failure to localize a product in the cuticle was unexpected however, given the fact that this protein was originally identified by extrinsic iodination with a relatively non-invasive technique (Sutanto *et al.*, 1985). A possible analogous situation has been described in *T. spiralis*, in which surface-labelled proteins of 48 kDa and 50–55 kDa have been localized to the stichocytes (secretory cells), the gut, and the surface of the cuticle by immunoperoxidase staining (Silberstein and Despommier, 1984). It is possible that the proteins are produced in the stichocytes and/or the gut, and passively absorbed onto the cuticle following secretion.

The concept of secreted antigens as targets of immunity is an old one, prompted originally by the observation of precipitin reactions around body openings and in the intestine of nematodes incubated in immune serum or removed from the tissues of immune hosts (Taliaferro and Sarles, 1939). More recently, high levels of protection have been achieved with secreted products, most notably with purified preparations of the 48 kDa and 50–55 kDa proteins of *T. spiralis* (Silberstein and Despommier, 1984).

ANTIGENICITY AND CROSS-REACTIVITY OF CUTICULAR PROTEINS

Although the cuticular collagens are inaccessible to the immune system in intact worms, immunoprecipitation analyses show that people with lymphatic filarial infection possess serum antibodies to the full complement of proteins (Figure 2.7a). This is to be expected, as the body will be exposed to these proteins following moulting, attrition of parasites via immune responses to other targets and natural death. Individuals infected with other species of filarial worms (*Wuchereria*, *Onchocerca*, *Mansonella* and *Loa*) produce antibodies which cross-react with cuticular collagens of *B. malayi*, whereas people with non-filarial nematodiases (hookworm and *Toxocara* infection) display little or no cross-reacting antibody (Figure 2.7a). Antibody binding is weak in comparison with the non-cuticular surface proteins (Figure 2.7b), and this is consistent with the general weak antigenicity of collagens (Timpl, 1982).

Antibodies raised to human type IV (basement membrane) collagen react specifically with at least one filarial collagen of 35 kDa relative mass, whereas antibodies raised to other collagen types isolated from bovine skin, tendon and cartilage were uniformly negative (Selkirk *et al.*, 1989). A few individuals with lymphatic filariasis can be demonstrated to possess elevated titres of autoreactive antibodies to type IV collagen, but there is no significant correlation with disease. Indeed, antibodies which cross-react with vertebrate β-tubulin (Helm *et al.*, 1989) and myosin heavy chains (unpublished observations) can be readily detected in lymphatic filariasis patients, but there is generally no tissue damage evident that could be attributed to autoantibodies to such broadly distributed structural and contractile elements.

The pattern of apparent conservation of at least some aspects of cuticular protein structure within the filariae, indicated by immunological cross-reactivity between collagens, has been reported previously for the non-collagenous surface proteins (Maizels *et al.*, 1983, 1985), and is represented in Figure 2.7b. Again, no cross-reactivity is observed with sera taken from people with other nematode infections. In this respect, cuticular proteins display a greater degree of immunological specificity than somatic extracts, yet this result serves to illustrate why it has proven difficult to discriminate serologically between *B. malayi* and *Wuchereria bancrofti*, or *Onchocerca volvulus* and the less pathogenic human- or animal-infective filarids. As both onchocercal and lymphatic filarial infection often exist in individuals without overt clinical symptoms, accurate diagnosis has been problematic.

Gp29 HOMOLOGUES ARE PRESENT ON ALL SPECIES OF LYMPHATIC FILARIAE

Gp29 is present on all species of lymphatic filariae (Maizels *et al.*, 1983, 1989; Morgan *et al.*, 1986), is highly immunogenic and elicits an antibody response relatively early in infection (Maizels *et al.*, 1986). Adult *Loa loa* possess a dominant

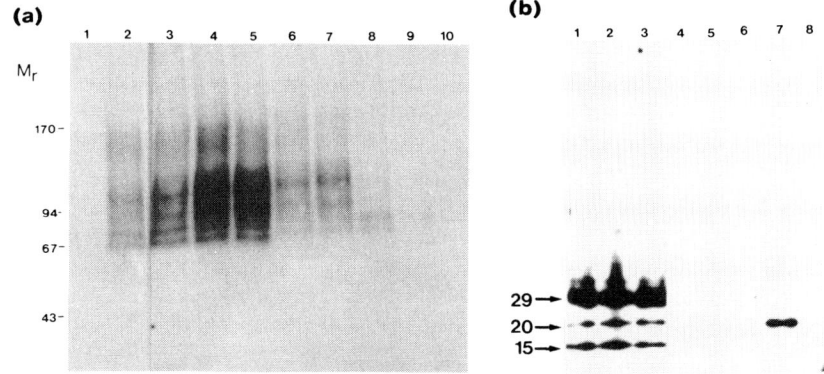

Figure 2.8. Cross-reactivity of *B. malayi* surface proteins (from Maizels *et al.*, 1985, Selkirk *et al.*, 1989). (a) Immunoprecipitation of cuticular collagens with human infection sera: (1) *B. malayi*; (2) *W. bancrofti*; (3) *M. perstans*; (4) *L. ioa*; (5) *O. volvulus*; (6) *S. stercoralis*; (7) *T. spiralis*; (8) *N. americanus*. (b) Immunoprecipitation of non-collagenous surface proteins with human infection sera: (1) *B. malayi*; (2) *B. timori*; (3) *W. bancrofti*; (4) *S. stercoralis*; (5) *O. volvulus*; (6) *T. spiralis*.

surface 29–31 kDa glycoprotein which cross-reacts with sera from *Brugia* infected patients (Egwang *et al.*, 1988), and the predominant surface-labelled protein of adult *Onchocerca* is somewhat smaller (20–22 kDa), but also possesses two N-linked oligosaccharide chains (Maizels *et al.*, 1987; Bradley *et al.*, 1989). Interestingly, a number of human anti-*Onchocerca* antibodies react with gp29 from *B. malayi*, but preferentially react with a 20 kDa breakdown product of it suggesting that breakdown exposes a shared epitope normally presented by native *Onchocerca* antigen (Figure 2.8a,b) (Maizels *et al.*, 1985).

It is thus possible that all filariae express homologues of gp29. The 15 kDa protein also bears epitopes which are cross-reactive between filariae (Maizels *et al.*, 1985), and possible homologues of similar molecular weight have been recorded in numerous species (Betschart *et al.*, 1985; Morgan *et al.*, 1986; Tamashiro *et al.*, 1986; Taylor *et al.*, 1986). In the next few years, it will be interesting to deduce the structure and biological function of these molecules, in addition to their potential as immunoprophylactic agents.

ACKNOWLEDGEMENTS

This work was supported by the Medical Research Council, the Wellcome Trust, the Filariasis component of the UNDP/World Bank/WHO Special Programme for Research and Training in Tropical Diseases and the Commission of the European Communities. I would like to thank Bill Gregory and Rosalind Jenkins for their permission to include unpublished data, and Rick Maizels for providing the schematic illustration of the cuticle.

REFERENCES

Anya, A. O., 1966, The structure and chemical composition of the nematode cuticle. Observations on some oxyuroids and *Ascaris*, *Parasitology*, **56**, 179–98.

Bradley, J. E., Grieve, R. B., Bowman, D. D. and Glickman, L. T., 1987, Immune-mediated adherence of eosinophils to *Toxocara canis* infective larvae: the role of excretory-secretory antigens, *Parasite Immunology*, **9**, 133–43.

Bailey, A. J., 1971, Comparative studies on the nature of the cross-links stabilising the collagen fibres of invertebrates, cyclostomes and elasmobranchs, *FEBS Letters*, **18**, 154–8.

Betschart, B. and Jenkins, J. M., 1987, Distribution of iodinated proteins in *Dipetalonema viteae* after surface labelling, *Molecular and Biochemical Parasitology*, **22**, 1–8.

Betschart, B., Rudin, W. and Weiss, N., 1985, The isolation and immunogenicity of the cuticle of *Dipetalonema viteae* (Filarioidea), *Zeitschrift für Parasitenkunde*, **71**, 87–95.

Betschart, B., Glaser, M., Keifer, R., Rudin, W. and Weiss, N., 1987, Immunochemical analysis of the epicuticle of parasitic nematodes, *Journal of Cellular Biochemistry*, **11A**(Suppl.), 164.

Bird, A. F., 1971, *The Structure of Nematodes*, New York: Academic Press.

Bradley, J. E., Gregory, W. F., Bianco, A. E. and Maizels, R. M., 1989, Biochemical and immunochemical characterisation of a 20-kilodalton complex of surface-associated antigens from adult *Onchocerca gutturosa* filarial nematodes, *Molecular and Biochemical Parasitology*, **34**, 197–208.

Chen, S. N. and Howells, R. E., 1981, The uptake of monosaccharides, disaccharides and nucleic acid precursors by adult *Dirofilaria immitis*, *Annals of Tropical Medicine and Parasitology*, **75**, 329–34.

Collin, W. K., 1971, Ultrastructural morphology of the oesophageal region of the infective larva of *Brugia pahangi* (Nematoda: Filarioidea), *Journal of Parasitology*, **57**, 449–68.

Cox, G. N., Kusch, M. and Edgar, R. S., 1981a, Cuticle of *Caenorhabditis elegans*: its isolation and partial characterization, *Journal of Cell Biology*, **90**, 7–17.

Cox, G. N., Staprans, S. and Edgar, R. S., 1981b, The cuticle of *Caenorhabditis elegans*. II. Stage specific changes in ultrastructure and protein composition during postembryonic development, *Developmental Biology*, **86**, 456–70.

Cox, G. N., Kramer, J. M. and Hirsh, D., 1984, Number and organisation of collagen genes in *Caenorhabditis elegans*, *Molecular and Cellular Biology*, **4**, 2389–95.

Devaney, E., 1987, Preliminary studies on the characterisation of the M_r 30 000 surface antigen of *Brugia pahangi*, *Parasite Immunology*, **9**, 401–5.

Devaney, E., 1988, The biochemical and immunochemical characterisation of the 30 kilodalton surface antigen of *Brugia pahangi*, *Molecular and Biochemical Parasitology*, **27**, 83–92.

Egwang, T. G., Akue, J.-P., Dupont, A. and Pinder, M., 1988, The identification and partial characterisation of an immunodominant 29–31 kilodalton surface antigen expressed by adult worms of the human filaria *Loa loa*, *Molecular and Biochemical Parasitology*, **31**, 263–72.

Eyre, D. R., Paz, M. A. and Gallop, P. M., 1984, Cross-linking in collagen and elastin, *Annual Reviews of Biochemistry*, **53**, 717–48.

Fattah, D. I., Maizels, R. M., McLaren, D. J. and Spry, C. J. F., 1986, *Toxocara canis*: interaction of human eosinophils with the infective larvae, *Experimental Parasitology*, **61**, 421–31.

Fujimoto, D., 1968, Isolation of collagens of high hydroxyproline, hydroxylysine and carbohydrate content from muscle layer of *Ascaris lumbricoides* and pig kidney, *Biochimica et Biophysica Acta*, **168**, 537–43.

Fujimoto, D. and Kanaya, S., 1973, Cuticlin: a noncollagen structural protein from *Ascaris* cuticle, *Archives of Biochemistry and Biophysics*, **157**, 1–6.

Fujimoto, D., Horiuchi, K. and Hirama, M., 1981, Isotrityrosine, a new crosslinking amino acid isolated from *Ascaris* cuticle collagen, *Biochemical and Biophysical Research Communications*, **99**, 637–43.

Helm, R., Selkirk, M. E., Bradley, J. F., Burns, R. G., Hamilton, A. J., Croft, S. and Maizels, R. M., 1989, Localisation and immunogenicity of tubulin in the filarial nematodes *Brugia malayi* and *Brugia pahangi*, *Parasite Immunology*, **11**, 479–502.

Hessle, H. and Engvall, E., 1984, Type IV collagen. Studies on its localisation, structure and biosynthetic form with monoclonal antibodies, *Journal of Biological Chemistry*, **259**, 3955–61.

Howells, R. E. and Blainey, L. J., 1983, The moulting process and the phenomenon of intermoult growth in the filarial nematode *Brugia pahangi*, *Parasitology*, **87**, 493–505.

Howells, R. E. and Chen, S. N., 1981, *Brugia pahangi*: feeding and nutrient uptake *in vitro* and *in vivo*, *Experimental Parasitology*, **51**, 42–58.

Howells, R. E., Bray, P. G. and Allen, D., 1983a, An analysis of glucose uptake by *Brugia pahangi*, *Transactions of the Royal Society for Tropical Medicine and Hygiene*, **77**, 273.

Howells, R. E., Mendis, A. M. and Bray, P. G., 1983b, The mechanisms of amino acid uptake by *Brugia pahangi in vitro*, *Zeitschrift für Parasitenkunde*, **69**, 247–53.

Kennedy, M. W., Foley, M., Kusel, J. R. and Garland, P. B., 1987, Biophysical properties of the surface lipid of parasitic nematodes, *Molecular and Biochemical Parasitology*, **22**, 233–40.

Kramer, J. M., Cox, G. N. and Hirsh, D., 1982, Comparisons of the complete sequences of two collagen genes from *Caenorhabditis elegans*, *Cell*, **30**, 599–606.

Kramer, J. M., Cox, G. N. and Hirsh, D., 1985, Expression of the *Caenorhabditis elegans* collagen genes col-1 and col-2 is developmentally regulated, *Journal of Biological Chemistry*, **260**, 1945–51.

Kwan-Lim, G.-E., Gregory, W. F., Selkirk, M. E., Partono, F. and Maizels, R. M., 1989, Secreted antigens of filarial nematodes: a survey and characterisation of *in vitro* excreted/secreted products of adult *Brugia malayi*, *Parasite Immunology*, **11**, 629–54.

Lee, D. L., 1962, The distribution of esterase enzymes in *Ascaris lumbricoides*, *Parasitology*, **52**, 241–60.

Lee, D. L., 1970, Moulting in nematodes: the formation of adult cuticle during final moult in *Nippostrongylus brasiliensis*, *Tissue and Cell*, **2**, 139–53.

Lee, D. L., Wright, K. A. and Shivers, R. R., 1984, A freeze-fracture study of the infective stage larva of the nematode *Trichinella* (Nematoda), *Tissue and Cell*, **16**, 819–28.

Lee, D. L., Wright, K. A. and Shivers, R. R., 1986, A freeze–fracture study of the body wall of adult, *in utero* larvae and infective stage larvae of *Trichinella* (Nematoda), *Tissue and Cell*, **18**, 219–30.

Lehrach, H., Frischauf, A. M., Hanahan, D., Wozney, J., Fuller, F., Crkvenjakov, R., Boedtker, H. and Doty, P., 1978, Construction and characterisation of a 2.5 kilobase procollagen clone, *Proceedings of the National Academy of Sciences (USA)*, **75**, 5417–21.

Leushner, J. R. A., Semple, N. E. and Pasternak, J. P., 1979, Isolation and characterization of the cuticle from the free-living nematode *Panagrellus silusiae*, *Biochimica et Biophysica Acta*, **580**, 166–74.

Maizels, R. M. and Selkirk, M. E., 1988, Antigens of filarial parasites, *ISI Atlas of Science: Immunology*, **1**, 143–8.

Maizels, R. M., Philipp, M. and Ogilvie, B. M., 1982, Molecules on the surface of parasitic nematodes as probes of the immune response in infection, *Immunological Reviews*, **61**, 109–36.

Maizels, R. M., Partono, F., Oemijati, S., Denham, D. A. and Ogilvie, B. M., 1983, Cross-reactive surface antigens on three stages of *Brugia malayi*, *B. pahangi* and *B. timori*, *Parasitology*, **87**, 249–63.

Maizels, R. M., de Savigny, D. and Ogilvie, B. M., 1984, Characterisation of surface and excretory–secretory antigens of *Toxocara canis* infective larvae, *Parasite Immunology*, **6**, 23–37.

Maizels, R. M., Sutanto, I., Gomez-Priego, A., Lillywhite, J. and Denham, D. A., 1985, Specificity of surface molecules of adult *Brugia* parasites: cross-reactivity with antibody from *Wuchereria, Onchocerca* and other human filarial infections, *Tropical Medicine and Parasitology*, **36**, 233–7.

Maizels, R. M., Selkirk, M. E., Sutanto, I. and Partono, F., 1987, Antibody responses to human lymphatic filarial parasites, in *Filariasis: Ciba Foundation Symposium No. 127*, pp. 189–202, Chichester: John Wiley.

Maizels, R. M., Bianco, A. E., Burke, J., Flint, J. E., Gregory, W. F., Kennedy, M. W., Lim, G. E., Robertson, B. D. and Selkirk, M. E., 1987, Glycoconjugate antigens from parasitic nematodes, *UCLA Symposia on Molecular and Cellular Biology*, **59**, 267–79.

Maizels, R. M., Gregory, W. F., Kwan-Lim, G.-E. and Selkirk, M. E., 1989, Filarial surface antigens: the major 29 kilodalton glycoprotein and a novel 17–200 kilodalton complex from adult *Brugia malayi* parasites, *Molecular and Biochemical Parasitology*, **32**, 213–28.

Marshall, E. and Howells, R. E., 1985, An evaluation of different methods for labelling the surface of the filarial nematode *Brugia pahangi* with ^{125}iodine, *Molecular and Biochemical Parasitology*, **15**, 295–304.

Martinez-Palomo, A., 1978, Ultrastructural characterization of the cuticle of *Onchocerca volvulus* microfilaria, *Journal of Parasitology*, **64**, 127–36.

McBride, O. W. and Harrington, W. P., 1967, *Ascaris* cuticle collagen: on the disulfide cross-linkages and the molecular properties of the subunits, *Biochemistry*, **6**, 1484–98.

Morgan, T. M., Sutanto, I., Purnomo, Sukartono, Partono, F. and Maizels, R. M., 1986, Antigenic characterisation of adult *Wuchereria bancrofti* filarial nematodes, *Parasitology*, **93**, 559–69.

Ouazana, R., Herbage, D. and Godet, J., 1984, Some biochemical aspects of the cuticle collagen of the nematode *Caenorhabditis elegans*, *Comparative Biochemistry and Physiology*, **77B**, 51–6.

Pappas, P. W. and Read, C. P., 1975, Membrane transport in helminth parasites: a review, *Experimental Parasitology*, **37**, 469–530.

Peczon, B. D., Venable, J. H., Beams, C. G., Jr, and Hudson, B. G., 1975, Intestinal basement membrane of *Ascaris suum*. Preparation, morphology and composition, *Biochemistry*, **14**, 4069–75.

Philipp, M., Davis, T. B., Storey, N. and Carlow, C. K. S., 1988, Immunity in filariasis: perspectives for vaccine development, *Annual Reviews in Microbiology*, **42**, 685–716.

Politz, J. C. and Edgar, R. S., 1984, Overlapping stage-specific sets of numerous small collagenous polypeptides are translated *in vitro* from *Caenorhabditis elegans* RNA, *Cell*, **37**, 853–60.

Politz, S. M., Politz, J. C. and Edgar, R. S., 1986, Small collagenous proteins present during the moult in *Caenorhabditis elegans*, *Journal of Nematology*, **18**, 303–10.

Pritchard, D. I., McKean, P. G. and Rogan, M. T., 1988, Cuticle preparations from *Necator americanus* and their immunogenicity in the infected host, *Molecular and Biochemical Parasitology*, **28**, 275–84.

Proudfoot, L., Kusel, J. R., Smith, H. V., Harnett, W., Worms, M. J. and Kennedy, M. W., 1990, The surface lipid of parasitic nematodes: organisation and modifications during transition to the mammalian host environment, *Acta Tropica*, **47**, 323–30.

Rogers, R., Denham, D. A. and Nelson, G. S., 1974, Studies with *Brugia pahangi*. 5. Structure of the cuticle, *Journal of Helminthology*, **48**, 113–17.

Runnegar, B., 1985, Collagen gene construction and evolution, *Journal of Molecular Evolution*, **22**, 141–9.

Rutherford, T. A. and Webster, J. M., 1974, Transcuticular uptake of glucose by an entomophilic nematode, *Mermis nigrescens*, *Journal of Parasitology*, **60**, 804–8.

Samuelson, J. C., Caulfield, J. P. and David, J. R., 1980, *Schistosoma mansoni*: post-transformational changes in schistosomulae grown *in vitro* and in mice, *Experimental Parasitology*, **50**, 369–83.

Sayers, G., MacKenzie, C. D. and Denham, D. A., 1984, Biochemical surface components of *Brugia pahangi* microfilariae, *Parasitology*, **89**, 425–34.

Selkirk, M. E., Denham, D. A., Partono, F., Sutanto, I. and Maizels, R. M., 1986, Molecular characterisation of antigens of lymphatic filariae, *Parasitology*, **91**, S15–38.

Selkirk, M. E., Nielsen, L., Kelly, C., Partono, F., Sayers, G. and Maizels, R. M., 1989, Identification, synthesis and immunogenicity of cuticular collagens from the filarial nematodes *Brugia malayi* and *Brugia pahangi*, *Molecular and Biochemical Parasitology*, **32**, 229–46.

Selkirk, M. E., Gregory, W. F., Yazdanbakhsh, M., Jenkins, R. E. and Maizels, R. M., 1990, Cuticular localisation and turnover of the major surface protein (gp29) of adult *Brugia malayi*, *Molecular and Biochemical Parasitology*, **42**, 31–44.

Silberstein, D. S. and Despommier, D. D., 1984, Antigens from *Trichinella spiralis* that induce a protective response in the mouse, *Journal of Immunology*, **132**, 898–904.

Smith, H. V., Quinn, R., Kusel, J. R. and Girdwood, R. W. A., 1981, The effect of temperature and antimetabolites on antibody binding to the outer surface of second stage *Toxocara canis* larvae, *Molecular and Biochemical Parasitology*, **4**, 183–93.

Stoner, R. D. and Hankes, L. V., 1958, *In vitro* metabolism of DL-tyrosine-2-*C*-14 and DL-tryptophan-2-*C*-14 by *Trichinella spiralis* larvae, *Experimental Parasitology*, **7**, 145–51.

Sutanto, I., Maizels, R. M. and Denham, D. A., 1985, Surface antigens of a filarial nematode: analysis of adult *Brugia pahangi* surface components and their use in monoclonal antibody production, *Molecular and Biochemical Parasitology*, **15**, 203–14.

Takahashi, Y., Uno, T., Nishiyama, T., Yamada, S. and Araki, T., 1988, Immunocytolocalization study of the external covering of *Trichinella spiralis* muscle larva, *Journal of Parasitology*, **74**, 270–4.

Taliaferro, W. H. and Sarles, M. P., 1939, The cellular reactions in the skin, lungs and intestines of normal and immune rats after infection with *Nippostrongylus muris*, *Journal of Infectious Diseases*, **64**, 157–92.

Tamashiro, W. K., Ehrenberg, J. P., Levy, D. A. and Scott, A. L., 1986, Antigenic peptides on the surface of *Dirofilaria immitis* microfilariae, *Molecular and Biochemical Parasitology*, **18**, 369–76.

Taylor, D. W., Goddard, J. M. and McMahon, J. E., 1986, Surface components of *Onchocerca volvulus*, *Molecular and Biochemical Parasitology*, **18**, 283–300.

Timpl, R. T., 1982, Antibodies to collagens and procollagens, in *Methods in Enzymology*, Vol. **82**, *Structural and Contractile Proteins. Part A; Extracellular Matrix*, pp. 472–98, London: Academic Press.

Vincent, A. L., Ash, L. R. and Frommes, S. P., 1975, The ultrastructure of adult *Brugia malayi* (Brug, 1929), (Nematoda: Filarioidea), *Journal of Parasitology*, **61**, 499–512.

Wright, K. A., 1987, The nematode's cuticle — its surface and the epidermis: function, homology, analogy — a current consensus, *Journal of Parasitology*, **73**, 1077–83.

3. The surface antigens of the filarial nematode *Brugia*, and the characterization of the major 30 kDA component

E. Devaney

INTRODUCTION

The lymphatic-dwelling filarial parasites of man, *Wuchereria bancrofti*, *Brugia malayi* and *Brugia timori*, are a major cause of disease throughout the tropics with an estimated 90 million people infected and a further 900 million at risk of infection (WHO Expert Committee on Filariasis, 1984). The parasites enter the mammalian host as infective third stage larvae (L3) via the wound made by the mosquito vector and develop to adults within the lymphatics. Following mating the female worm produces an abundance of first-stage larvae (or microfilariae) which circulate in the blood and which may be ingested by the mosquito vector. The adult parasite can survive for many years in the infected host, apparently unaffected by the host immune response.

Lymphatic filariasis gives rise to a wide spectrum of clinical manifestations which are thought to reflect the intensity and the nature of the immune response to parasite antigens (Ottessen, 1984). Filarial parasites are complex multicellular organisms containing many potential antigens. In the past few years parasite surface antigens have been the subject of considerable interest because the recognition of surface exposed molecules by the host immune system might contribute to the development of an effective immune response. In some parasites such as *Plasmodium falciparum* and *Schistosoma mansoni* the application of innovative techniques in cell and molecular biology has greatly increased our understanding of the biosynthesis and expression of the surface antigens. In contrast, the surface of filarial parasites remains something of an enigma, perhaps because in the past the filarial cuticle was regarded as an antigenically inert exoskeleton. It now appears that antigenic molecules are expressed upon the surface of the cuticle and recent studies are directed at defining the immunological and functional roles of these antigens in the parasite life cycle.

The developmental cycle of filarial nematodes is characterized by a series of four moults at each of which the entire cuticle is shed. In some nematodes, e.g. *Trichinella spiralis* the surface antigens expressed at any stage in the life cycle are strictly stage specific (Philipp *et al.*, 1981). This is not the case with *Brugia* species where antibodies raised against one stage in the life cycle will precipitate surface-labelled antigens of a different stage (Maizels *et al.*, 1983).

In the adult filarial nematode *Brugia pahangi* the major antigen seen after labelling with ^{125}I, detergent extraction and immunoprecipitation with serum from infected hosts, has a relative molecular mass (M_r) of 29–30 kDa (see also Chapter 1). The 30 kDa antigen of adult *Brugia* is highly conserved amongst the lymphatic filarial parasites and contains epitopes which are cross-reactive with other species of filarial parasite (Maizels *et al.*, 1983) and with other stages in the life cycle (Philipp *et al.*, 1986). The 30 kDa antigen has been the subject of extensive study both because of its immunological characteristics and because it was, until recently, generally accepted to be localized to the worm surface. However, at the time of writing, some doubt exists as to the exact localization of this antigen because, in contrast to the biochemical studies, immunocytochemical studies have failed to localize the 30 kDa antigen to the epicuticle (Devaney *et al.*, 1990).

In this chapter I review the data on the biochemical and immunochemical characterization of the 30 kDa antigen in adult *B. pahangi*.

THE FILARIAL EPICUTICLE

The cuticle of adult *Brugia* species, in common with other nematodes, is an acellular matrix composed of collagens and other structural proteins cross-linked by disulphide bonding (Selkirk *et al.*, 1989). The outermost layer of the cuticle, the epicuticle, forms the interface between parasite and host, yet surprisingly little is known of its composition or structure. It has a characteristic appearance at the ultrastructural level (Figure 3.1), the trilaminate-like appearance presumably reflecting the presence of lipids in this layer (see Chapter 1).

Biochemical studies with the free-living nematode *Caenorhabditis elegans* have shown that the epicuticle consists of covalently cross-linked non-collagenous proteins which are resistant to standard methods of solubilization (Cox *et al.*, 1981). In the parasitic nematode *Ascaris lumbricoides* the epicuticle contains a highly insoluble non-collagenous protein, called 'cuticlin' (Fujimoto and Kanaya, 1973).

A similar phenomenon has been demonstrated in the filarial parasite *Acanthocheilonema* (= *Dipetalonema*) *viteae* by Betschart *et al.* (1985). Progressive solubilization of the adult parasite in sodium dodecylsulphate (SDS) and 2-mercaptoethanol (2ME) resulted in the cortical zone, including the epicuticle, remaining intact. In *B. pahangi* the epicuticle is also resistant to

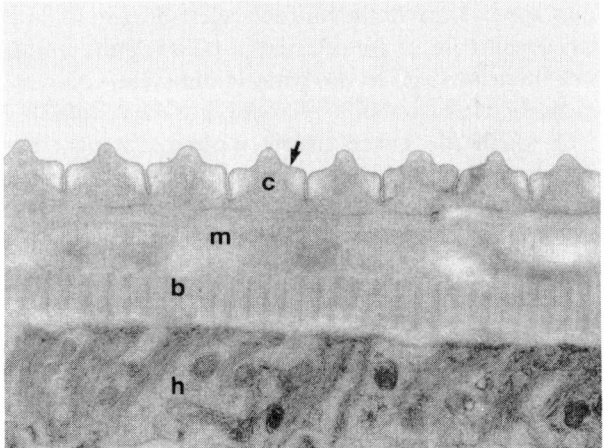

Figure 3.1. Transmission electron micrograph of adult *B. pahangi* showing the epicuticle (arrowed), the three layers of the cuticle, cortical (c), median (m) and basal (b) and the underlying hypodermis (h) with membranous infoldings. The worms were fixed in 3 per cent cacodylate buffered glutaraldehyde, post-fixed in osmium, embedded in epon/araldite and stained *en bloc* with uranyl acetate. Sections (approximately 50 nm thick) were stained in uranyl acetate and lead citrate and photographed on a Philips CM-10 operated at 80 kV. Magnification × 35 000.

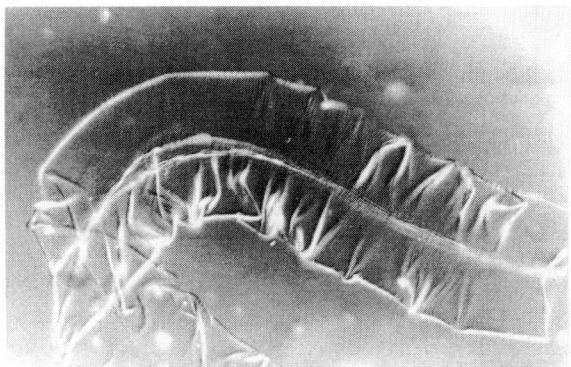

Figure 3.2. Purification of the epicuticle of adult *B. pahangi*. Female worms, cut into pieces, were stirred overnight in 1.0 ml of SDS_1 buffer (Betschart *et al.*, 1985). The supernatant was removed and the pieces of worm further extracted in SDS_2 buffer followed by three extractions in SDS/2ME overnight at 4°C. Magnification × 275.

solubilization (Selkirk *et al.*, 1989; Devaney *et al.*, 1990), although it appears to be less highly cross-linked in this species than in *A. viteae* (Figure 3.2).

The nature of the insoluble components of the epicuticle remains unclear, largely because of the difficulties inherent in analysing such material. Cox *et al.* (1981) and Maizels (unpublished, cited in Maizels *et al.*, 1989) have demonstrated the partial degradation of the insoluble fraction by elastase and other hydrolytic proteases. An alternative approach adopted by Betschart and co-workers (Betschart *et al.*, 1985; Kiefer *et al.*, 1986) is to raise antibodies to the insoluble fraction. Although mice immunized with the insoluble cuticular fraction of *A. viteae* produced antibodies which reacted with the cuticle, no labelling of the outermost surface of the worm was detected by immunofluorescence or by immunogold labelling. The reason for the lack of reactivity with the worm surface remains unclear. Kiefer *et al.* (1986) suggested that the epicuticle may have to be processed in the immunized mouse and that only 'breakdown' products would produce antibody which reacted with the worm surface. Alternatively, the insoluble epicuticle may be poorly antigenic, the mice responding only to the antigenic stimulus of the matrix of the cortical zone, which is also present in the insoluble cuticular preparation used as antigen. On the other hand, the harsh denaturing conditions required to produce the purified epicuticle might destroy the antigenicity of some epitopes. Similar results have been obtained in this laboratory where rabbits immunized with purified insoluble material from cuticles of *B. pahangi* failed to produce antibodies which bound to the worm surface (Devaney, unpublished).

Further support for the highly cross-linked nature of the epicuticle in filarial nematodes was provided by the results of Kennedy *et al.* (1987a). These workers observed that fluorescent lipid probes inserted into the epicuticle showed minimal lateral mobility, in contrast to the situation with classical plasma membranes which are characterized by the fluidity of the lipid bilayer (but see Chapter 1 for a detailed discussion).

As these studies illustrate, it is now clear that the epicuticle bears little resemblance to a classical eukaryotic plasma membrane. Our understanding of the components which comprise the bulk of the epicuticle and the manner in which they may be organized remains scanty. Whether the insoluble components of the epicuticle are actually exposed upon the worm surface and whether they are antigenic in the naturally infected host remains to be determined.

THE SOLUBLE ANTIGENS OF THE CUTICLE

Despite the fundamental differences in the nature of the surfaces concerned, much of the recent information concerning the proteins of the nematode cuticle has come from methods developed for the study of eukaryotic plasma membrane proteins, e.g. surface labelling with $Na^{125}I$ followed by detergent extraction of soluble antigens (Parkhouse *et al.*, 1981). When adult male or female *B. pahangi* are labelled by means of the Iodogen or Chloramine-T methods (both of which preferentially label tyrosine residues), detergent extracted and analysed by polyacrylamide gel electrophoresis in the presence of SDS (SDS-PAGE) and autoradiography, a restricted set of labelled polypeptides is seen. The major labelled polypeptide in *B. pahangi* and in each of the other species of *Brugia* is of M_r 29–30 kDa (Maizels *et al.*, 1983).

Iodination via the Bolton–Hunter reagent (a lysine directed method) labels a much wider range of polypeptides (Marshall and Howells, 1985; Maizels *et al.*,

1989) and it is now widely accepted that this finding reflects the permeability of the filarial cuticle to this reagent. Radioactive iodine permeates the cuticle of all filarial species to some extent (Marshall and Howells, 1985; Betschart and Jenkins, 1987) and labelling of a polypeptide with ^{125}I should never be accepted as proof of a surface location. Confirmation of surface exposure by alternative methods is, therefore, imperative. With classical plasma membrane proteins, the sensitivity of a polypeptide to proteolytic cleavage or its accessibility to antibody binding *in situ* is generally accepted as evidence of surface exposure.

In the case of adult *B. pahangi*, only one polypeptide, the 30 kDa, appears to fulfil these criteria. Serum from experimentally infected jirds contains antibodies to the 30 kDa antigen and incubation of live adult *B. pahangi* with infected jird serum at 4 °C has been shown to adsorb most of the antibody activity against the 30 kDa antigen (Devaney, 1987) (Figure 3.3, lane B). More recently, Maizels *et al.* (1989) have shown that incubation of ^{125}I labelled *Brugia* with 100 µg trypsin ml^{-1} for 60 min cleaves a 2 kDa fragment from the 30 kDa polypeptide, suggesting that at

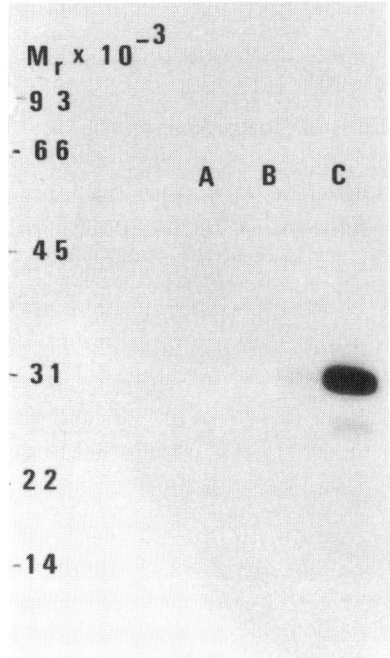

Figure 3.3. Immunoprecipitation of ^{125}I labelled polypeptides of *B. pahangi* with pre-adsorbed jird serum. Adult female worms were iodinated via the Iodogen method and solubilized in 1 per cent DOC. For the pre-adsorption, 150 live worms were incubated for 2 h at 4 °C in 200 µl of PBS (pH 7.2) containing 5 µl of infected jird serum. 800 000 trichloroacetic acid (TCA) precipitable counts per minute (cpm) were mixed with an equal volume of control jird serum (lane A), pre-adsorbed infected jird serum (lane B) or infected jird serum (lane C). Antigen/antibody complexes were formed overnight at 4 °C and immune complexes were precipitated with 10 mg Protein A Sepharose complexed to rabbit anti-mouse immunoglobulin G (IgG). Equal volumes of each precipitate were analysed by SDS-PAGE on a 10 per cent gel followed by autoradiography. (Reproduced by permission of Blackwell Scientific Publications.)

least a small domain of this molecule is surface exposed. The alternative interpretation of this data would necessitate the epicuticle being permeable to macromolecules of the size of trypsin (23.5 kDa). This explanation seems unlikely, because in the experiments concerned only the 30 kDa antigen was cleaved and not the cuticular collagens. However, the recent results of Alvarez *et al.* (1989), demonstrated that all the Iodogen-labelled polypeptides in adult *B. malayi* were degraded by prolonged (overnight) exposure to trypsin. This would imply that trypsin can progressively degrade the surface and it could be that the cleavage of the 30 kDa polypeptide merely reflects its greater proximity to the surface than other polypeptides, but not that it is exposed. A time-course experiment ought to resolve this discrepancy.

OTHER SURFACE ASSOCIATED COMPONENTS

Although the 30 kDa polypeptide is the major labelled component in detergent extracts of iodinated worms, a number of minor components have been identified

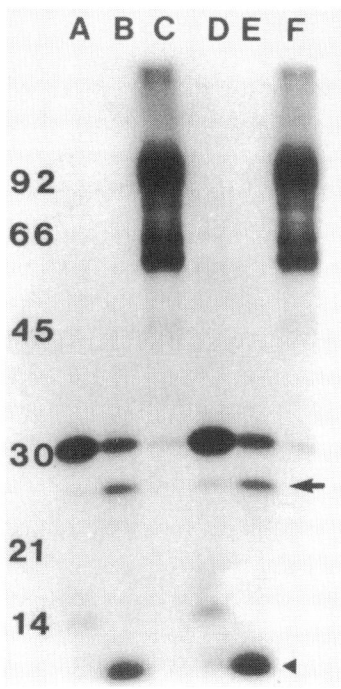

Figure 3.4. Profile ^{125}I labelled polypeptides of adult female. *B. pahangi* by Iodogen labelling. Worms were homogenized in 10 mM Tris containing protease inhibitors (lanes A and D). The Tris insoluble pellet was solubilized with 1 per cent DOC (lanes B and E) and the DOC insoluble pellet boiled with SDS-PAGE sample cocktail containing 8M urea (lanes C and F). 3×10^4 cpm per lane analysed on a 10 per cent SDS polyacrylamide gel. Arrow denotes 28 kDA component. Arrowhead denotes fast migrating material released by DOC and presumed to be lipid. (Reproduced by permission of Elsevier Publications.)

at 15, 17, 20 and 51 kDa. The 20 kDa polypeptide shows some structural similarities to the 30 kDa antigen and it has been suggested that it is a partial breakdown product of the 30 kDa antigen (Maizels *et al.*, 1989). In this laboratory an additional labelled component is routinely observed in detergent extracts of Iodogen-labelled adult *B. pahangi* at 27/28 kDa (see, for example, Figure 3.4, arrow).

A novel set of polypeptides ranging in M_r from 17 to 200 kDa has recently been described, following labelling of adult *Brugia* with Bolton–Hunter reagent and incubation of intact worms in 2ME (Maizels *et al.*, 1989). The rapid shedding of this 'ladder' of polypeptides may be indicative of a localization close to the worm surface, but to date there is no conclusive evidence for surface exposure of any of these components (see Chapter 2).

BIOCHEMICAL CHARACTERIZATION OF THE 30 kDa POLYPEPTIDE

The 30 kDa polypeptide has been extensively characterized and for detailed accounts of these studies, including methods employed, the reader is referred to Devaney (1988) and Maizels *et al.* (1989). The experiments in this laboratory on the biochemical characterization of the 30 kDa antigen were initiated with the aim of defining the interaction of this polypeptide with the epicuticle. The 30 kDa polypeptide was followed through the different analyses using the ^{125}I label as a marker. An immune rabbit serum (IRS) raised against the polypeptide excised from polyacrylamide gels (Devaney, 1987) was used as an additional probe in these studies.

THE SOLUBILITY CHARACTERISTICS OF THE 30 kDa ANTIGEN

One of the most striking features of the 30 kDa antigen is its solubility in aqueous solution, a feature which contrasts markedly with the insoluble nature of the bulk of epicuticular components. The 30 kDa polypeptide is effectively released from ^{125}I labelled *B. pahangi* by homogenization and extraction in one of a number of detergents. No consistent differences have been observed in the radioactivity associated with the 30 kDa polypeptide released by homogenization and extraction in an anionic (deoxycholate (DOC)), cationic (cetyltrimethylammonium bromide, CTAB) or non-ionic (Triton X-114, TX-114) detergent.

Detergent is not essential for the release of the 30 kDa polypeptide and its hydrophilic nature has been confirmed by the extraction of adult worms in the detergent TX-114 and the subsequent partitioning of the 30 kDa antigen into the aqueous phase (Devaney, 1988). The proportion of ^{125}I-labelled 30 kDa antigen released by homogenization in aqueous buffers (e.g. Tris and phosphate buffered saline) has been estimated by densitometric scanning of autoradiographs (see Figure 3.4, for example). A total of 60–70 per cent of the ^{125}I-labelled 30 kDa polypeptide is released into the aqueous extract (lane A), with most of the remainder in the DOC extract (lane B), although the percentage varies from experiment to experiment and, in part, reflects the quality of homogenization. The high M_r polypeptides released by treatment with 8 M urea (lane C) are the cuticular collagens (Sutanto *et al.*, 1985) which are heavily labelled despite being inaccessible on the worm surface (Selkirk *et al.*, 1989). Even after this harsh final solubilization step, a significant proportion of the radioactivity associated with the worm remains in the insoluble pellet (Betschart and Jenkins, 1987; Selkirk *et al.*, 1989), associated with the epicuticle and the cortical zone of the cuticle.

Although the 30 kDa polypeptide can be readily released from the worm following disruption by homogenization, it is more difficult to release from intact worms. Incubation of whole worms in detergent solutions (1 per cent DOC), in high salt (500 mM KCl), in chelating agents (10 mM ethylenediaminetetraacetic acid or ethylene glycol-bis (β-aminoethyl ether) N,N,N',N',-tetraacetic acid (EDTA)(EGTA)) or under conditions of high pH (100 mM Na_2CO_3, pH 11.0) releases a minimal amount of radioactivity associated with the 30 kDa polypeptide (Devaney, 1988; unpublished). However, incubation of intact parasites in 2ME does release a significant proportion of the 30 kDa polypeptide (Maizels *et al.*,

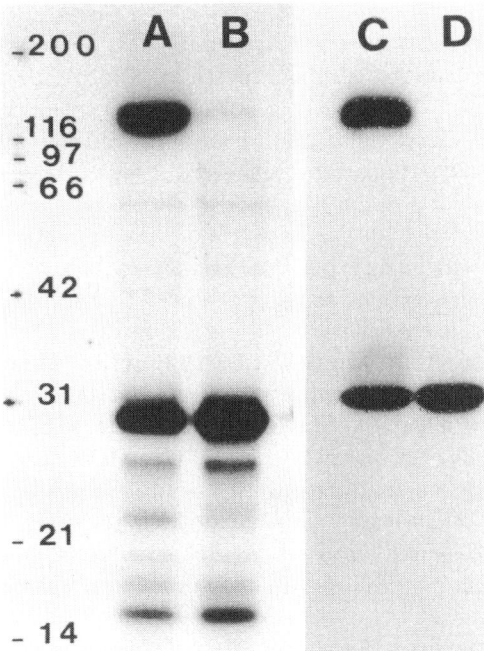

Figure 3.5. Autoradiograph of a DOC extract of ^{125}I-labelled polypeptides of *B. pahangi* analysed by SDS-PAGE without boiling (lane A) and with boiling (lane B). For lanes C and D, 500 000 cpm of a DOC extract of ^{125}I-labelled adult parasites was immunoprecipitated with 20 μl of the IRS. Immune complexes were precipitated with Protein A Sepharose and prepared for SDS-PAGE by incubation in sample cocktail at room temperature (lane C) or by boiling in sample cocktail (lane D). Note that presence of the high M_r component (*c.* 120 kDa) in lanes A and C.

1989), suggesting an interaction with the cuticular collagens or with other as yet unidentified components of the cuticle.

Analysis by SDS-PAGE and autoradiography of a DOC soluble extract under reducing and non-reducing conditions reveals a slight shift in the mobility of the polypeptide from 30 kDa under reducing conditions to 28 kDa when non-reduced, indicating the presence of intramolecular disulphide bonds (Devaney, 1988). No evidence was obtained for high M_r aggregates of the 30 kDa polypeptide analysed under non-reducing conditions, although a high M_r form of the polypeptide (120 kDa) is seen when the boiling step is omitted from the sample preparation prior to SDS-PAGE (Figure 3.5, lane A). The high M_r form can also be immuno-precipitated from extracts of ^{125}I-labelled parasites (Figure 3.5, lane C) suggesting that the 30 kDa antigen may exist in a tetrameric conformation *in situ*.

THE 30 kDa ANTIGEN IS NOT ANCHORED BY A LIPID LINKAGE

The solubility characteristics of the 30 kDa polypeptide bear some resemblance to that of another well characterized parasite surface antigen, the variant surface glycoprotein (VSG) of *Trypanosoma brucei*. VSG and a number of other eukaryotic plasma membrane proteins have been shown to be anchored in the membrane via a covalent linkage between the 1,2-diacylglycerol moiety of phosphatidylinositol and the polypeptide chain (Low *et al.*, 1986). VSGs are released in a soluble form during lysis of the cell by the action of an endogenous phospholipase C which cleaves the hydrophobic domain from the glycolipid. It seems unlikely, however, that the 30 kDa *Brugia* antigen is anchored in the epicuticle via a lipid linkage. First, there is no evidence for the existence of a hydrophobic form of the 30 kDa antigen as might otherwise be expected. Extraction of iodinated parasites in boiling SDS (to inactivate any enzyme activity) followed by the addition of TX-114 and phase separation, still results in a hydrophilic form of the 30 kDa polypeptide (Devaney, unpublished). In addition, treatment of live ^{125}I-labelled *Brugia* with a variety of phospholipase C enzymes (from *Bacillus cerus* or from *Clostridium perfringens*) fails to release a significant amount of 30 kDa antigen (Devaney, unpublished; Maizels *et al.*, 1989).

Rabbit antiserum prepared against the soluble form of one VSG contains antibodies which cross-react with heterologous VSGs, the cross-reacting determinant (CRD) being localized to the glycan component. The anti-CRD antibody binds to soluble VSG and to the soluble form of some other proteins with a similar membrane anchor (reviewed in Ferguson and Williams, 1988). The soluble 30 kDa antigen of *B. pahangi* is not recognized by anti-CRD antibody (kindly provided by Dr M.J. Turner) in immunoprecipitation or immunoblotting experiments (Figure 3.6, lane D), although a number of other *Brugia* components of different M_r are recognized.

To date, therefore, there is no direct evident for the 30 kDa *Brugia* antigen being integrated in the epicuticle via hydrophobic interactions with lipid. Lipids are

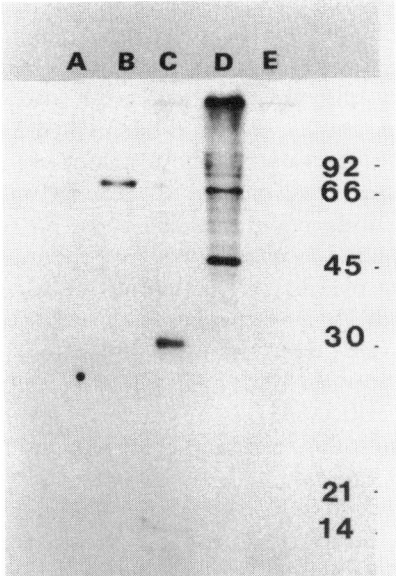

Figure 3.6. Immunoblot of a DOC extract of adult *B. pahangi* with the anti-CRD serum. Lanes C, D and E contain 70 μg of *B. pahangi* extract probed with 1/50 dilution of the IRS (lane C), 1/25 dilution of the anti-CRD serum (lane D) or 1/25 dilution of control rabbit serum (lane E). As a positive control, lanes A and B contain 5 μg of membrane form VSG and 5 μg of soluble VSG, respectively, probed with 1/25 dilution of anti-CRD serum. Bound IgG was visualized with [125]I Protein A (2.5 × 10[5] cpm/lane) followed by autoradiography.

undoubtedly present in the epicuticle of filarial nematodes (Kennedy *et al.*, 1987a), but they remain largely uncharacterized. In detergent extracts of *B. pahangi* analysed using SDS-PAGE and autoradiography, the fast migrating material at the dye front is presumed to be lipid as it is absent from parasites extracted in aqueous buffers (compare lanes A and B (arrowhead) in Figure 3.4).

THE 30 kDa ANTIGEN IS GLYCOSYLATED

Another feature of the 30 kDa polypeptide which distinguishes it from a typical eukaryotic plasma membrane protein relates to the organization of the oligosaccharide side-chains. Some 3–4 kDa of the 30 kDa polypeptide are carbohydrate in nature as revealed by digestion with glycopeptidase F. The carbohydrate appears to be organized as two side-chains each of 1.7–2.0 kDa (Maizels *et al.*, 1989) which are typical *N*-linked high mannose oligosaccharides, as revealed by their lectin binding characteristics. There is no evidence for the presence of *O*-linked oligosaccharides. The solubilized 30 kDa antigen binds to concanavalin A (Con-A) sepharose, but fluorescein isothiocyanate conjugated (FITC) Con-A does not bind to the worm surface. These results were interpreted to suggest that, contrary to the situation with classical membrane glycoproteins where carbohydrate residues are always exposed on the extracytoplasmic face of the membrane, the carbohydrate residues of the 30 kDa polypeptide are not accessible *in situ* (Devaney, 1988). A similar organization of glycoprotein antigens in the epicuticle of *T. spiralis* has also been observed (Parkhouse *et al.*, 1981; Ortega-Pierres *et al.*, 1984). In the case of the *Brugia* 30 kDa antigen, these results may have to be reassessed once the precise localization of the 30 kDa antigen has been determined.

While these biochemical studies have shed some light upon the possible interaction of the 30 kDa antigen with the cuticle, the precise means by which it is integrated into the cuticle and by which it is secreted remain unclear. In addition, at the present time the exact localization of the 30 kDa antigen remains ambiguous. Its accessibility to antibody binding and to proteolytic cleavage clearly support its surface localization, but more recent immunocytochemical studies at the electron microscope level (Devaney *et al.*, 1990) have failed to confirm its surface exposure. Inevitably these results (discussed below in more detail) have necessitated a reevaluation of the existing data and studies in progress are aimed at clarifying these apparent contradictions.

The presence of both soluble components, e.g. the 30 kDa antigen, and insoluble components in the epicuticle would, however, explain some of the other apparently paradoxical observations in the literature. A number of studies with different nematode species and stages of the life cycle have demonstrated the 'dynamic' nature of the epicuticle (for a review see Philipp and Rumjaneck, 1984) while other studies have suggested that the epicuticle has little capacity for dynamic change (Howells and Blainey, 1983). The apparent immobility of the surface lipid of parasitic nematodes could also be said to support the latter argument although it need not preclude other dynamic properties (Kennedy *et al.*, 1987b). These observations need not be mutually exclusive, however, but might represent heterogenicity in the composition of the nematode surface, as suggested, for example, by the possible existence of lipid domains in the epicuticle (Chapter 1).

THE BIOSYNTHESIS AND SECRETION OF THE 30 kDa ANTIGEN

The first indication that the 30 kDa antigen might be released from the adult parasite came from the studies of Kaushal *et al.* (1982) and Maizels *et al.* (1985a). Both groups identified excretory/secretory (ES) antigens released by adult *Brugia* with the same M_r (15 and 29 kDa) as surface-labelled molecules, although the excretory–secretory antigens were not definitively proven to be surface-derived. In contrast, a subsequent study by Marshall and Howells (1986) could find no release of the ^{125}I-labelled 30 kDa antigen from adult *B. pahangi* over a 7-day period *in vivo*. The apparent discrepancy in these results probably relates to the sensitivity of the methods employed.

[35S]Methionine labelling of adult *B. pahangi* has shown that the 30 kDa polypeptide is continuously synthesized in both male and female parasites. In the original experiments (Devaney, 1988), adult *B. pahangi* were labelled over a prolonged period *in vitro* (48–60 h), extracted and immunoprecipitated with the IRS raised against the 30 kDa polypeptide. A doublet at 30 kDa was immunoprecipitated plus a number of high M_r components (over 66 kDa). Attempts were made to confirm the surface localization of the newly synthesized 30 kDa polypeptide by incubating ^{35}S-labelled parasites with serum from infected jirds, solubilizing the parasites and precipitating immune complexes with protein A sepharose coupled to rabbit anti-mouse IgG. These experiments failed to demonstrate unequivocally that the newly synthesized 30 kDa polypeptide was associated with the worm surface. The possibility that it instead represented an intracellular pool of 30 kDa antigen could not be ruled out.

It was, however, possible to demonstrate the presence of ^{35}S-labelled 30 kDa antigen in the spent culture fluid of adult parasites by immunoprecipitation with the IRS. The identity of the 30 kDa component was confirmed by limited proteolysis of the immunoprecipitated material with *Staphylococcus aureus* V8 protease. The resulting peptide maps were identical to those obtained using ^{125}I-labelled 30 kDa antigen from the worm surface. These experiments were difficult to carry out because of the small amounts of ^{35}S-labelled excretory–secretory antigen available. To confirm the results obtained by metabolic labelling, excretory–secretory antigens were collected from short-term (24 h) bulk cultures of male and female *B. pahangi*. The excretory–secretory antigen was concentrated, iodinated and analysed by immunoprecipitation with the IRS followed by peptide mapping, as above. This confirmed that the secreted 30 kDa component was identical to the 30 kDa surface antigen (Devaney, 1988; Figure 3.7). Additional confirmation that the surface and secreted 30 kDa antigens are identical has also been provided by Kwan-Lim *et al.* (1989) using comparative two-dimensional electrophoresis.

These experiments are always open to the criticism of having been carried out *in vitro* under conditions of culture which may compromise the integrity of the worm surface. This explanation can never be completely excluded, although ultra-structural examination of female parasites cultured under identical conditions has failed to demonstrate any gross morphological changes associated with the worm surface (Devaney, unpublished).

PRECURSORS OF THE 30 kDa ANTIGEN

More recent pulse and chase studies (Devaney and Jecock, unpublished) have attempted to define the stability and half-life of the newly synthesized 30 kDa polypeptide. The original ^{35}S-labelling experiments suggested that the synthesis and release of the 30 kDa antigen from the adult parasite occurred at rather a slow rate, as demonstrated by the small amounts of ^{35}S-labelled 30 kDa antigen released by the adults over a prolonged period of culture. In the pulse and chase experiments worms were labelled with [35S]methionine over a 30-min period and then chased in medium containing no radioisotope for various periods up to 4 h. The worms were then extracted in boiling SDS solution, to avoid proteolysis, prior

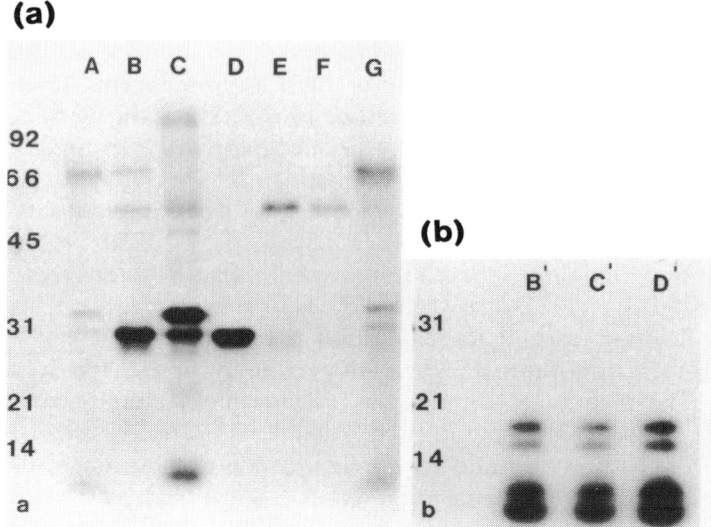

Figure 3.7. (a) Profile of ^{125}I-labelled excretory–secretory products of adult male and female *B. pahangi*. Excretory–secretory antigen was collected from three cultures each containing approximately 75 male and female *B. pahangi*, concentrated by ultrafiltration in an Amicon cell with a YM10 membrane and labelled via the Iodogen method. Lanes A and G: total ^{125}I-excretory–secretory material. I-labelled excretory–secretory material was immunoprecipitated with the IRS (lane B), infected jird serum (lane C), control rabbit serum (lane E), or control jird serum (lane F) overnight at 4°C. Immune complexes were precipitated with Protein A Sepharose (rabbit sera) or with Protein A Sepharose complexed to rabbit anti-mouse IgG (jird sera) and resolved on a 10 per cent SDS polyacrylamide gel. For comparison, a DOC extract of ^{125}I-labelled female *B. pahangi* immunoprecipitated with the IRS is shown in lane D. (b) Peptide map of 30 kDa components immunoprecipitated from excretory–secretory antigen by the IRS (B'), infected jird serum (C') and from ^{125}I-labelled adult parasites precipitated with the IRS (D'). Components resolving at 30 kDa were cut out of the first gel and further analysed by limited proteolysis with 5 μg *S. aureus* V8 protease in a 12.5 per cent second gel using the method of Cleveland (1983). Peptides resolving at M_r 19, 17, 13 and 11 kDa. (Reproduced by permission of Elsevier Publications.)

to immunoprecipitation with the IRS. The 30 kDa antigen was labelled after a short pulse of 30 min and during subsequent chase periods (up to 4 h) there was no significant decrease in the radioactivity associated with the 30 kDa antigen, suggesting that it is stable for at least 4 h after synthesis. Selkirk *et al.* (1988) have suggested that the 30 kDa antigen is synthesized as a higher M_r precursor prior to export to the cuticular matrix. In our pulse and chase studies, immunoprecipitation with the IRS has revealed components at approximately 200 kDa and 32 kDa present at time 0 and at 30 min of chase, but to date we have no evidence of a precursor–product relationship between these components and the 30 kDa antigen.

Synthesis of the 30 kDa polypeptide has also been studied in a cell free system, using total RNA isolated from adult parasites to direct protein synthesis (Jecock and Devaney, unpublished). Immunoprecipitation of ^{35}S-labelled *in vitro* translation products with the IRS raised against the 30 kDa antigen yields one major band on the autoradiograph with an estimated M_r of 31 kDa and a minor band at 28 kDa

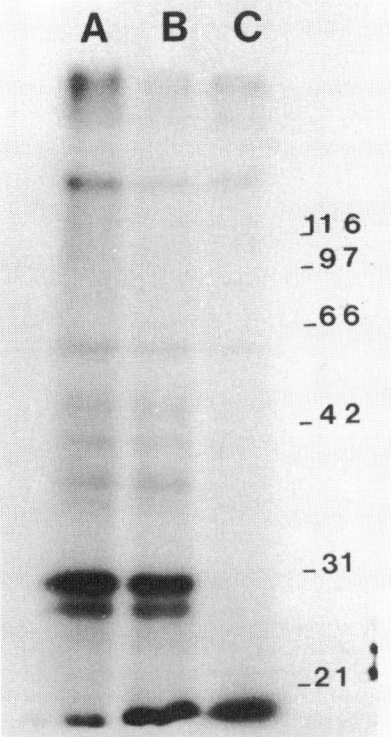

Figure 3.8. *In vitro* translation of total RNA from adult male and female *B. pahangi*. 22 µg total RNA was translated in a rabbit reticulocyte lysate system with 120 µCi[³⁵S]methionine. Translated products (1.2 × 10⁶ cpm) were immunoprecipitated with 25 µl of IRS raised against the 30 kDa antigen (lanes A and B) of 25 µl of control rabbit serum (lane C). Immune complexes were precipitated with Protein A Sepharose and resolved on a 10 per cent SDS polyacrylamide gel and fluorographed by immersion in Amplify.

(Figure 3.8). The discrepancy in M_r between the immunoprecipitated translation products and the native protein may be explained by post-translational processing events, such as the cleavage of a signal sequence and/or glycosylation.

THE LOCALIZATION OF THE 30 kDa ANTIGEN AND ITS ROUTE OF SECRETION

A major question arising from these experiments which also relates to the structure and molecular organization of the cuticle is the means by which the 30 kDa antigen is secreted. There is no morphological evidence for the presence of pores or other structures in the filarial cuticle through which it might be released. Immunocytochemical studies at the light-microscope level using the IRS raised against the 30 kDa antigen showed labelling throughout the cuticle (Devaney, 1987). At the higher resolution of the electron microscope, immunogold labelling

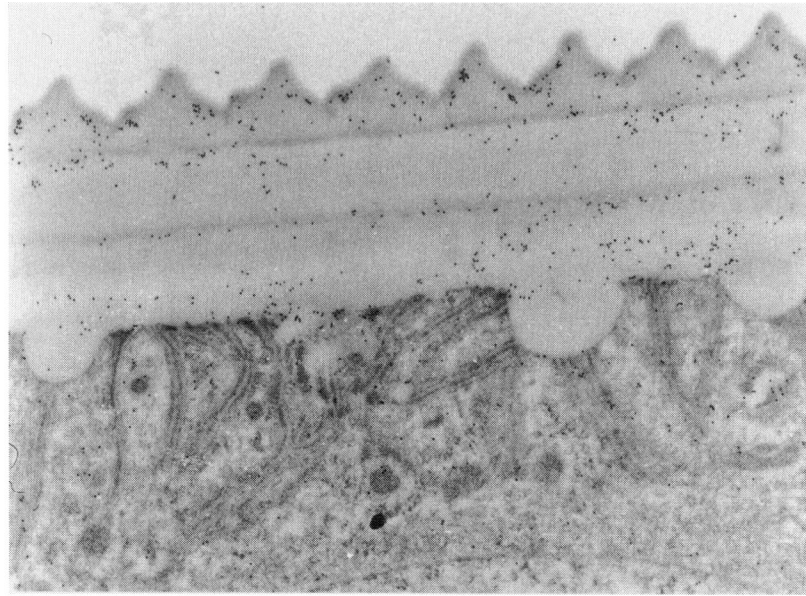

Figure 3.9. The immunocytochemical localization of the 30 kDa antigen in adult female *B. pahangi*. Worms were fixed in 0.5 per cent phosphate buffered glutaraldehyde and embedded in Lowicryl K4M. Sections were incubated with a 1:1000 dilution of the IRS raised against the 30 kDa antigen followed by goat anti-rabbit Ig coupled to 10 nm gold particles. The most intensive labelling was observed in the cortical zone with some labelling in the median and basal zones. Labelling of the epicuticular surface was not observed. Gold particles were also observed in the hypodermis, around the folds of the hypodermal membrane and over vesicles. Sections incubated with control rabbit sera showed almost no labelling of the cuticle but some labelling of the hypodermis (data not shown). Magnification × 30 000.

has failed to localize the 30 kDa antigen to the worm surface. Gold particles are distributed throughout the cuticle, though at a higher density in the external cortical layer (Figure 3.9) (Devaney *et al.*, 1990).

The immunocytochemical studies have also shown that the IRS labels vesicular bodies in the hypodermis and the folds of the hypodermal membrane. In common with all the components of the cuticle, the 30 kDa antigen must be synthesized in the hypodermis, following which it may be transported in vesicles to the hypodermal membrane where it may be inserted and from which it may be secreted. If the 30 kDa antigen is secreted across the cuticle it may be transiently expressed at the parasite surface. However, the means by which the 30 kDa antigen might reach its eventual site of expression remain entirely speculative. The permeability of the filarial cuticle to low molecular weight compounds such as glucose and amino acids is well established (Howells, 1980) and Weiss (unpublished, cited in Betschart and Jenkins, 1987) has observed penetration of fluorescein–isothiocyanate (molecular weight 389 kDa) down to the hypodermal membrane of *A. viteae* within a few seconds. Whether the cuticle might also be permeable to macromolecular compounds is presently under investigation in this laboratory.

IMMUNOLOGICAL CHARACTERISTICS OF THE 30 kDa ANTIGEN

In addition to its potential role as a 'molecular marker' for the epicuticle, the 30 kDa antigen has been the subject of interest because of its immunological characteristics. It is the immunodominant surface antigen in each of the species of *Brugia* examined (Maizels *et al.*, 1983) and in *W. bancrofti* (Morgan *et al.*, 1986), as defined by labelling with ^{125}I and immunoprecipitation with serum from infected hosts. The 30 kDa antigen of adult *B. pahangi* also contains epitopes which cross-react with antibodies elicited by earlier stages in the life cycle, e.g. in cats experimentally infected with L3 of *B. pahangi*, antibodies which precipitate the 30 kDa adult antigen are seen as early as 14 days post-infection, some 10–14 days before the fourth stage larva (L4) moults to the adult stage (Philipp *et al.*, 1986). The immature larval stages must therefore contain epitopes which cross-react with the adult 30 kDa antigen. Marshall and Howells (1986) could not, however, detect a 30 kDa component in the infective larva (L3) of *B. pahangi* labelled with ^{125}Iodosulphanilic acid, although a component of M_r 30 kDa was present in the L4.

In addition, the 30 kDa *Brugia* antigen cross-reacts with antibodies present in the serum of patients infected with each of the species of lymphatic filariae and with *Loa loa* and *Mansonella perstans*. Serum from *Onchocerca volvulus* infected individuals preferentially recognizes the 20 kDa antigen of *B. pahangi*, which may be a partial breakdown product of the 30 kDa antigen (Maizels *et al.*, 1989). The cross-reactive epitopes appeared to be restricted to the filarial nematodes as serum from non-filarial infections did not immunoprecipitate the 30 kDa antigen of *B. pahangi* (Maizels *et al.*, 1985b).

Recent studies with adult *Loa loa* (Egwang *et al.*, 1988) have also shown a major ^{125}I-labelled component at 29–31 kDa which was recognized by serum from 'resistant' and amicrofilaraemic individuals, but not by serum from microfilaraemic individuals. In *Loa loa* it appeared that carbohydrate determinants of the 29–31 kDa antigen were important in the reactivity with human serum.

In contrast to the situation with *Loa loa*, no correlation has been observed between antibody levels to the *Brugia* 30 kDa antigen and disease status (Maizels *et al.*, 1987). Indeed, in that study, it was observed that individuals resident in an endemic area, but with no history or clinical signs of filariasis, made antibodies particularly to the 30 kDa antigen. Whether the presence of antibody to this antigen might be indicative of exposure to the parasite remains unclear.

Although no correlation has so far been found between clinical status and levels of antibody to the 30 kDa antigen in *Brugia* infection, it is likely that the availability of more defined reagents will improve the resolution of such analyses. The pattern of isotype and subclass-specific responses to defined filarial antigens requires further study (Hussain *et al.*, 1981; Ottesen *et al.*, 1985). In addition, the mapping of the antigenic epitopes of filarial antigens and the analysis of humoral responses to defined peptide epitopes is now feasible using a combination of monoclonal antibodies and recombinant antigens, as has been done for the major 65 kDa antigen of *Mycobacterium leprae* (Mehra *et al.*, 1986).

In the case of the *Brugia* 30 kDa antigen, the immunodominant epitopes recognized by infected human serum have not been fully characterized. The nature of the humoral response to the 30 kDa antigen suggest that it might contain epitopes which dominate the B cell response and which could serve to deflect the antibody response from more biologically relevant, but immunorecessive, epitopes. Alternatively, the immunodominant epitopes of the 30 kDa may elicit antibodies of an inappropriate isotype for effector function.

CONCLUDING REMARKS

To date, only the humoral response to the 30 kDa *Brugia* antigen has been considered; methods are now available for investigating T cell recognition of filarial antigens (Nutman *et al.*, 1984) and for identifying T cell determinants of protein antigens (Lamb *et al.*, 1987). Mapping of the B and T cell epitopes of the 30 kDa antigen will resolve the molecular basis of the cross-reactivity and may provide clues to the possible function of this molecule.

Such functional analyses are severely hampered by the paucity of parasite material available, but several laboratories have sought to overcome this problem by the use of recombinant DNA technology. Complementary DNA (cDNA) libraries have been prepared from adult *Brugia* (see, for example, Selkirk *et al.*, 1986) and screened with either human infection serum or with more specific reagents, e.g. antibodies to gel purified 30 kDa antigen (Selkirk *et al.*, 1988; Jecock and Devaney, unpublished). The identification and characterization of the gene encoding the 30 kDa antigen will provide information on its functional role as well as providing insight into the control of expression of this highly conserved protein in the filarial life cycle.

ACKNOWLEDGEMENTS

These studies have been supported by the Wellcome Trust and the Medical Research Council. I would like to thank Drs M. J. Turner and A. Gurnett of Merck, Sharpe and Dohme for the gift of the anti-CRD antibody and the VSG, Professor R. E. Howells, Dr M. L. Chance and Ms R. M. Jecock for critically reading the manuscript. Special thanks are also due to Drs B. Betschart and W. Rudin of the Swiss Tropical Institute for permission to use unpublished results and for their helpful comments, and to Janis Carter for help with the word processing.

REFERENCES

Alvarez, R. M., Henry, R. W. and Weil, G. J., 1989, Use of Iodogen and sulfosuccinimidobiotin to identify and isolate cuticular proteins of the filarial parasite *Brugia malayi*, *Molecular and Biochemical Parasitology*, **33**, 183–90.

Betschart, B. and Jenkins, J. M., 1987, Distribution of iodinated proteins in *Dipetalonema viteae* after surface labelling, *Molecular and Biochemical Parasitology*, 22, 1–8.

Betschart, B., Rudin, W. and Weiss, N., 1985, The isolation and immunogenicity of the cuticle of *Dipetalonema viteae* (Filarioidea), *Zeitschrift für Parasitenkunde*, 71, 87–95.

Cleveland, D. W., 1983, Peptide mapping in one dimension by limited proteolysis of sodium dodecylsulphate solubilised proteins, *Methods of Enzymology*, 96, 222–9.

Cox, G. N., Kusch, M. and Edgar, R. S., 1981, Cuticle of *Caenorhabditis elegans*: Its isolation and partial characterization, *Journal of Cell Biology*, 90, 7–17.

Devaney, E., 1987, Preliminary studies on the characterization of the M_r 30 000 surface antigen of *Brugia pahangi*, *Parasite Immunology*, 9, 401–5.

Devaney, E., 1988, The biochemical and immunochemical characterisation of the 30 kilodalton surface antigen of *Brugia pahangi*, *Molecular and Biochemical Parasitology*, 27, 83–92.

Devaney, E., Betschart, B. and Rudin, R., 1990, The analysis of the 30 kDa antigen of *Brugia pahangi* and its interaction with the cuticle, *Acta Tropica*, 47, 365–72.

Egwang, E. G., Akue, J.-P., Dupont, A. and Pinder, M., 1988, The identification and partial characterization of an immunodominant 29–31 kilodalton surface antigen expressed by adult worms of the human filaria *Loa loa*, *Molecular and Biochemical Parasitology*, 31, 263–72.

Ferguson, M. A. J. and Williams, A. F., 1988, Cell-surface anchoring of proteins via glycosylphosphatidylinositol structures, *Annual Review of Biochemistry*, 57, 285–320.

Fujimoto, D. and Kanaya, S., 1973, Cuticlin: a noncollagen structural protein from *Ascaris* cuticle, *Archives of Biochemistry and Biophysics*, 157, 1–6.

Howells, R. E., 1980, Filariae: dynamics of the surface, in Bossche, H. V. D. (Ed.) *The Host Invader Interplay*, pp. 69–84, Amsterdam: Elsevier/North-Holland Biomedical Press.

Howells, R. E. and Blainey, L. J., 1983, The moulting process and the phenomenon of intermoult growth in the filarial nematode *Brugia pahangi*, *Parasitology*, 87, 493–505.

Hussain, R., Hamilton, R. G., Kumaraswami, V., Adkinson, Jr, N. F. and Ottesen, E. A., 1981, IgE responses in human filariasis. I. Quantitation of filaria-specific IgE, *Journal of Immunology*, 127, 1623–29.

Kaushal, N. A., Hussain, R., Nash, T. E. and Ottesen, E. A., 1982, Identification and characterization of excretory–secretory products of *Brugia malayi* adult filarial parasites, *Journal of Immunology*, 129, 338–43.

Kennedy, M. W., Foley, M., Knox, K., Harnett, W., Worms, M. J., Kusel, J. R., Birmingham, J. and Garland, P. B., 1987a, Are the biophysical properties of the surface lipid of filariae different from other parasitic nematodes? in MacInnes, A. J. (Ed.) *Molecular Paradigms for Eradicating Helminthic Parasites*, pp. 289–300, New York: Alan R. Liss.

Kennedy, M. W., Foley, M., Kuo, Y.-M., Kusel, J. R. and Garland, P. B., 1987b, Biophysical properties of the surface lipid of parasitic nematodes, *Molecular and Biochemical Parasitology*, 22, 233–40.

Kiefer, E., Rudin, W., Betschart, B., Weiss, N. and Hecker, H., 1986, Demonstration of anti-cuticular antibodies by immuno-electron microscopy in sera of mice immunized with cuticular extracts and isolated cuticles of adult *Dipetalonema viteae* (Filarioidea), *Acta Tropica*, 43, 99–112.

Kwan-Lim, G. E., Gregory, W. F., Selkirk, M. E., Partono, F. and Maizels, R. M., 1989, Secreted antigens of filarial nematodes: a survey and characterization of *in vitro* excreted/secreted products of adult *Brugia malayi*, *Parasite Immunology*, 11, 629–54.

Lamb, J. R., Ivanyi, J., Rees, A. D. M., Rothbard, J. B., Howland, K., Young, R. A. and Young, D. B., 1987, Mapping of T cell epitopes using recombinant antigens and synthetic peptides, *EMBO Journal*, 6, 1245–9.

Low, M. G., Ferguson, M. A. J., Futerman, A. H. and Silman, I., 1986, Covalently attached phosphatidylinositol as a hydrophobic anchor for membrane proteins, *Trends in Biochemical Sciences*, 11, 212–15.

Maizels, R. M., Partono, F., Oemijati, S., Denham, D. A. and Ogilvie, B. M., 1983, Cross-reactive surface antigens on three stages of *Brugia malayi*, *B. pahangi* and *B. timori*, *Parasitology*, **87**, 249–63.

Maizels, R. M., Denham, D. A. and Sutanto, I., 1985a, Secreted and circulating antigens of the filarial parasite *Brugia pahangi*: analysis of *in vitro* related components and detection of parasite products *in vivo*, *Molecular and Biochemical Parasitology*, **17**, 277–88.

Maizels, R. M., Sutanto, I., Gomez-Priego, A., Lillywhite, J. and Denham, D. A., 1985b, Specificity of surface molecules of adult *Brugia* parasites: cross-reactivity with antibody from *Wuchereria*, *Onchocerca* and other human filarial infections, *Tropical Medicine and Parasitology*, **36**, 233–7.

Maizels, R. M., Selkirk, M. E., Sutanto, I. and Partono, F., 1987, Antibody responses to human lymphatic filarial parasites, in *Filariasis*, Ciba Foundation Symposium No. 127, Chichester: Wiley.

Maizels, R. M., Gregory, W. F., Kwan-Lim, G.-E. and Selkirk, M. E., 1989, Filarial surface antigens: the major 29 kilodalton glycoprotein and a novel 17–200 kilodalton complex from adult *Brugia malayi* parasites, *Molecular and Biochemical Parasitology*, **32**, 213–28.

Marshall, E. and Howells, R. E., 1985, An evaluation of different methods for labelling the surface of the filarial nematode *Brugia pahangi* with ^{125}Iodine, *Molecular and Biochemical Parasitology*, **15**, 295–304.

Marshall, E. and Howells, R. E., 1986, Turnover of the surface proteins of adult and third and fourth stage larval *Brugia pahangi*, *Molecular and Biochemical Parasitology*, **18**, 17–24.

Mehra, V., Sweetser, D. and Young, R. A., 1986, Efficient mapping of protein antigenic determinants, *Proceedings of the National Academy of Sciences, (USA)*, **83**, 7013–17.

Morgan, T. M., Sutanto, I., Purnomo, Sukartono, Partono, F. and Maizels, R. M., 1986, Antigenic characterization of adult *Wuchereria bancrofti* filarial nematodes, *Parasitology*, **93**, 559–69.

Nutman, T. B., Ottesen, E. A., Fauci, A. S. and Volkman, D. J., 1984, Parasite antigen-specific human T cell lines and clones: Major histocompatibility complex restriction and B cell helper function, *Journal of Clinical Investigation*, **73**, 1754–62.

Ortega-Pierres, G., Chayen, A., Clark, N. W. T. and Parkhouse, R. M. E., 1984, The occurrence of antibodies to hidden and exposed determinants of surface antigens of *Trichinella spiralis*, *Parasitology*, **88**, 359–69.

Ottesen, E. A., 1984, Immunological aspects of lymphatic filariasis and onchocerciasis in man, *Transactions of the Royal Society of Tropical Medicine and Hygiene*, **78**, 9–17.

Ottesen, E. A., Skvaril, F., Tripathy, S. P., Poindexter, R. W. and Hussain, R., 1985, Prominence of IgG4 in the IgG antibody response to human filariasis, *Journal of Immunology*, **134**, 2707–12.

Parkhouse, R. M. E., Philipp, M. and Ogilvie, B. M., 1981, Characterization of surface antigens of *Trichinella spiralis* infective larvae, *Parasite Immunology*, **3**, 339–52.

Philipp, M. and Rumjaneck, F. D., 1984, Antigenic and dynamic properties of helminth surface structures, *Molecular and Biochemical Parasitology*, **10**, 245–68.

Philipp, M., Taylor, P. M., Parkhouse, R. M. and Ogilvie, R. M., 1981, Immune response to stage-specific surface antigens of the parasitic nematode *Trichinella spiralis*, *Journal of Experimental Medicine*, **154**, 210–15.

Philipp, M., Maizels, R. M., McLaren, D. J., Davies, M. W., Suswillo, R. and Denham, D. A., 1986, Expression of cross-reactive surface antigens by microfilariae and adult worms of *Brugia pahangi* during infections in cats, *Transactions of the Royal Society of Tropical Medicine and Hygiene*, **80**, 385–93.

Selkirk, M. E., Denham, D. A., Partono, F., Sutanto, I. and Maizels, R. M., 1986, Molecular characterization of antigens of lymphatic filarial parasites, *Parasitology*, **91**, S15–38.

Selkirk, M. E., Yazdanbakhsh, M., Blaxter, M., Gregory, W. and Maizels, R. M., 1988, Organization, synthesis and structure of cuticular proteins of *Brugia* sp., *Transactions of the Royal Society of Tropical Medicine and Hygiene*, **82**, 820.

Selkirk, M. E., Nielsen, L., Kelly, C., Partono, F., Sayers, G. and Maizels, R. M., 1989, Identification, synthesis and immunogenicity of cuticular collagens from the filarial nematodes *Brugia malayi* and *Brugia pahangi*, *Molecular and Biochemical Parasitology*, **32**, 229–46.

Sutanto, I., Maizels, R. M. and Denham, D. A., 1985, Surface antigens of a filarial nematode: analysis of adult *Brugia pahangi* surface components and their use in monoclonal antibody production, *Molecular and Biochemical Parasitology*, **15**, 203–14.

WHO Expert Committee on Filariasis, 1984, Lymphatic filariasis, *WHO Technical Report Series 702*, Geneva: WHO.

4. Collagen genes in *Ascaris*

I. B. Kingston

INTRODUCTION

Collagens constitute the major component of the extracellular cuticle in *Ascaris* and they have also been isolated from the intestinal basement membrane and the muscle layer of these nematodes. Collagens have not, however, been studied widely in the invertebrates and in discussing the collagens of *Ascaris* it is informative also to consider the extensively studied vertebrate proteins. In addition, the detailed work on the collagens and collagen genes in the free-living, soil-dwelling nematode *Caenorhabditis elegans* provides an important and useful background to the studies in *Ascaris*.

COLLAGEN STRUCTURE

Collagens are a heterogeneous group of structural proteins which have been found in all multicellular animals in which they have been sought. Collagens are distinguished by an unusual and characteristic triple-helical structure formed from regions of three polypeptide chains, the α-chains, which have a repeating amino acid sequence (Gly–X–Y) in which X and Y can be any amino acid but are frequently proline and hydroxyproline (for reviews, see Bornstein and Sage, 1980; Miller and Gay, 1987). Lysine residues which occur in the Y position may also be hydroxlated. This structure is shown schematically in Figure 4.1. Each of the α-chains has a slightly twisted left-handed helical conformation and three of these α-chains are coiled about each other to form a triple-stranded right-handed super helix. Every third residue must be glycine because this is the only amino acid small enough to occupy the interior of the triple helix. This helical conformation is one of the few into which proline residues can fit and their presence imparts rigidity to the structure.

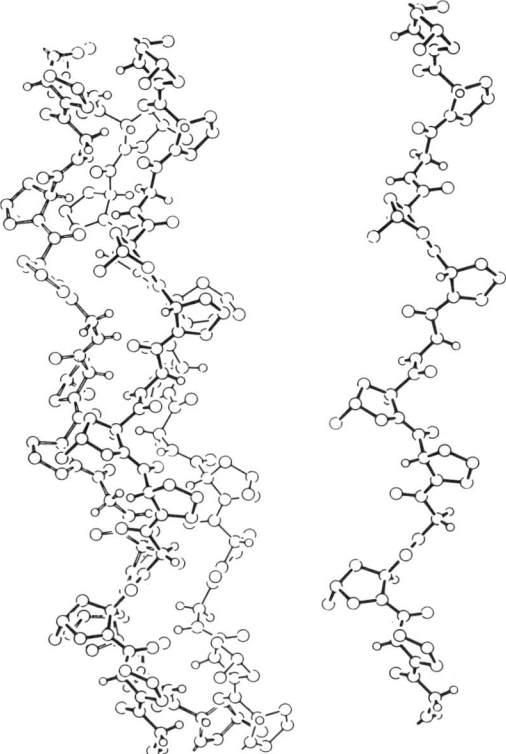

Figure 4.1. A model structure for the collagen molecule represented by the repeating sequence –Gly–Pro–γOH Pro–. The three-stranded coil is shown on the left and a single strand is shown on the right. (From Fraser *et al.*, 1979).

VERTEBRATE COLLAGENS

Collagens constitute some 25 per cent of the total body protein of vertebrates and provide the extracellular framework of almost every organ and tissue. At least 13 different collagen types occur, each with distinct structural properties and tissue distributions. The different types are classified according to their α-chain compositions. The most abundant is the fibre-forming type I collagen found in skin, tendons and bones. This collagen has served as a reference for the other vertebrate fibrillar collagens, types II, III, V and XI. Each of the α-chains in type I collagen has a relative molecular mass (M_r) of approximately 95 000 and each contains an uninterrupted triple-helical domain composed of approximately 1000 amino acids. A schematic drawing of a type I collagen is shown in Figure 4.2. This figure represents the pro-protein and shows the presence of extension peptides at both the *N*-terminal and *C*-terminal ends. These are later cleaved off, but they have an important role in ensuring the correct alignment of the α-chains in the triple helix and of preventing fibre formation within the fibroblasts in which the collagen

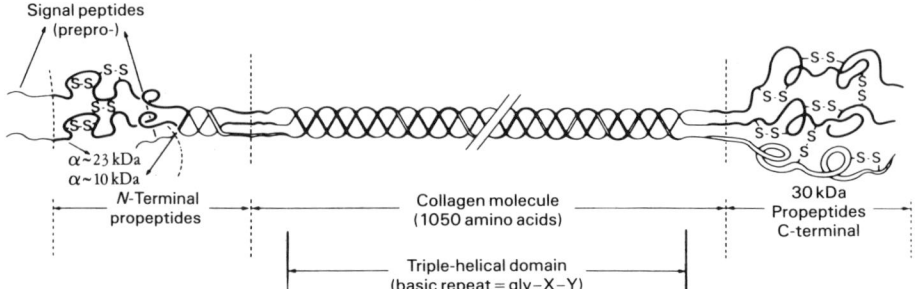

Figure 4.2. Schematic diagram of a type I collagen protein composed of two α1 subunits and one α2 subunit. The dashed lines indicate cleavage sites for specific proteases during the biosynthesis of collagens. (From Vogeli *et al.*, 1980.)

is synthesized. This synthetic process is complex and involves a number of post-translational modifications including the hydroxylation of proline and lysine residues and the establishment of cross-links. Cross-links derived from aldehyde derivatives of lysine are typical in vertebrates but disulphide cross-links also occur.

The other vertebrate collagen types, for example type IV, do not appear to form fibrils or fibres and characteristically have triple-helical regions which are interrupted by amino acid sequences which are incompatible with helix formation. The non-fibrillar collagens can be considered as belonging to one of two groups, one containing those collagens with α-chains similar in size to those found in type I collagen and another containing collagens with α-chains shorter than those found in type I collagen. This latter group is typified by type IX and type X collagen (Olsen *et al.*, 1985).

It is estimated that the multigene family encoding the α-chains of the different vertebrate collagen types consists of a minimum of 20 different genes (Miller and Gay, 1987) and the co-ordinate expression and regulation of these is believed to be important in vertebrate development and differentiation (Adamson, 1982). The genes for a number of these vertebrate collagens have been cloned and their organization and nucleotide sequences have been determined, those of the fibrillar collagens being the best characterized. They are all large genes in the region of 20 to 40 kilobases (kb) long, containing a remarkable number of intervening sequences. For example, the coding sequence of the chicken pro-α2 (type I) gene spans almost 39 kb and is interrupted by 51 introns (Ohkubo *et al.*, 1980; Wozney *et al.*, 1981; Boedtker *et al.*, 1985). Most of the exons code for the triple-helical domain of the chicken pro-α2 (I) polypeptide and these appear to be related to a basic 54 base pair (bp) coding unit; over half are 54 bp long and the remainder are all approximate multiples of 54, being 44, 99, 108 or 162 bp (Boedtker *et al.*, 1985). A similar organization has been found for other vertebrate fibrillar collagen genes and this has led to the hypothesis that the triple-helical coding regions of the fibrillar collagen genes have evolved from an ancestral 54 bp coding unit (Yamada *et al.*, 1980). It now appears likely that only the fibrillar collagen genes have evolved from this ancestral unit as none of the other vertebrate collagen genes has yet been found to exhibit a similar organization.

COLLAGENS AND COLLAGEN GENES IN THE NEMATODE *CAENORHABDITIS ELEGANS*

In contrast to the situation regarding vertebrate collagens, relatively little is known about the structure of invertebrate collagens or about the genes which encode them (Adams, 1978; Murray *et al.*, 1982). Of the invertebrates, the collagens and collagen genes of the free-living, soil dwelling nematode *C. elegans* are currently the best characterized. Like all nematodes, *C. elegans* has a cuticle which completely covers the outside of the animal. This cuticle is a multilayered complex structure, composed principally of proteins, about 80 per cent of which are collagens with smaller amounts of lipid and carbohydrate (McBride and Harrington, 1967; Bird, 1971; Evans *et al.*, 1976; Leushner *et al.*, 1979; Cox *et al.*, 1981a) and it is replaced at each of the four post-embryonic moults which characterize nematode development. The cuticle of each developmental stage, which may have a distinctive morphology and distinctive protein components (Bird, 1971; Cos *et al.*, 1981c), is elaborated by underlying hypodermis that becomes activated for the synthesis of cuticle components at each moult (Kan and Davey, 1968; Singh and Sulston, 1978).

Detailed studies of the synthesis of cuticle proteins during post-embryonic development have been carried out in *C. elegans* and also in *Panagrellus silusiae*, another free-living nematode. In *P. silusiae*, total collagen synthesis during post-embryonic development was found to be discontinuous and a peak of collagen production preceded each moult that was studied (Leushner and Pasternak, 1975; Pasternak and Leushner, 1975). It was postulated that this discontinuous synthesis reflected a quantitive regulation of the production of cuticular collagen during development. More detailed studies of cuticular protein synthesis in *C. elegans* have been in complete accord with this postulate (Cox *et al.*, 1981c; Cox *et al.*, 1981b). It appears, therefore, that during development in these nematodes there is an oscillating pattern of cuticle synthesis with high rates during moulting and reduced rates in intermoult periods.

In addition, the detailed studies of *C. elegans* have shown that the cuticles of the first and second larval stages (L1 and L2) and of the dauer larvae and adult stages differ substantially from one another in ultrastructure and collagen composition and work on the expression of cuticle collagen genes supports the notion that the different collagens present in these developmental stages arise as a result of differential collagen gene expression. Thus, during post-embryonic development in *C. elegans*, cuticle synthesis involves not only a discontinuous synthesis of cuticle components synchronized to the moulting stages but also a programme of genetic switches leading to the elaboration of some cuticle collagens only at specific stages (Cox and Hirsh, 1985). The cuticular collagens of *C. elegans* have also been linked to the development of normal body shape. Genetic analysis of a number of body-shape mutants which have demonstrable anatomical alterations in their cuticle has shown that these mutants probably arise as a result of mutations in collagen genes (Kramer *et al.*, 1988; von Mende *et al.*, 1988).

The collagen genes encoding the cuticular collagens of *C. elegans* have received considerable attention. One of the remarkable findings has been that *C. elegans* has at least 40 and possibly as many as 150 distinct collagen genes dispersed throughout its genome. The reason for this high number of collagen genes is not known, but it may be related to the diversity of the architectural features of the cuticles of the different developmental stages or it may reflect the manner in which the temporal expression of the collagen genes is regulated or the need to produce large amounts of cuticle collagens in relatively short periods of time (Hirsh *et al.*, 1985).

The nucleic acid sequences for a number of *C. elegans* cuticular collagen genes have been determined and they have been found to be quite distinct from vertebrate collagen genes. They encode comparatively small proteins with M_rs of 30 000 to 40 000. They contain only one or two introns, show no evidence of a 54 bp basic unit, and their predicted triple-helical coding regions are interrupted by stretches coding for 2 to 18 amino acids that do not conform to a Gly–X–Y repeating structure (Kramer *et al.*, 1982; Cox *et al.*, 1989).

ASCARIS COLLAGENS AND COLLAGEN GENES

Collagens have been identified in *Ascaris* cuticle, intestinal basement membranes and muscle layer. Those present in the cuticle have been the most extensively studied and various estimates have been made of their molecular weights. The earlier studies of Evans *et al.* (1976) were consistent with the cuticle containing only three different collagen chains each encoded by a different gene but each having a molecular weight of approximately 52 000. The more recent work of Winkfein *et al.* (1985) has not supported this finding and these authors have reported a more heterogeneous composition of not less than seven collagenous components with molecular weights ranging from 173 to 16 kDa.

The collagenous components of basement membrane have been investigated by Hudson and co-workers who have shown that they occur as triple-helical monomers and dimers joined end to end by disulphide bonds (Noelken *et al.*, 1986; Hung *et al.*, 1981). Two polypeptides with molecular weights of 185 000 and 179 000 were identified which appear to be similar to vertebrate type IV collagen. The muscle layer collagens have been the least studied and only preliminary data is available (Fujimoto *et al.*, 1969).

DETECTION AND ISOLATION OF MULTIPLE *ASCARIS SUUM* COLLAGEN GENES

Ascaris collagen genes are now under detailed investigation (Kingston *et al.*, 1989), the findings to date being summarized below. The results of Southern blot hybridization analysis of *Ascaris suum* genomic DNA are shown in Figure 4.3. Two different probes, which were derived from *C. elegans* and *A. suum* collagen genes, were used. The cloned *C. elegans* gene (provided by D. Hirsh, G. N. Cox and C. Fields) was contained within a plasmid (pJC130) which was derived from a genomic clone (λJC43) and contained the collagen gene *col-*18 which is expressed mainly in the adult *C. elegans* (C. Fields, personal communication; Cox and Hirsh, 1985). pJC130 was cleaved with *XhoI* and the fragments derived from the insert were subcloned into M13mp8. Partial sequence analysis of these subclones together with Southern blot analysis showed that all those containing a collagen-coding region hybridized to *A. suum* genomic DNA and gave a common pattern of bands in the Southern blot. The intensity of the bands varied between the subclones; however, pJC130.2.5 giving the strongest signal. This subclone was used in all further hybridization experiments. Partial sequence analysis of the only *C. elegans* subclone which did not hybridize to *A. suum* DNA indicated that it was unlikely to contain a collagen-encoding region; this same subclone hybridized to only a single

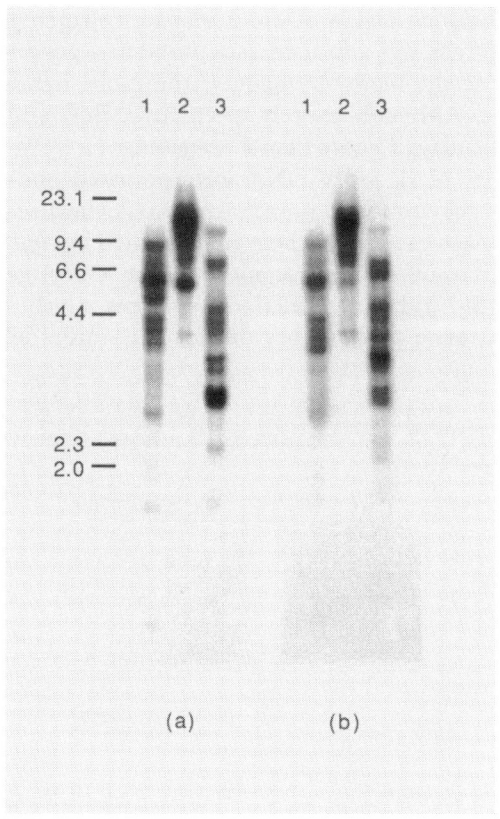

Figure 4.3. Genomic Southern blots of *A. suum* DNA. *A. suum* DNA was digested with *Hind*III (track 1), *Bam* HI (track 2) and *Eco* RI (track 3) and fractionated by electrophoresis on a 0.5 per cent agarose gel. After transfer to nitrocellulose, the blot was hybridized to ^{32}P-labelled probes under conditions of low stringency (55°C, 4 × SET). Two probes were used: (a) a *C. elegans* collagen gene probe derived from *col-*18 (see text); (b) an *A. suum* collagen gene probe corresponding to the sequence from nucleotides 1 to 210 in Figure 4.6. The numbers on the left are lengths (in kb) of restriction fragment markers. (61 × SET is 0.15 M NaCl, 2mM EDTA, 0.03 M Tris-HCl pH 7.4). The probes were single-strand DNA probes derived from M13 clones. Probe preparation and hybridization conditions were as described by Kingston and Anderson (1986). (Data from Kingston *et al.*, 1989.)

band in a genomic Southern blot of *C. elegans* DNA (not shown). This result supports the notion that the hybridization of the *C. elegans* probe pJC130.2.5 is due to similarity to the triple-helical coding region of the *C. elegans* collagen gene and that there is little or no similarity between the flanking region of this *C. elegans* gene and the flanking regions of *A. suum* genes.

At low stringency, the *C. elegans* collagen gene fragment hybridized to approximately 15 bands with varying intensity (Figure 4.3a). At a higher stringency, the probe hybridized strongly to only one or two of these bands (not shown). When the *A. suum* collagen gene was used as a probe, a rather similar result was obtained (Figure 4.3b). Both probes appeared to hybridize largely to the same set of restriction fragments. The two probes also hybridized at similar intensities suggesting that the *A. suum* genes are no more similar to each other than they are to the subcloned region of the *C. elegans* collagen gene. The *C. elegans* collagen gene fragment was also used as a probe in a genomic Southern blot of *C. elegans* DNA. At the higher stringency, the probe hybridized to at least 20 bands (not shown). Although entirely speculative, this result may indicate that *A. suum* has fewer collagen genes than *C. elegans*.

Genomic libraries of *A. suum* DNA were prepared and screened using the *C. elegans* collagen gene probe. A number of clones hybridized with the probe and simple restriction maps of seven of these are presented in Figure 4.4. These appear to represent seven non-overlapping segments of the *Ascaris* genome. The stippled regions in Figure 4.4 indicate restriction fragments which hybridized with the *C. elegans* collagen gene probe. They do not necessarily delineate the collagen

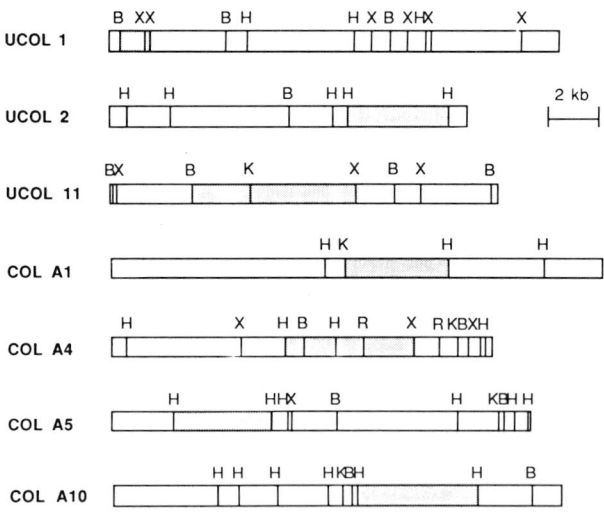

Figure 4.4. Restriction maps of recombinant bacteriophage containing *A. suum* collagen genes. Only the *A. suum* DNA inserts are shown. The hatched regions indicate those fragments which hybridized with the *C. elegans* collagen gene probe. Only two *Eco* RI sites (R) are shown in COLA4, but at least nine others are present: B, *Bam* HI; X, *Xho* I; H, *Hind* III; and K, *Kpn* I. The vector used was λ2001 (Karn *et al.*, 1984). (From Kingston and Pettitt, 1990.)

genes but represent only the closest set of restriction enzyme sites used in the mapping experiments. In most cases the collagen genes appear to be located within approximately 4 to 6 kb segments of DNA.

DNA SEQUENCE ANALYSIS OF TWO *A. SUUM* COLLAGEN GENES

Two recombinant bacteriophages, COLA4 and UCOL1, have been isolated from genomic libraries of *A. suum* DNA and subjected to nucleic acid sequence analysis. The nucleic acid sequence of the entire 4.5 kb at the 5′ end of the insert in clone UCOL1 was determined and part of this sequence is presented in Figure 4.5. The sequence contains an open reading frame of 684 bp which encodes a repeating, but discontinuous Gly–X–Y sequence. Three triple-helical domains containing glycyl residues in every third position are interrupted by stretches of amino acids which depart from the Gly–X–Y sequence. The average proline content of the triple-helical domains is 34 per cent. The nucleic acid sequences at the 5′ and 3′ ends of the open reading frame encode non-triple-helical amino acid sequences and at the 3′ end there is an in-frame termination codon. Potential polyadenylation signals, AATAAA (Proudfoot and Brownlee, 1976), are present 356 bp and 1963 bp downstream from the termination codon. A potential intron/exon acceptor splice site has been identified. This site closely resembles a consensus intron/exon boundary sequence (Mount, 1982) and upstream from this site there are numerous termination codons in all three reading frames. No region upstream from this site contains a sequence which appears to correspond to an exon/intron donor site and this site must, therefore, occur upstream from the *A. suum* DNA sequence present in this clone.

A 3 kb *Eco*RI fragment which hybridized with the *C. elegans* probe was isolated from COLA4. This *Eco*RI fragment was the only *Eco*RI fragment that appeared to hybridize in an initial Southern blot with the *C. elegans* probe. Subsequent more detailed restriction mapping, however, revealed that the region of hybridization in this clone was more extensive (Figure 4.4). The region in which this additional hybridization was detected contains five *Eco*RI sites (not shown) that give rise to small fragments, which presumably failed to hybridize or gave only a weak or undetectable signal in the original Southern blot.

The nucleotide sequence of the 3 kb *Eco*RI fragment from clone COLA4 is presented in Figure 4.6. There are two putative collagen-coding regions (denoted by an overbar): one, which is interrupted by a seven amino acid stretch that departs from the Gly–X–Y sequence, extends from the 5′ end of the fragment to base 181 and the other extends from base 742 to base 813. They contain glycine as every third amino acid and have a high proline content (28 per cent). The sequence between these two collagen-coding regions is predicted to be an intron. This prediction is based on the observation that there are numerous termination codons in all three potential reading frames of the sequence and that it is bounded by sequences which closely resemble consensus exon/intron and intron/exon boundary sequences

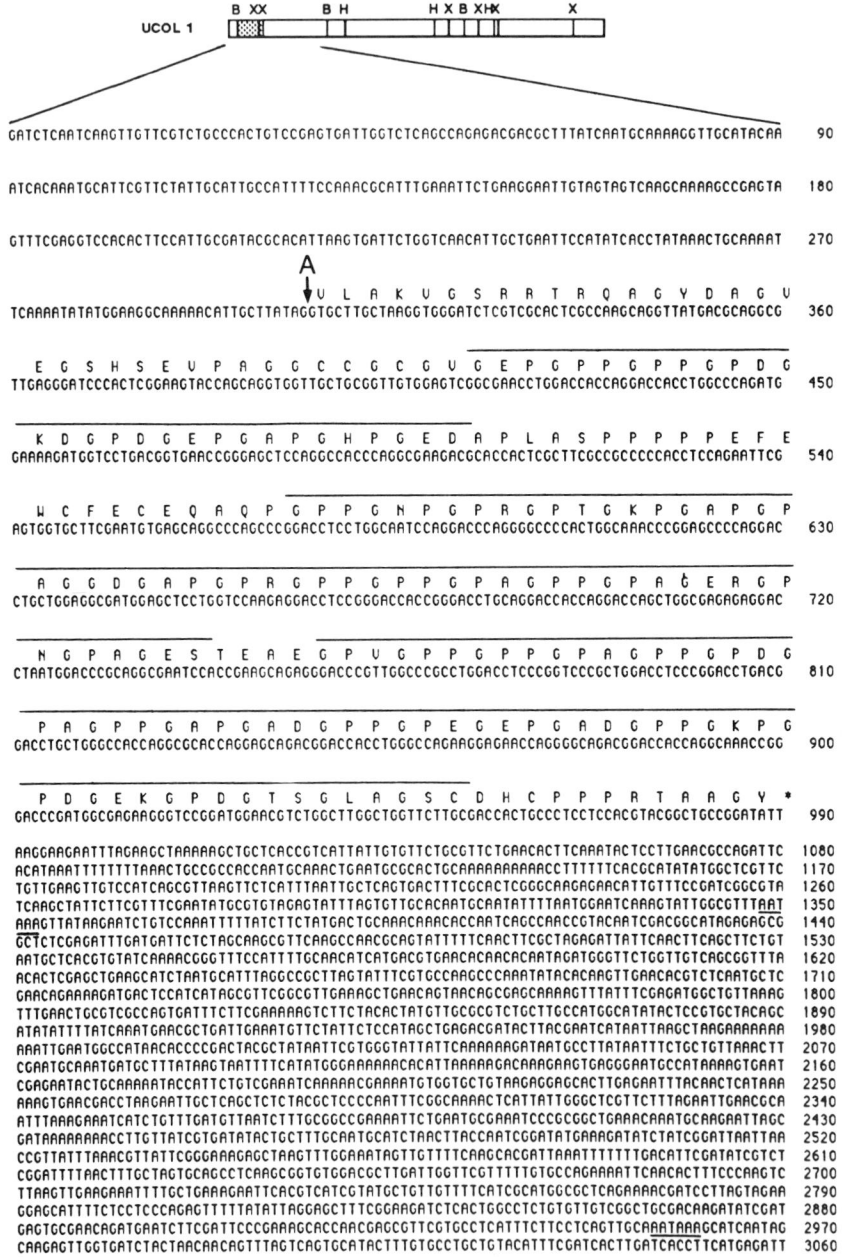

Figure 4.5. Restriction map of clone UCOL1 and the nucleotide sequence of the 5′ end of the clone. The derived amino acid sequence spanning the collagen-coding region is shown above the nucleotide sequence. Potential triple-helical regions containing glycine as every third amino acid are denoted by an overbar. An arrow indicates a potential intron/exon (acceptor) splice junction. Potential polyadenylation signals, AATAAA, are underlined. In the restriction map, B, *Bam* HI; X, *Xho* I; H, *Hind* III. The sequence was determined by the chain termination method (Bankier *et al.*, 1988). (From Kingston *et al.*, 1989.)

(Mount, 1982). The intron is 560 bp long. There is an in-frame termination codon at the end of the last segment and there are potential polyadenylation signals (AATAAA) at positions 1911 and 2480. The 5′ collagen-encoding region is not preceded by an initiation codon and, in agreement with the restriction map data, this putative gene must begin further upstream.

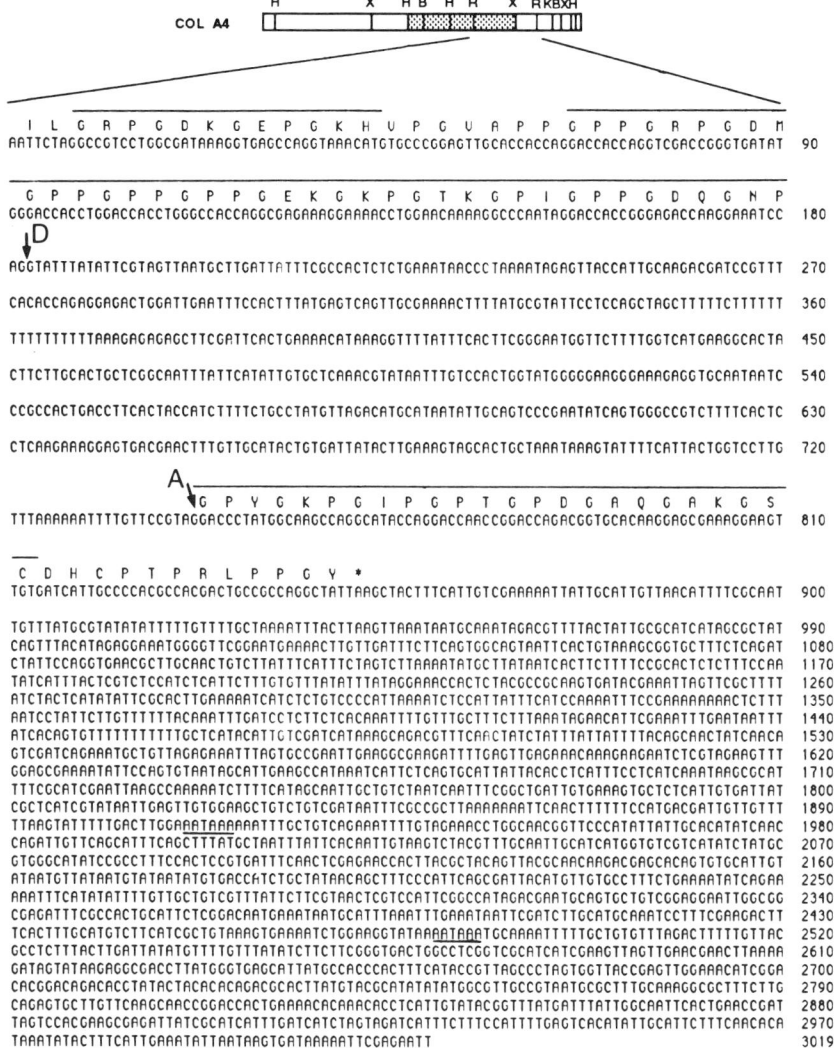

Figure 4.6. Restriction map of clone COLA4 and the nucleotide sequence of a 3 kb Eco RI fragment from the clone. The derived amino acid sequence spanning the collagen-coding regions are shown above the nucleotide sequence. Potential triple-helical regions containing glycine as every third amino acid are denoted by an overbar. The asterisk indicates an in-frame termination codon. Arrows indicate potential exon/intron (donor) and intron/exon (acceptor) splice junctions. Potential polyadenylation signals, AATAAA, are underlined. In the restriction map: B, *Bam* HI: X, *Xho*, I; H, *Hind* III; K, *Kpn* I. Only two *Eco* RI sites (R) are shown but at least nine others are present (see text). (From Kingston *et al.*, 1989.)

EXPRESSION OF *A. SUUM* COLLAGEN GENES

In order to obtain preliminary information about the expression of *A. suum* collagen genes, Northern blot hybridization analysis of messenger RNA isolated from total body-wall tissue of *A. suum* was carried out. Four different probes were used. Two were prepared from COLA4, one being the complement of the sequence from position 1 to position 210 in Figure 4.6 and the other being the complement of the sequence from position 976 to position 1175. The latter would correspond to the non-translated 3′ end of the gene. One probe was prepared from UCOL1 and was the complement of the sequence from approximately position 120 to position 546 in Figure 4.5. The fourth probe was the *C. elegans* probe pJC130.2.5 described above.

The probe derived from UCOL1 gave rise to two strongly hybridizing bands (Figure 4.7) whose sizes, estimated from the positions of the 18S and 28S rRNA

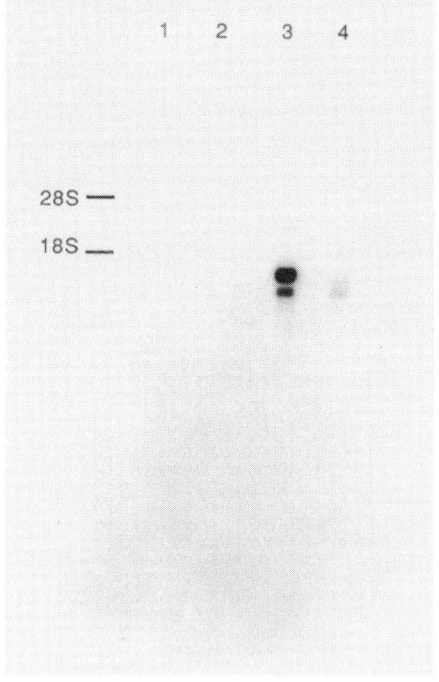

Figure 4.7. Northern blot of *A. suum* RNA. Four samples of RNA were denatured using glyoxal and fractionated on a 1.1 per cent agarose gel. Four different probes were used (see text). In track 1, the probe corresponded to the complement of the sequence from the nucleotides 1 to 210 in the COLA4 gene (Figure 4.6). In track 2, the probe corresponded to the complement of the sequence from nucleotide 976 to 1175 in the COLA4 gene (Figure 4.6). In track 3, the probe corresponded to the complement of the sequence from approximately nucleotide 120 to 546 in the UCOL1 gene (Figure 4.5). In track 4, the probe was the *C. elegans* clone pJC130.2.5. The probes were single-strand DNA probes from M13 clones. Probe preparation and hybridization conditions were as described by Kingston and Anderson (1986). The final wash was in 0.15 × SCP/0.1 per cent sodium dodecyl sulphate (SDS) at 65 °C (1 × SCP is 0.1 M NaCl, 0.03 M NaH$_2$ PO$_4$, 1 mM ethylene glycol-bis(β-aminoethyl ether) N, N, N', N'-tetraacetic acid (EDTA), pH 6.2). (From Kingston *et al.*, 1989.)

bands, were 1100 and 1400 bp, respectively. Transcripts of rather similar size were detected much more weakly by the *C. elegans* probe and very much more weakly (and visible only after prolonged exposure of the autoradiograph) by the probe derived from the triple-helical coding region of COLA4. This result suggests that the UCOL1 collagen gene is expressed in adult *A. suum* body-wall tissue and that the COLA4 collagen gene is probably not. The sizes of the transcripts which hybridized to the UCOL1 probe support the notion that the genomic sequence shown in Figure 4.5 contains almost the entire collagen coding sequence and they are consistent with this gene encoding a protein of approximately M_r 30 000.

COMPARISON OF THE TWO *A. SUUM* COLLAGEN GENES WITH EACH OTHER AND WITH OTHER COLLAGEN GENES

The nucleic acid sequences of two collagen genes, *col*-1 and *col*-2, from *C. elegans* have been reported by Kramer *et al.* (1982). The *col*-1 gene encodes a protein of 296 amino acids and contains two short introns of 102 nucleotides and 52 nucleotides. The *col*-2 gene encodes a protein of 301 amino acids and contains a single short intron of 47 nucleotides. The proteins encoded by these genes show a remarkable similarity in the position of triple-helical regions and in the position of cysteine and lysine residues (Kramer *et al.*, 1982). A comparison of the organization and coding potential of the *A. suum* collagen genes with each other and with these two *C. elegans* collagen genes is shown in Figure 4.8. A striking feature of this comparison is the almost identical size and position of the triple-helical domains in all four deduced polypeptides. The gene in clone UCOL1 encodes a total of 51 Gly–X–Y repeats in the three triple-helical domains shown in Figure 4.8, and the two *C. elegans* genes both encode 50 such repeats. When the sequences are aligned in this manner, there is also a striking near identity in the location of cysteine residues and some of these may, therefore, form the disulphide cross-links which are known to be important in nematode cuticle collagens (McBride and Harrington, 1967; Leushner *et al.*, 1979; Cox *et al.*, 1981a; Ouazana and Herbage, 1981). In the *C. elegans* proteins, the position of lysine residues also show a close correspondence (Kramer *et al.*, 1982) but only one of these, within the 30 residue triple-helical region, is found in a corresponding position in the UCOL1 gene product.

Although the exons in each of the four collagen genes depicted in Figure 4.8 encode similar polypeptides, the size and position of their introns differ. The introns present in the *A. suum* genes are larger than those found in *C. elegans*. The COLA4 gene stands out from the others in that it contains an intron of 560 bp which interrupts a triple-helical coding region. The UCOL1 gene contains an intron in a similar position to that found in the *C. elegans* gene *col*-1 but, even though the sequence is incomplete, the *A. suum* intron is clearly longer.

The *A. suum* collagen gene in clone UCOL1 was compared with the *C. elegans* genes *col*-1 and *col*-2 using the computer program DIAGON (Staden, 1982), ALN3 (Gotoh, 1986) and ALIGN (PIR, National Biomedical Research Foundation,

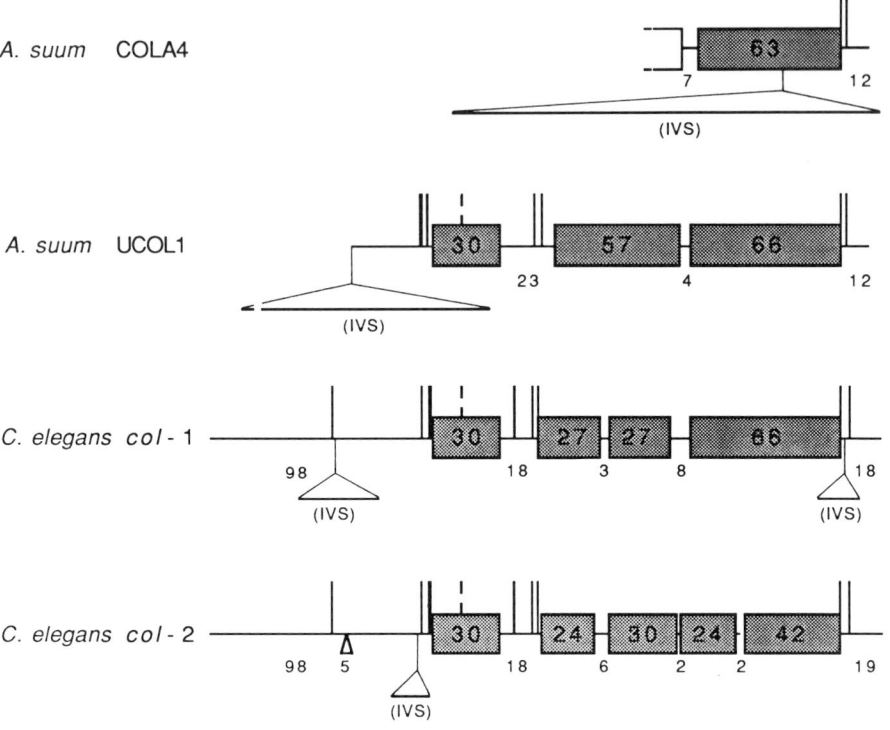

Figure 4.8. Comparison of the *A. suum* collagen genes in clones UCOL1 and COLA4 and the *C. elegans* collagen genes *col*-1 and *col*-2 (Kramer *et al.*, 1982). The shaded boxes represent triple-helical coding regions which encode glycine as every third amino acid. Non-helical regions are represented by the horizontal line. The numbers of amino acids in each region are indicated. Cysteine residues are indicated by solid vertical lines and the position of a single lysine residue is indicated by a broken vertical line. The position of intervening sequences (IVS) is also shown. For maximum alignment of *col*-1 and *col*-2, a five amino acid coding region (indicated by the triangle) has been looped out of *col*-2 between the amino terminus and the first triple-helical region and a one amino acid gap has been made in *col*-2 immediately in front of the fifth triple-helical region (Kramer *et al.*, 1982). From (Kingston *et al.*, 1989.)

Georgetown University Medical Centre, Washington, DC, USA). The result of a comparison of the UCOL1 gene with *col*-1 using ALIGN is shown in Figure 4.9. The regions compared were restricted to the triple-helical coding regions and their associated interruptions, i.e. in UCOL1, bases 405 to 953 (Figure 4.4), and in *col*-1, bases 391 to 936 (Kramer *et al.*, 1982). This corresponds to the sequences between the common cysteine residues immediately preceding the 30 amino acid triple-helical region and the common cysteine residue at the end of the C-terminal triple-helical region (Figure 4.8). The constraint of having a glycine codon as every third codon in the triple-helical coding regions increases their overall level of similarity. After introducing three gaps into the UCOL1 gene sequence and one into the *col*-1 sequence, there are 313 identical bases out of 540 possible matches. A comparison of the UCOL1 gene with *col*-2 gave a similar result and the programme ALN3 produced a similar proportion of matches (not shown).

```
TGCCTTCCAGGACCACCAGGACCAGCTGGAGCCCCAGGAAAGCCAGGAAAGCCAGGACGT      450
**         **    *** ******  *  *** *  ** **           *****    **  *  *
TGTGGAGTCGGCGAACCTGGACCACCAGGACCACCTGGCCCAGATGGAAAAGATGGTCCT      464

CCAGGAGCACCAGGAACTCCAGGAACCCCAGGAAAGCCACCAGTTGCCCCATGTGAGCCA      510
**  *  *** *** *******   ******   *      **      * * *     ***
GACGGTGAACCGGGAGCTCCAGGCCACCCAGGCGAAGACGCACCACTCGCTTC---GCCG      521

ACTACTCCACCACCATGCAAGCCATGCCCACAAGGACCACCAGGACCACCAGGACCACCA      570
*  *  *** **  ** ** * **    ***   ***   *  *  *  ** ***** ***
CCCCCACCTCCAGAATTCGAGTGGTGCTTCGAATGTGAGCAGGCCCAGCCCGGACCTCCT      581

GGAGCACCAGGAGACCCGGGAGAGGCTGGAACCCCAGGACGCCCAGGGACCGATGCCGCC      630
**   ******  *  ***    ****  ** ** *** ******   *  *  **   * *
GGCAATCCAGGACCCAGGGGCCCCACTGGCAAACCCGGAGCCCCAGGACCTGCTGGAGGC      641

CCAGGATCCCCAGGACCACGTGGACCACCAGGACCAGCTGGAGAGGCCGGAGCCCCAGGA      690
 *** * ** **  ** *  * *****  **  ******  *  *** **  *** * ******
GATGGAGCTCCTGGTCCAAGAGGACCTCCGGGACCACCGGGACCTGCAGGACCACCAGGA      701

CCAGCCGGAGAGCCAGGAACCCCAGCTATTTCCGAGCCACTCACCCCAGGAGCACCAGGA      750
*****  ** ***   **** **    *       *      * ** *  ** ** * ****      ***
CCAGCTGGCGAGAGAGGA--CCTAATGGACCCGCAGGCGAATCCACCGA---AGCAGAGGGA  758

GAGCCAGGAGACTCCGGACCACCAGGACCACCAGGACCACCAGGAGCACCAGGAAACGAC      810
  *  *        *  * ***** **  ** **   *  ***** ** *** *     ***  *
CCCGTTGGCCCGCCTGGACCTCCCGGTCCCGCTGGACCTCCCGGACCTGACGGACCTGCT      818

GGACCGCCAGGACCACCAGGACCAAAGGGAGCCCCAGGACCAGACGGACCACCAGGAGCC      870
** **  *****   ********* **  * *** *  * *    ***** ***   ****** **
GGGCCACCAGGCGCACCAGGAGCAGACGGACCACCTGGGCCAGAAGGAGAACCAGGGGCA      878

GACGGACAATCCGGACCACCAGGACCACCAGGA----------CCAGCTGGAACCCCAGGA      921
******** *  *  **   *** *****   **          ** * *******   *  **
GACGGACCACCAGGCAAACCGGGACCCGATGGCGAGAAGGGTCCGGATGGAACGTCTGGC      938

GAGAAGGGAATCTGT                                                936
 *   **    **
TTGGCTGGTTCTTGC                                                953
```

Figure 4.9. Comparison of the nucleotide sequences of the *A. suum* collagen gene in clone UCOL1 and the *C. elegans* gene *col*-1 (Kramer *et al.*, 1982). The regions compared are those spanning the triple-helical regions (see text). The upper sequence is part of the *A. suum* gene and the lower one is part of the *C. elegans* gene. The sequences are aligned to maximize similarity and dashes (—) indicate where compensatory gaps have been introduced. An asterisk (*) indicates where the sequences are identical. The boxed nucleotide sequences are potential triple-helical coding regions.

Although the glycine and proline codons in the two *Ascaris* genes show some preference for adenine in the third position they do not show the extreme preference exhibited by the *C. elegans* collagen genes (Kramer *et al.*, 1982). In the UCOL1 gene, 30 of the 57 glycine codons end with adenine as do 26 of the 63 proline codons. In the COLA4 gene, 14 of 27 glycine codons and 18 of 28 proline codons end with adenine.

The *A. suum* collagen genes show little resemblance to those studied in two other invertebrates *Drosophila melanogaster* and *Strongylocentrotus purpuratus*. The analysis of a collagen gene from *D. melanogaster* is incomplete (Monson *et al.*, 1982), but a segment of the gene contains two long triple-helical coding regions of

682 bp and 725 bp interrupted by a single short (62 bp) intron. The analysis of a collagen gene from *S. purpuratus* is similarly incomplete but this also appears quite distinct from the *A. suum* collagen gene. The *S. purpuratus* gene hybridized to a 9 kb mRNA and contained multiple (at least 15) short (200–400 bp) exons interrupted by intervening sequences that ranged in size from 400 to 1300 bp (Venkatesan *et al.*, 1986). None of the vertebrate collagens described to date correspond very closely to the *A. suum* collagens. The most similar are the short-chain collagens type IX and type XII (Gordon *et al.*, 1987).

ROLE OF THE TWO *A. SUUM* COLLAGEN GENE PRODUCTS

The very strong similarity between the deduced *A. suum* proteins and the *C. elegans* proteins suggests that they share a common and specific structural role. There is strong evidence that the *C. elegans* proteins are cuticular collagens (Kramer *et al.*, 1982, 1985) and thus it is likely that the *Ascaris* genes described here also encode cuticular collagens. This notion is supported by the Northern blot result, in which the UCOL1 gene hybridized to mRNA species almost identical in size to those detected when *col*-1 or *col*-2 were used to probe *C. elegans* mRNA (Cox *et al.*, 1984). The *C. elegans* genes hybridized to abundant transcripts of 1.2 kb and 1.4 kb and the *A. suum* UCOL1 gene hybridized to transcripts of 1.1 kb and 1.4 kb (Figure 4.6).

The transcripts detected by the UCOL1 gene would be expected to encode a polypeptide of approximately M_r 30 000. Although various estimates have been made of the sizes of *Ascaris* cuticular collagens, none has identified a major cuticle collagen of this size (McBride and Harrington, 1967; Evans *et al.*, 1976; Winkfein *et al.*, 1985). Winkfein *et al.* (1985) concluded that 95 per cent of the cuticular collagens were accounted for by polypeptides with molecular weights of 99, 90 and 65 kDa, the rest being accounted for by polypeptides with molecular weights of 173, 157, 28 and 16 kDa. These polypeptides were identified after polyacrylamide gel electrophoresis (PAGE) of reduced and carboxymethylated cuticle proteins. Although lysine cross-links have not been detected in *Ascaris* cuticle proteins (Bailey, 1971), their presence remains a possibility, particularly in view of the occurrence of a lysine residue in the protein encoded by the UCOL1 gene whose position matches that of a common lysine in the proteins encoded by *col*-1 and *col*-2. It has been suggested that lysine cross-links may occur in *C. elegans* cuticle collagens (Kramer *et al.*, 1985) and it is tempting to speculate that in *Ascaris* too, lysine cross-links may occur at a very low frequency so that the sizes of collagen polypeptides detected by sodium dodecyl sulphate (SDS) gel electrophoresis after full reduction and alkylation of cysteine residues may not represent single gene products. Although we do not yet know whether the UCOL1 gene product is a major cuticle component, it is notable that the abundant proteins detected by Winkfein *et al.* (1985) had molecular weights which are approximate multiples of the predicted approximately M_r 30 000 protein encoded by the UCOL1 gene.

CONCLUDING REMARKS

It is clear that, despite overall similarities, the structure of the two *Ascaris* putative cuticular collagen genes described above show important differences from those of *C. elegans*. It remains to be established, however, whether or not these differences reflect disparities in their function and developmental control. The observed multiplicity of collagen genes in *C. elegans* has not been found to apply to *Ascaris* but further studies are required in order to assess more confidently the number of *Ascaris* collagen genes. There is no reason to suppose, for instance, that the cuticle of *Ascaris* in all its developmental stages is any simpler than that of *C. elegans* and that *Ascaris* consequently requires a smaller set of collagen proteins. A final point would be that the sheer size of many parasitic species of nematode, *Ascaris* being an extreme example, might have imposed structural requirements which do not apply to *C. elegans*. This might, therefore, have led to the evolution of cuticular collagens in large parasites which fall outside the classes and families which have been described in their smaller free-living relatives.

ACKNOWLEDGMENTS

Work in the author's laboratory was supported by a grant from The Wellcome Trust.

REFERENCES

Adams, E., 1978, Invertebrate collagens, *Science*, **202**, 591–8.
Adamson, E. D., 1982, The effect of collagen on cell division, cellular differentiation and embryonic development, in Weiss, J. B. and Jayson, M. I. V. (Eds) *Collagen in Health and Disease*, pp. 218–43, Edinburgh: Churchill Livingstone.
Bailey, A. J., 1971, Comparative studies on the nature of the cross-links stabilizing the collagen fibres of invertebrates, cyclostomes and elasmobranchs, *FEBS Letters*, **18**, 154–8.
Bankier, A. T., Weston, K. M. and Barrell, B. G., 1988, Random cloning and sequencing by the M13/dideoxynucleotide chain termination method, *Methods in Enzymology*, **155**, 51–93.
Bird, A. F., 1971, *The Structure of Nematodes*, New York: Academic Press.
Boedtker, H., Finer, M. and Aho, S., 1985, The structure of the chicken α2 collagen gene, *Annals of the New York Academy of Science*, **460**, 85–116.
Bornstein, P. and Sage, H., 1980, Structurally distinct collagen types, *Annual Reviews in Biochemistry*, **49**, 957–1003.
Cox, G. N. and Hirsh, D., 1985, Stage-specific patterns of collagen gene expression during development of *Caenorhabditis elegans*, *Molecular and Cellular Biology*, **5**, 363–72.
Cox, G. N., Kusch, M. and Edgar, R. S., 1981a, Cuticle of *Caenorhabditis elegans*: its isolation and partial characterization, *Journal of Cell Biology*, **90**, 7–17.
Cox, G. N., Kusch, M., De Nevi, K. and Edgar, R. S., 1981b, Temporal regulation of cuticle synthesis during development of *Caenorhabditis elegans*, *Developmental Biology*, **84**, 277–85.

Cox, G. N., Staprans, S. and Edgar, R. S., 1981c, The cuticle of *Caenorhabditis elegans*, *Developmental Biology*, **86**, 456–70.

Cox, G. N., Kramer, J. M. and Hirsh, D., 1984, Number and organization of collagen genes in *Caenorhabditis elegans*, *Molecular and Cellular Biology*, **4**, 2389–95.

Cox, G. N., Fields, C., Kramer, J. M., Rozenzweig, B. and Hirsh, D., 1989, Sequence comparisons of developmentally regulated collagen genes of *Caenorhabditis elegans*, *Gene*, **76**, 331–44.

Evans, H. J., Sullivan, C. E. and Piez, K. A., 1976, The resolution of *Ascaris* cuticle collagen into three chain types, *Biochemistry*, **15**, 1435–9.

Fraser, R. D. B., MacRae, T. P. and Suzuki, E., 1979, Chain conformation in the collagen molecule, *Journal of Molecular Biology*, Series C **129**, 463–81.

Fujimoto, D., Ikeuchi, T. and Nozawa, S., 1969, Intermolecular disulphide cross-linkages in the collagen from the muscle layer of *Ascaris lumbricoides*, *Biochemica et Biophysica Acta*, **188**, 295–301.

Gordon, M. K., Gerecke, D. R. and Olsen, B. R., 1987, Type XII collagen: distinct extracellular matrix component discovered by cDNA cloning, *Proceedings of the National Academy of Sciences (USA)*, **84**, 6040–4.

Gotoh, O., 1986, Alignment of three biological sequences with an efficient traceback procedure, *Journal of Theoretical Biology*, **121**, 327–37.

Hirsh, D., Cox, G. N., Kramer, J. M., Stinchcombe, D. and Jefferson, R., 1985, Structure and expression of the collagen genes of *Caenorhabditis elegans*, *Annals of New York Academy of Science*, **460**, 163–71.

Hung, C.-H., Noelken, M. E. and Hudson, B. G., 1981, Intestinal basement membrane of *Ascaris suum*, physical properties of the collagenous domain, *Journal of Biological Chemistry*, **256**, 3822–6.

Kan, S. P. and Davey, M. G., 1968, Moulting in a parasitic nematode, *Phocanema decipiens*. III. The histochemistry of cuticle deposition and protein synthesis, *Canadian Journal of Zoology*, **46**, 723–7.

Karn, J., Matthes, H. W. D., Gait, M. J. and Brenner, S., 1984, A new selective phage cloning vector, 2001, with sites for *Xba*I, *Bam*HI, *Hind*.III, *Eco*RI, *Sst*I and *Xho*I, *Gene*, **32**, 217–24.

Kingston, I. B. and Anderson, S., 1986, Sequences encoding two trypsin inhibitors occur in strikingly similar genomic environments, *Biochemical Journal*, **233**, 443–50.

Kingston, I. B. and Pettitt, J., 1990, Structure and expression of *Ascaris suum* collagen genes: a comparison with *Caenorhabditis elegans*, *Acta Tropica*, **47**, 283–7.

Kingston, I. B., Wainwright, S. M. and Cooper, D., 1989, Comparison of collagen gene sequences in *Ascaris suum* and *Caenorhabditis elegans*, *Molecular and Biochemical Parasitology*, **37**, 137–46.

Kramer, J. M., Cox, G. N. and Hirsh, D., 1982, Comparisons of the complete sequences of two collagen genes from *Caenorhabditis elegans*, *Cell*, **30**, 599–606.

Kramer, J. M., Cox, G. N. and Hirsh, D., 1985, Expression of the *Caenorhaditis elegans* collagen genes *col*-1 and *col*-2 is developmentally regulated, *Journal of Biological Chemistry*, **260**, 1945–51.

Kramer, J. M., Johnson, J. L., Edgar, R. S., Basch, C. and Roberts, S., 1988, The *sqt*-1 gene of *C. elegans* encodes a collagen critical for organismal morphogenesis, *Cell*, **55**, 555–65.

Leushner, J. R. A. and Pasternak, J. P., 1975, Programmed synthesis of collagen during postembryonic development of the nematode *Panagrellus silusiae*, *Developmental Biology*, **47**, 68–80.

Leushner, J. R. A., Semple, N. E. and Pasternak, J. P., 1979, Isolation and characterization of the cuticle from the free-living nematode *Panagrellus silusiae*, *Biochimica et Biophysica Acta*, **580**, 166–74.

McBride, O. W. and Harrington, W. F., 1967, *Ascaris* cuticle collagen: on the disulphide cross-linkages and the molecular properties of the subunits, *Biochemical Journal*, **6**, 1484–94.

Miller, E. J. and Gay, S., 1987, The collagens: an over view and update, *Methods in Enzymology*, **144**, 3–41.

Monson, J. M., Natzle, J., Friedman, J. and McCarthy, B. J., 1982, Expression and novel structure of a collagen gene in *Drosophila*, *Proceedings of the National Academy of Sciences (USA)*, **79**, 1761–5.

Mount, S. M., 1982, A catalogue of splice junction sequences, *Nucleic Acid Research*, **10**, 459–72.

Murray, L. W., Waite, J. H., Tanzer, M. L. and Houschka, P. V., 1982, Preparation and characterization of invertebrate collagens, *Methods in Enzymology*, **82**, 65–96.

Noelken, M. E., Wisdom, B. J., Dean, D. C., Hung, C.-H. and Hudson, B. G., 1986, Intestinal basement membrane of *Ascaris suum*, *Journal of Biological Chemistry*, **261**, 4706–14.

Ohkubu, H., Vogeli, G., Mudryi, M., Avvedimento, V. E., Sullivan, M., Pasten, I. and de Crombrugghe, B., 1980, Isolation and characterization of overlapping genomic clones covering the chicken α2 (type I) collagen gene, *Proceedings of the National Academy of Sciences (USA)*, **77**, 7059–63.

Olsen, B. R., Ninomiya, Y., Lanzano, G., Komomi, H., Gordon, M., Green, G., Parson, J., Seyer, J., Thompson, H. and Vasios, G., 1985, Short-chain collagen genes and their expression in cartilage, *Annals of the New York Academy of Sciences*, **460**, 141–53.

Ouazana, R. and Herbage, D., 1981, Biochemical characterization of the cuticle collagen of the nematode *Caenorhabditis elegans*, *Biochimica et Biophysica Acta*, **669**, 236–43.

Pasternak, J. P. and Leushner, J. R. A., 1975, Programmed collagen synthesis during postembryonic development of the nematode *Panagrellus silusiae*: effect of transcription and translation inhibitors, *Journal of Experimental Zoology*, **194**, 519–28.

Proudfoot, N. J. and Brownlee, G. G., 1976, 3′ Non-coding region sequences in eukaryotic messenger RNA, *Nature (London)*, **263**, 211–14.

Singh, R. N. and Sulston, J. E., 1978, Some observations on moulting in *Caenorhabditis elegans*, *Nematologica*, **24**, 63–71.

Staden, R., 1982, An interactive graphics program for comparing and aligning nucleic acid and amino acid sequences, *Nucleic Acid Research*, **10**, 2951–61.

Venkatesan, M., de Pablo, F., Vogeli, G. and Simpson, R. T., 1986, Structure and developmentally regulated expression of a *Strongylocentrotus pupuratus* collagen gene, *Proceedings of the National Academy of Sciences (USA)*, **83**, 3351–5.

Vogeli, G., Ohkubo, H., Avvedimento, V. E., Sullivan, M., Yamack, Y., Mudryi, M., Pastan, I. and DeCrombrugghe, B., 1980, A repetitive structure in the chick 2-collagen gene, *Cold Spring Harbor Symposia on Quantitative Biology*, **45**, 777–83.

von Mende, N., Bird, D. Mck., Albert, P. S. and Riddle, D. L., 1988, *dpy*-13: a nematode collagen gene that affects body shape, *Cell*, **55**, 567–76.

Winkfein, R. J., Pasternak, J., Mundry, T. and Martin, L. H., 1985, *Ascaris lumbricoides*: characterization of the collagenous components of the adult cuticle, *Experimental Parasitology*, **59**, 197–203.

Wozney, J., Hanahan, D., Tate, V., Boedtker, H. and Doty, P., 1981, Structure of the pro-α2 (I) collagen gene, *Nature (London)*, **294**, 129–35.

Yamada, T., Avvedimento, V. E., Mudryi, M., Ohkubo, H., Vogeli, G., Irani, M., Pastan, I. and de Crombrugghe, B., 1980, The collagen gene: evidence for its evolutionary assembly by amplification of a DNA segment containing an exon of 54 bp, *Cell*, **22**, 887–92.

5. Synthesis and replacement of nematode cuticle components

C. M. Preston-Meek and D. I. Pritchard

INTRODUCTION

Despite the wealth of information which has been accrued on nematode cuticle ultrastructure, the dynamic nature of the nematode surface has only recently received attention. The suggestion that the cuticle retains a capacity for growth and differentiation after formation was proposed almost 30 years ago (Lee, 1965; Watson, 1965), but it is only during the last 10 years that the relevance and potential importance of this phenomenon has been realized.

Several studies have shown the presence of various stage-specific and intramoult surface antigens in a number of animal parasitic species (reviewed by Maizels *et al.*, 1982; Philipp and Rumjaneck, 1984; Maizels and Selkirk, 1988). Only limited studies, however, have been made to determine the source, ontogeny and mechanisms which regulate the transport, expression and release of these surface molecules. The cuticle comprises the primary interface between a nematode and its environment and, as such, it is important to understand the processes involved in the determination of turnover and replacement of surface components. An important consideration is the dynamism of the nematode surface and whether turnover of surface molecules could comprise an immune evasion stratagem. The existence of surface dynamism of the nematode cuticle raises two important questions: How are surface molecules transferred from the site of synthesis to the cuticle surface? and How is the synthesis of surface molecules regulated and controlled.

SURFACE DYNAMICS

The active shedding of cuticle surface components *in vitro* has been demonstrated in a number of studies (Vetter and Klaver-Wesseling, 1978; Smith *et al.*, 1981; Maizels *et al.*, 1983; Philipp and Rumjaneck, 1984). This has been graphically demonstrated in scanning electron microscopy of *Toxocara canis* larvae incubated with immune serum and peritoneal exudate cells (Badley *et al.*, 1987). This showed a layer sloughing off the surface, and consisted of surface antigens and antibody complexes. In spite of these demonstrations of surface shedding, the actual mechanisms of surface antigen release remains a matter of conjecture. Pritchard (1986) suggested two possible sources of replacement surface antigens. Firstly, the adsorption of excretory–secretory products onto the cuticle surface. Secondly, surface antigens could be synthesized within the inner cuticle or cellular hypodermis and then transported to the cuticle surface. The presence of RNA and a number of enzymes associated with protein synthesis within the cuticle and hypodermis (Anya, 1966; Lee, 1977) would seem to provide circumstantial evidence to support the hypothesis of replacement of surface antigens by a transcuticular route.

A MODEL FOR SURFACE DAMAGE AND REPAIR

The development of post-embedding immunostaining using colloidal gold probes has enabled the location of antigens in sectioned material to be determined with much greater precision than that possible using peroxidase techniques. In this chapter, we describe the use of immunogold staining to determine the kinetics and route of resynthesis and transcuticular transport of surface molecules. Here, the stimulus for this is detergent stripping of surface material and we use the free-living stage of *Nippostrongylus brasiliensis* as a model system.

The epicuticle of a number of nematode species has proved resistant to enzymatic and organic solvent degradation (Murrell and Graham, 1982; Dunphy and Webster, 1987). The use of the cationic detergent cetyltrimethylammonium bromide (CTAB) has, however, enabled the selective and controlled removal of surface antigens (Pritchard *et al.*, 1985; Wright and Hong, 1988). A free-living stage was selected for this study in order to minimize any complications which may arise from the long-term *in vitro* incubation of parasitic stages or species. It is hoped that, once the mechanisms of surface turnover have been elucidated in free-living stages, this information can be applied to parasitic stages to determine if similar events and mechanisms occur in parasitic stages *in vivo* and thus comprise immune evasion strategies.

SURFACE ANTIGENS CAN BE REPLACED AFTER REMOVAL BY DETERGENTS

Of a range of detergents, enzymes, solvent and salt treatments tested, the cationic detergents CTAB and cetylpyridinium chloride (CPC) were by far the most effective at removing surface molecules (Table 5.1). Both CTAB and CPC removed 40 per cent of the surface antigens expressed by *N. brasiliensis* infective stage larvae. However, larvae did not survive exposure to CPC and subsequent studies were, therefore, confined to events following cuticle stripping with CTAB.

Table 5.1. Percentage reduction in antibody binding assessed by IFAT following treatment of infective stage larvae with detergents, organic solvents or enzyme digestion for 1 h at 37°C. The primary antibody was raised in mice and directed at CTAB-stripped molecules of *N. brasiliensis* infective stage larvae.

Extraction agent	Reduction in fluorescence[a] (%)
CTAB (0.25%)	43.7 ± 2.0[b]
CPC (0.25%)	40.2 ± 1.7[b]
SDS (1%)	22.6 ± 1.8[c]
Nonidet-P40 (1%)	9.3 ± 0.5[c]
Triton-X 100 (1%)	8.1 ± 1.7[d]
Zwittergent 3–10 (1%)	NDD
Sodium deoxycholate (Na DOC; 1%)	NDD
Protease (500 µg ml^{-1})	7.4 ± 0.5 NS
Papain (400 µg ml^{-1})	NDD
β-Glucosidase (100 µg ml^{-1})	4.1 ± 0.5 NS
α-Mannosidase (500 µg ml^{-1})	NDD
Lipase (1 mg ml^{-1})	4.3 ± 0.3 NS
Neuraminidase (1 unit/ml)	NDD
Chloroform/methanol/water (14:6:1)	4.6 ± 0.7 NS
Isobutanol	6.8 ± 0.2 NS
Diethylether	6.4 ± 0.4 NS
Acetone	5.1 ± 0.3 NS
KCl (500 mM)	NDD

[a] NS = not significant by Student's *t*-test. NDD, no detectable difference with untreated controls. Values are mean ± standard error of 20 replicates.
[b] Significant at the 0.1 per cent level.
[c] Significant at the 1 per cent level.
[d] Significant at the 5 per cent level.

Indirect fluorescent antibody tests (IFAT) carried out at intervals following detergent treatment of free-living *N. brasiliensis* demonstrated recovery and resynthesis of surface antigens, the time course and temperature dependence of this is summarized in Figure 5.1. Maximum recovery was achieved after 48 h and no greater fluorescence levels were recorded with longer periods of recovery. Maintenance of larvae at 4°C following detergent treatment resulted in fluorescence of only 20 per cent of those of controls even after 48 h recovery. Full recovery to control levels was never observed following CTAB or sodium dodecyl sulphate (SDS) in any of the replicate experiments, the maximum recovery assessed by antibody binding being 90 per cent. Repeated CTAB stripping followed by 48 h recovery periods on one batch of larvae revealed an IFAT of > 85 per cent following the third detergent strip compared with non-detergent treated controls.

INHIBITION OF ANTIGEN REPLACEMENT

The extent of recovery of surface antigens following CTAB solubilization was reduced when the larvae were incubated in the presence of the metabolic inhibitors sodium fluoride, sodium iodoacetamide, dinitrophenol and rotenone (Table 5.2). Significant reductions in antigen resynthesis were also achieved by incubation of

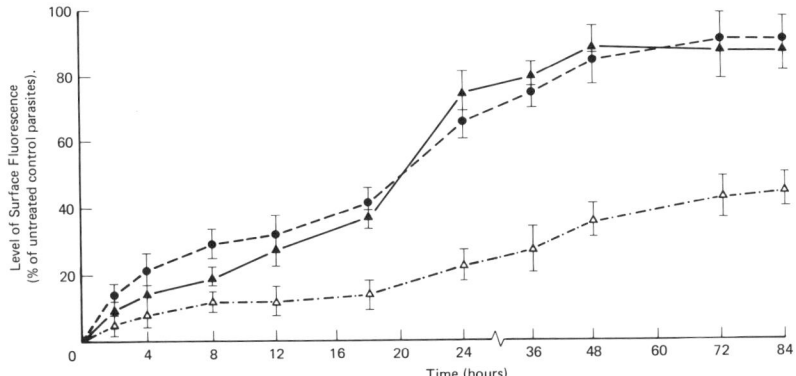

Figure 5.1. The time course of reappearance of surface antigens after detergent stripping. Infective *N. brasiliensis* larvae were incubated in 0.25 per cent CTAB (▲ , △) or 1 per cent SDS (●), allowed to recover at 4°C (△) or 25°C (▲), then IFAT tested. This graph indicates the recovery of fluorescence intensity over a period of 84 h expressed as a percentage of the pretreatment level. Points are means ± standard errors of 30 replicates.

larvae in the presence of transcription, translation and glycosylation inhibitors (actinomycin D, cycloheximide and tunicamycin, respectively).

Thus the free-living infective stage of *N. brasiliensis* is capable of repeatedly replacing surface molecules and this process was abrogated by inhibitors of metabolism, protein synthesis and/or intracellular transport.

A high degree of resistance to detergent, enzyme, solvent and high salt treatment is a feature of several nematode species (Murrell and Graham, 1982; Maizels *et al.*, 1983; Pritchard *et al.*, 1985; Devaney, 1988). Similar proportions of surface molecules were removed by CTAB from *Nematospiroides dubius* (Pritchard *et al.*, 1985) to those stripped from *N. brasiliensis*. The anionic detergent SDS solubilizes molecules from adult *Necator americanus* cuticle (Pritchard, 1986), although *Strongyloides ratti* larvae are resistant to this detergent (Murrell and Graham, 1982).

The ability of *T. canis* larvae to shed surface bound antibody is dependent on metabolic energy (Smith *et al.*, 1981). The present studies (see Table 5.1) indicate that the processes involved in the synthesis and expression of surface molecules of *N. brasiliensis* larvae are also dependent on metabolic energy, protein synthesis and microtubule-associated transport mechanisms. Of the metabolic inhibitors tested, sodium azide and dinitrophenol produced the most significant effects. Both of these compounds may inhibit translocation processes in addition to inhibiting oxidative phosphorylation. The general enzyme inhibitors sodium fluoride and sodium iodoacetamide also resulted in significant reduction in antigen replacement. In experiments involving inhibition of protein synthesis, recovery from CTAB stripping appeared to be dependent on translation of preformed mRNA rather than transcription of DNA because cyclohexamide inhibited resynthesis, but actinomycin D did not. The glycosylation inhibitor tunicamycin

Table 5.2. Following solubilization of surface antigens by CTAB, *N. brasiliensis* larvae were washed extensively then resuspended in non-lethal concentrations of inhibitors for 48 h at 25 °C. The level of replacement antigens is expressed as a percentage of fluorescence of control parasites allowed to replace antigens in the absence of any inhibitors.

Inhibitor	Target	Inhibition of antigen resynthesis[a] (%)	
		100 μM	250 μM
Sodium fluoride	General enzymes	3.5 ± 0.2 NS	6.5 ± 0.7[d]
Sodium iodoacetamide	General enzymes	3.8 ± 0.1 NS	8.5 ± 0.5[d]
Sodium azide	Oxidative phosphorylation	6.2 ± 0.4[d]	16.7 ± 0.7[c]
Dinitrophenol	Oxidative phosphorylation	1.1 ± 0.2 NS	10.7 ± 0.6[c]
Rotenone	Electron transport	2.0 ± 0.3 NS	8.1 ± 0.7[d]
Antimycin A	Electron transport	2.3 ± 0.4 NS	3.2 ± 0.1 NS
Salicylhydroxamic acid (SHAM)	Alternative electron transport	NDD	2.1 ± 0.2 NS
o-Hydroxphenol	Alternative electron transport	NDD	2.3 ± 0.4 NS
Actinomycin D	Transcription	NDD	7.5 ± 0.2[d]
Cycloheximide	Translation	NDD	17.0 ± 2.3[b]
Tunicamycin	Glycosylation	NDD	6.7 ± 0.3[d]
Cytochalasin B (5 μg ml^{-1})	Microfilament formation	7.4 ± 0.7[d]	
Colchicine (10 mM)	Microtubule formation	17.3 ± 1.0[b]	
Fenbendazole (5 μg ml^{-1}	Tubulin assembly	30.8 ± 0.7[b]	

[a] NS = not significant NDD, no detectable difference with untreated controls by Student's *t*-test. Values are mean ± standard error of 20 replicates.
[b] Significant at the 0.1 per cent level.
[c] Significant at the 1 per cent level.
[d] Significant at the 5 per cent level.

had a smaller, but still significant, effect on antigen resynthesis. Interpretation of the effects of the various inhibitors on the biochemical processes involved in surface antigen resynthesis is difficult as the concentration of these compounds at their internal sites of action is unknown. However, that significant inhibition of antigen resynthesis was obtained, even at low concentrations, might indicate that protein synthesis and glycosylation are of greater importance than these results at first suggest. The highly significant reduction in surface fluorescence during recovery in the presence of fenbendazole also suggests an important role of microtubule associated transport of surface destined molecules from internal sites.

SITE OF SYNTHESIS AND REPLACEMENT SURFACE ANTIGENS

The use of post-embedding immunogold electron microscopy has enabled the precise localization of newly synthesized surface molecules to be determined to resolutions equivalent to the autoradiographic tracking of proteins.

Figure 5.2. Opposite: Immunogold labelling of an infective stage *N. brasiliensis* larva showing the staining pattern in control untreated specimens (a). The outer surface (α) of the cuticle is uniformly labelled with both rabbit and mouse (insert) primary antisera raised against CTAB stripped proteins. The inner layers of the cuticle (cu) and the hypodermis (hy) are also stained. Larvae fixed immediately following CTAB stripping show a reduction in surface labelling (b) and immunostaining of inner cuticle layers and hypodermis are similar to that of untreated controls.

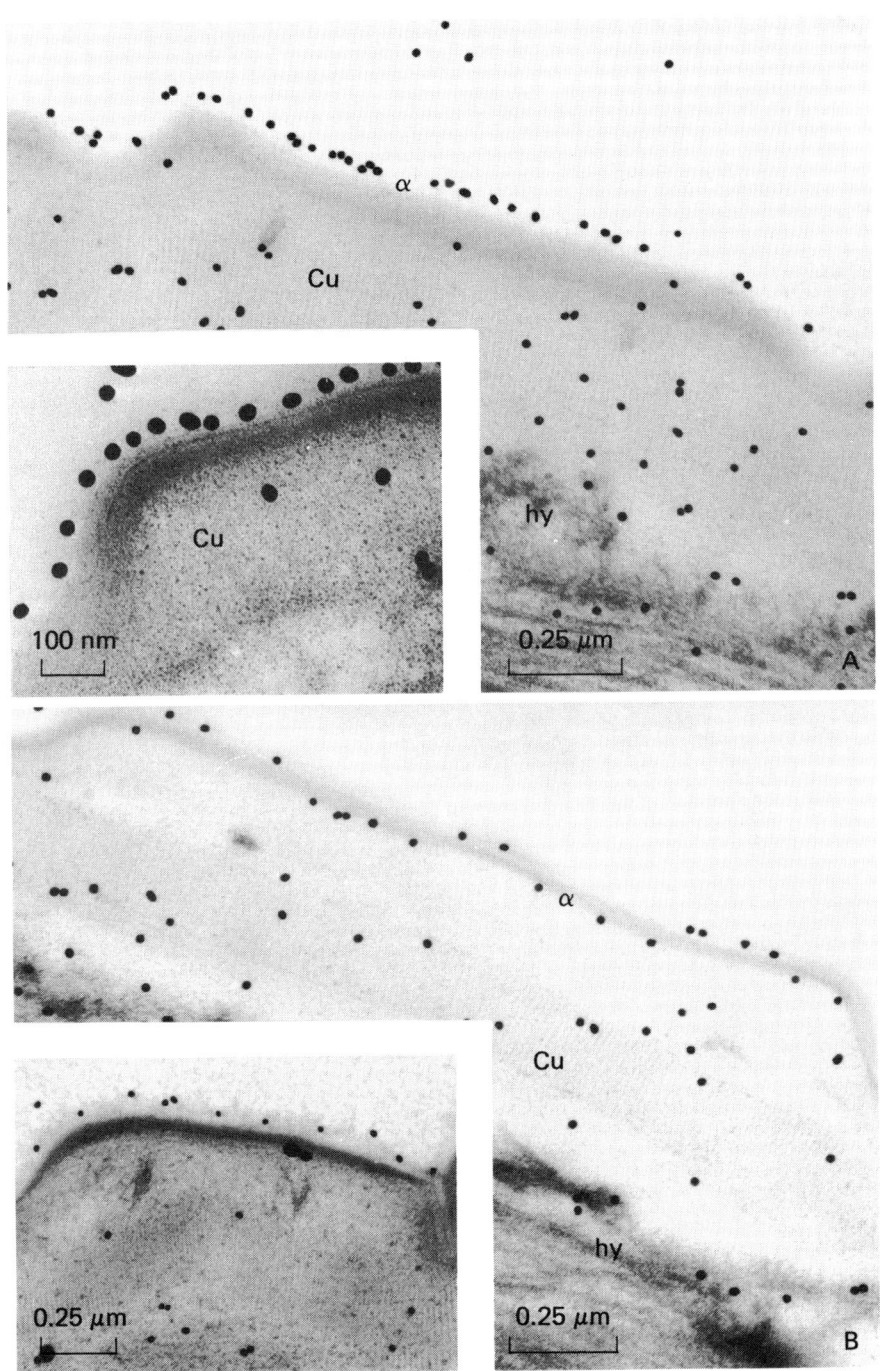

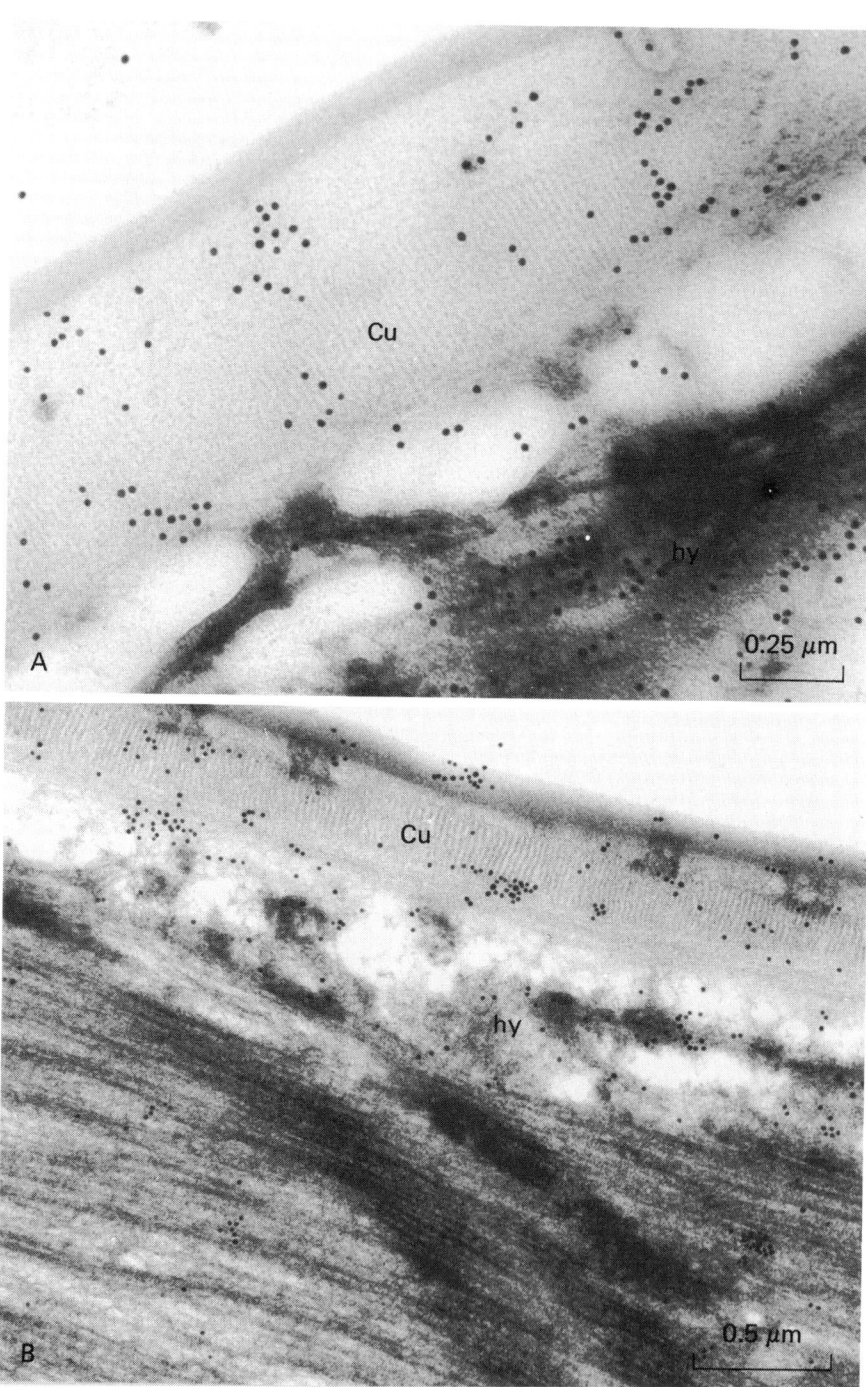

Figure 5.2a illustrates the immunogold staining pattern of control *N. brasiliensis* larvae. Monomeric gold particles were distributed evenly on the outer surface of the cuticle and appeared to be associated with a diffuse glycocalyx-like layer. Specific, but less intense, labelling or cortical and basal layers of the cuticle and the hypodermis was also evident in control larvae (Figure 5.3a).

Immunogold staining of larvae immediately following a 1-h incubation in CTAB revealed a reduction in surface labelling of approx 50 per cent to untreated controls (Figure 5.2b). At this stage the staining pattern of other regions of the cuticle and hypodermis was similar to control larvae.

Immunostaining 12 h after detergent stripping revealed a marked increase in the density of gold particles within the hypodermis (Figure 5.2a). Labelling of the cortical and basal layers appeared heterogeneous when compared with control sections (Figures 5.2a and 5.3a). This aggregation of labelling was further accentuated in 24-h recovery larvae (Figure 5.3b), at which time clusters of gold particles were present at intervals along the base of the cuticle and within the hypodermis. Surface immunostaining was patchy at 24-h recovery periods, but by 48 h gold particles were distributed evenly on the cuticle surface and in comparable numbers to that of controls (Figure 5.4a). Specificity of the staining reaction was confirmed in all batches of sections by the absence of significant labelling when normal mouse serum was used as the primary antibody (Figure 5.4b).

As yet it is uncertain whether precursors or fully formed molecules are released from the hypodermis for translocation across the cuticle. However, the immunogold staining within the hypodermis suggests that antigenic determinants of surface molecules are sufficiently developed to enable epitope recognition and binding by immune sera. This observation indicates that surface molecules are at least partly, if not fully, assembled whilst still within the hypodermis. Watson (1965) and Pritchard (1986) have suggested that surface molecules pass to the surface via cuticular pores, although the presence of such pores does not appear to be a common feature of nematode cuticles and none has been reported in *N. brasiliensis*. Wharton and Bone (1988) have described the formation of the first-stage cuticle of *Trichostrongylus colubriformis* and have given details of cuticle development which may help to explain the observations with *N. brasiliensis*. In *T. colubriformis* the striated basal zone is formed before the cortical zone, thus implying that the material which forms the cortical zone passes through the basal zone. Wharton and Bone (1988) suggested that either gaps exist in the basal zone or the entire layer is permeable to cortical zone components. The immunolabelling experiments with *N. brasiliensis* suggest that analogous mechanism(s) may be involved in the transport of new surface components through the basal and cortical layers of the cuticle following detergent stripping.

The ability of the nematode surface to renew surface components has also been indicated by Bird and Zuckerman (1989). Surface coats of dauer larvae of *Anguina*

Figure 5.3. Opposite: Following a 12-h recovery period the immunolabelling within the hypodermis is markedly more intense (a) compared with untreated larvae (see Figure 5.2a). The labelling at 12 and 24 h (b) is present as pockets of gold particles within the hypodermis (hy) and inner cuticle (cu).

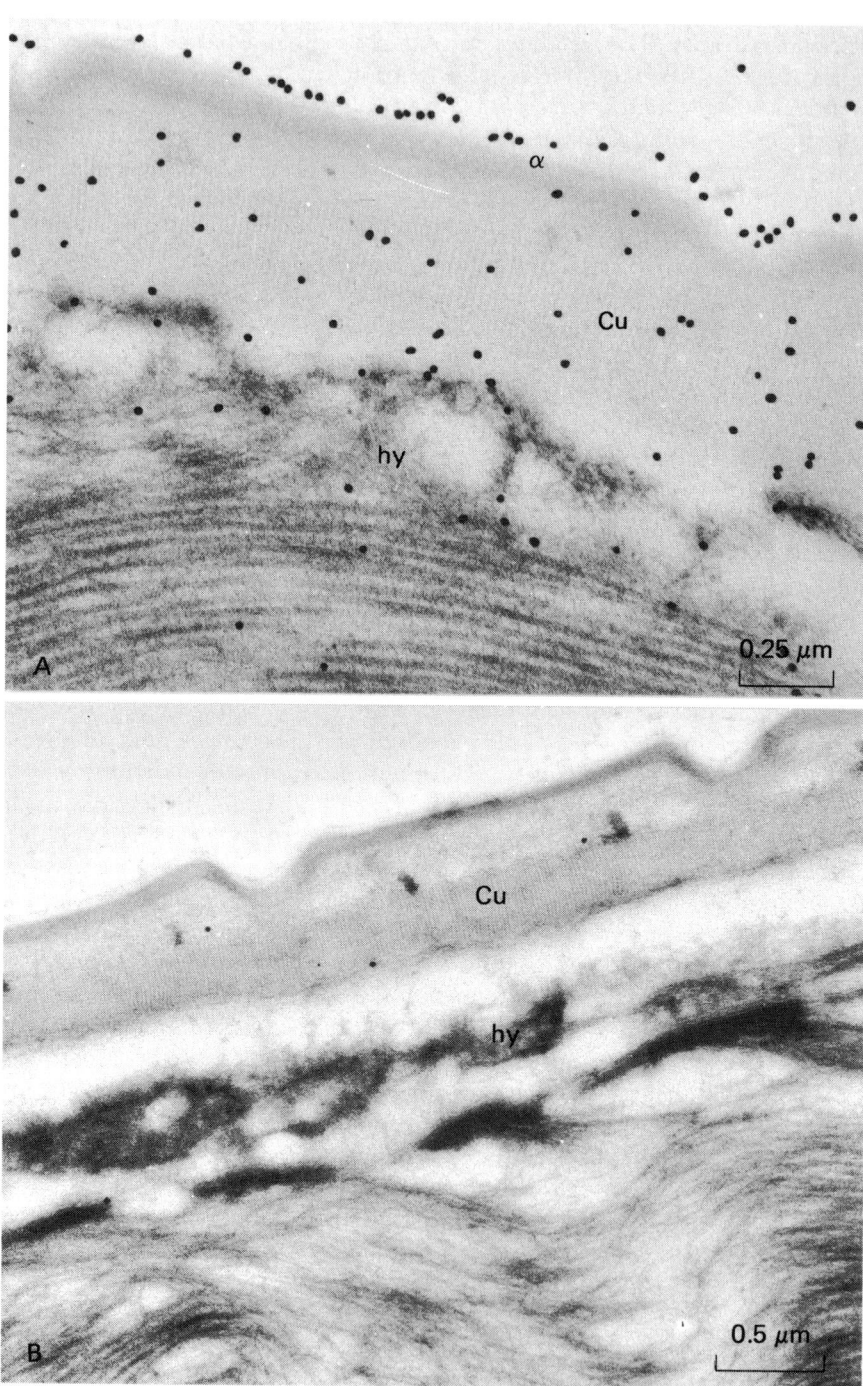

agrostis contain residues of *N*-acetyl-D-glucosamine and protein. Treatment with pepsin or trypsin inhibited the attachment of the corynebacterium *Clavibacter* sp. to the surface coat, indicating that proteins play a crucial role in the adhesion of bacteria to the nematode surface. Inhibition of attachment was reversed 18 h after removal of the larvae from enzymes, indicating that the nematode was capable of renewing and resynthesizing surface proteins.

Despite the increased immunostaining of the hypodermis of *N. brasiliensis* larvae during recovery from detergent stripping, very few, if any, organelles associated with protein synthesis were evident. Previous ultrastructural studies of the infective larvae (Bonner and Weinstein, 1972; Lee, 1977) revealed a thin hypodermal layer with few small mitochondria, ribosomes and microtubules. Investigations are currently in progress using amino acid autoradiography following detergent stripping to compare the sites and time course of protein synthesis with those of immunostaining. It is hoped that such studies will indicate whether antigen synthesis occurs throughout the hypodermis or in discrete regions with subsequent translocation from such sites.

CONCLUDING REMARKS

Future work must also include determination of the accumulation of newly synthesized RNA in infective larvae following detergent stripping. To measure new RNA synthesis, worms could be incubated in the presence of [^{32}P]orthophosphate and then the amount of ^{32}P incorporation and size distribution of the RNA determined. To obtain rates of RNA synthesis the ^{32}P is pulsed into aliquots of worms at intervals following detergent stripping.

The use of IFAT and immunogold staining following detergent stripping of larvae has provided additional support to the observations of surface turnover and release of surface components and reinforced the concept of the dynamic nature of the nematode cuticle during intermoult periods. In *N. brasiliensis* infective larvae, the source of replacement of surface molecules appears to be from the hypodermis with subsequent translocation through inner layers of the cuticle. Such a mechanism could explain the capacity for release of surface antigens over prolonged periods of *in vitro* cultivation in species such as *T. canis* (Smith *et al.*, 1981; Maizels *et al.*, 1983), and the replacement of surface proteins from enzyme treated larvae of *A. agrostis* (Bird and Zuckerman, 1989). The biological significance of this process remains to be determined, but could clearly be vital to nematode defences against immunological assaults in parasites and microbiological attack in free-living species or stages.

Figure 5.4. Opposite: Immunolabelling of *N. brasiliensis* infective larvae 48 h following CTAB stripping (a). The surface labelling is uniform and of a similar intensity to control specimens. Specificity of labelling is confirmed in sections and incubated in naive mouse serum (b). α, Outer surface of the cuticle; cu, inner layer of the cuticle; hy, hypodermis.

REFERENCES

Anya, A. O., 1966, Localization of ribonucleic acid in the cuticle of nematodes, *Nature*, **95**, 827–8.

Badley, J. E., Grieve, R. B., Rockey, J. H. and Glickman, L. T., 1987, Immune mediated adherence of eosinophils to *Toxocara canis* infective larvae: the role of excretory–secretory antigens, *Parasite Immunology*, **9**, 133–43.

Bird, A. F. and Zuckerman, B. M., 1989, Studies on the surface coat (glycocalyx) of the dauer larva of *Anguina agrostis*, *International Journal of Parasitology*, **19**, 235–40.

Bonner, T. P. and Weinstein, P. P., 1972, Ultrastructure of cuticle formation in the nematodes *Nippostrongylus brasiliensis* and *Nematospioides dubius*, *Journal of Ultrastructural Research*, **40**, 261–71.

Devaney, E., 1988, The biochemical and immunochemical characterisation of the 30 kilodalton surface antigen of *Brugia pahangi*, *Molecular and Biochemical Parasitology*, **27**, 83—92.

Dunphy, G. B. and Webster, J. M., 1987, Partially characterised components of the epicuticle of dauer juvenile *Steinernema feltiae* and their influence on hemocyte activity in *Galleria mellonella*, *Journal of Parasitology*, **73**, 584–8.

Lee, D. L., 1965, The cuticle of adult *Nippostrongylus brasiliensis*, *Parasitology*, **55**, 173–81.

Lee, D. L., 1977, The nematode epidermis and collagenous cuticle, its formation and ecdysis, *Symposia of the Zoological Society of London*, **39**, 145–70.

Maizels, R. M. and Selkirk, M. E., 1988, Immunobiology of nematode antigens, in Endlund, P. T. and Sher, A. (Eds) *The Biology of Parasitism*, pp. 285–308, New York: Alan R. Liss.

Maizels, R. M., Philipp, M. and Ogilvie, B. M., 1982, Molecules on the surface of parasitic nematodes as probes of the immune response in infection, *Immunological Review*, **61**, 109–36.

Maizels, R. M., Meghji, M. and Ogilvie, B. M., 1983, Restricted sets of parasite antigens from the surface of different stages and sexes of the nematode parasite *Nippostrongylus brasiliensis*, *Immunology*, **48**, 107–21.

Murrell, K. D. and Graham, C., 1982, Solubilization studies on the epicuticular antigens of *Strongyloides ratti*, *Veterinary Parasitology*, **10**, 191–203.

Philipp, M. and Rumjaneck, F. D., 1984, Antigenic and dynamic properties of helminth surface structures, *Molecular and Biochemical Parasitology*, **10**, 245–68.

Pritchard, D. I., 1986, Antigens of gastrointestinal nematodes, *Transactions of the Royal Society of Tropical Medicine and Hygiene*, **80**, 728–34.

Pritchard, D. I., Crawford, C. R., Duce, I. R. and Behnke, J. M., 1985, Antigen stripping from a nematode epicuticle using the cationic detergent cetyltrimethylammonium bromide (CTAB), *Parasite Immunology*, **7**, 575–85.

Smith, H. V., Quinn, R., Kusel, J. R. and Girdwood, R. W. A., 1981, The effect of temperature and antimetabolites on the antibody binding to the outer surface of the second stage *Toxocara canis* larvae, *Molecular and Biochemical Parasitology*, **4**, 183–93.

Vetter, J. C. M. and Klaver-Wesseling, J. C. M., 1978, IgG antibody binding to the outer surface of infective larvae of *Ancylostoma caninum*, *Zeitschrift für Parasitenkunde*, **58**, 91–6.

Watson, B. D., 1965, The fine structure of the body wall and the growth of the cuticle in the adult nematode *Ascaris lumbricoides*, *Quarterly Journal of Microbiol Science*, **106**, 83–91.

Wharton, D. A. and Bone, L. W., 1988, The formation of the first stage cuticle within the egg of *Trichostrongylus colubriformis*, *Parasitology*, **97**, 459–67.

Wright, K. A. and Hong, H., 1988, Characterization of the accessory layer of the cuticle of muscle larvae of *Trichinella spiralis*, *Journal of Parasitology*, **74**, 440–51.

6. *Toxocara canis*: secreted glycoconjugate antigens in immunobiology and immunodiagnosis

R. M. Maizels and B. D. Robertson

INTRODUCTION

One of the most widely distributed nematode parasites of mammals is the ascarid *Toxocara canis*, which infects the vast majority of canids by direct oral or indirect transplacental and colostorol routes. The ability of this parasite to infect *in utero* results in a near 100 per cent infection rate for new-born puppies (Scothorn *et al.*, 1965). Although *Toxocara* causes some morbidity in dogs, the major concern with this parasite is medical rather than veterinary, due to the common exposure of the human population to invasive larval forms, and the occasional permanent optical and neurological damage resulting from toxocaral infection in children and adult humans. However, the actual number of human infections, and the proportion who develop the symptoms of either visceral larva migrans (VLM) or ocular toxocariasis is not well defined. Moreover, while the immune responses of both definitive (canine) and paratenic (human and murine) hosts have been well documented (Lloyd, 1987), we still have no indication of how protective immunity may operate against this parasite.

The complex life cycle of *T. canis* was first fully described by Sprent (1958) and is illustrated in Figure 6.1. Mature adult worms in the canine gastrointestinal tract produce up to 200 000 eggs per day (Glickman *et al.*, 1979) which are voided with the faeces. These unembryonated eggs (Figure 6.2a) are not initially infective, but over the 2–3 weeks following deposition an embryonic larva forms which undergoes its first moult within the egg. Infective eggs (Figure 6.2b), containing fully formed parasites, hatch in the intestine when ingested by adult dogs to produce invasive second stage larvae (Figure 6.2c). In bitches, larvae migrate via the lungs and liver

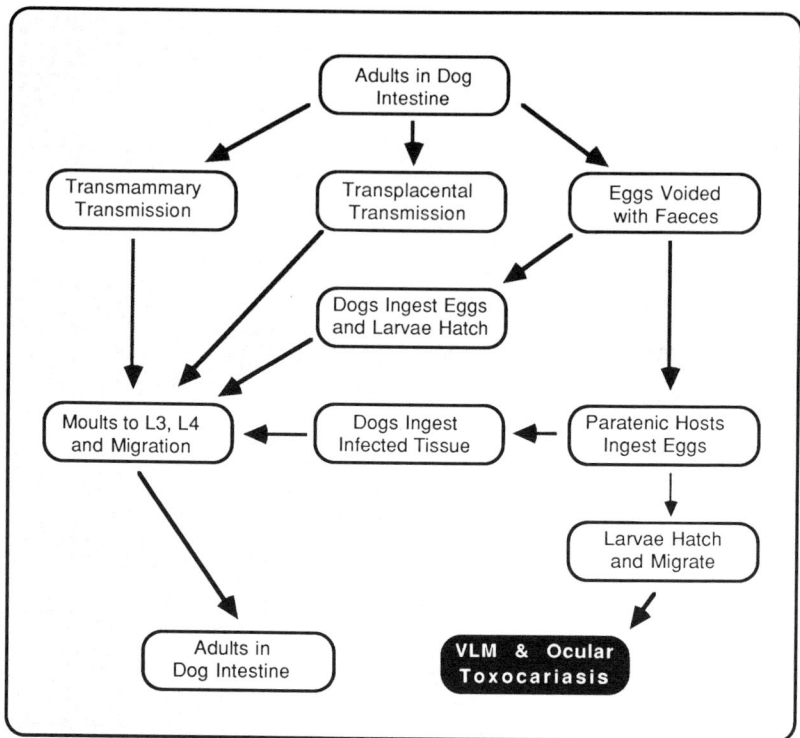

Figure 6.1. The life cycle of *T. canis*.

and disseminate throughout the body, remaining dormant until activated during pregnancy (Lloyd, 1987). At that time they can pass either to the fetus *in utero* via the placenta, which occurs in 98–100 per cent of experimental infections (Griesemer *et al.*, 1963; Burke and Roberson, 1985a), or post-parturition in milk to new-born puppies, although the occurrence of this is low in experimental infections (Scothorn *et al.*, 1965; Burke and Roberson, 1985a,b). The larvae then migrate via the lungs and trachea to the intestine where they develop to adults (Figure 6.2d).

The parasite may also be transmitted to paratenic hosts, commonly rodents and birds, which ingest embryonated *Toxocara* eggs. Humans are also susceptible to infection by the same route, and ocular toxocariasis or VLM may result. Larvae hatch within and invade paratenic hosts but do not mature: they remain in a state of arrested development analogous to that observed in the non-pregnant bitch. Although maturation is blocked, larvae migrate throughout the body and are known to accumulate in the musculature and central nervous system (Figure 6.2e) of infected mice (Bisseru, 1969; Kayes and Oaks, 1976; Dunsmore *et al.*, 1983). In mice, larvae show an affinity for white areas, and seem responsible for degeneration of fibre tracts, including the corpus callosum and areas of the cerebellum, perhaps reflecting the clear behavioural aberrations in mice infected with *Toxocara* larvae (Dolinsky *et al.*, 1981). Such larvae remain viable in the host tissues for at least two

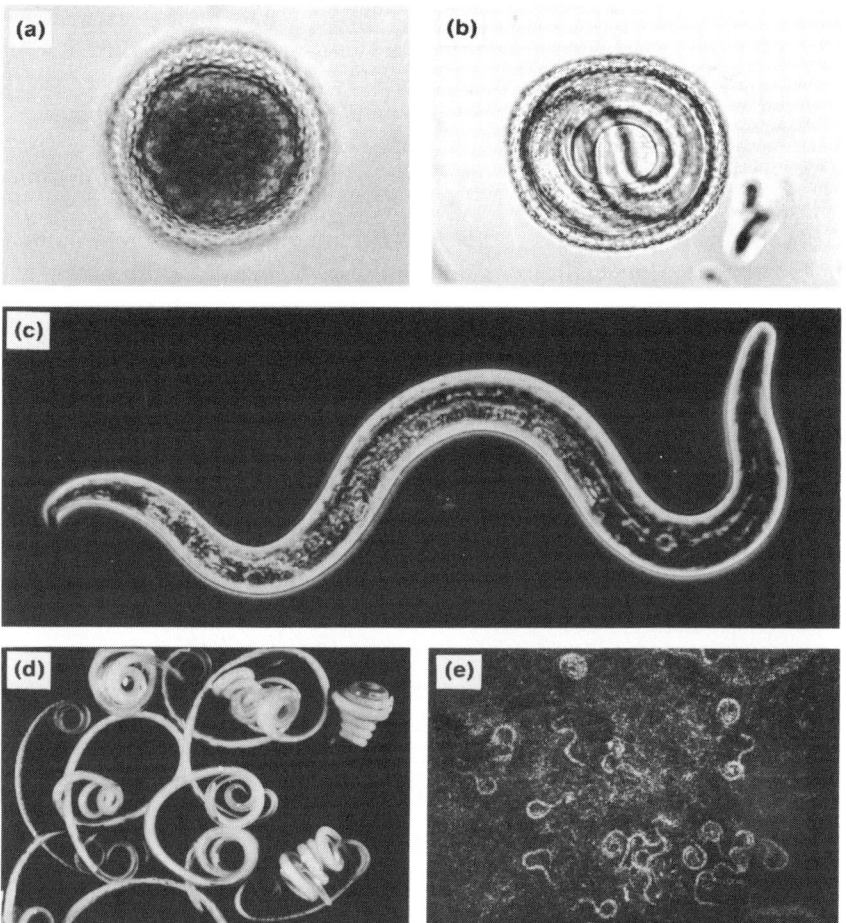

Figure 6.2. Successive stages of the parasite *T. canis* in definitive and paratenic hosts. (a) Unembryonated ovum. (b) Embryonated ovum. (c) Second stage larva (L2). (d) Adult worms from the intestine of a dog. (e) Larvae in the brain of a mouse.

years (Beaver, 1962) and can go on to infect patently canids which consume infected tissue (Sprent, 1958).

EPIDEMIOLOGY AND DISEASE

The close proximity of dogs and man means that this zoonosis poses a potentially widespread health hazard. The degree of soil contamination with *T. canis* infective eggs may be high, ranging between 10 and 20 per cent for public places in the UK and USA (Borg and Woodruff, 1973; Dada and Lindquist, 1979; Quinn *et al.*, 1980), and the proportion of children with *Toxocara* reactive antibody can be alarmingly high (see Table 6.1). Although the determination of seroprevalence is critically dependent upon the set of parasite antigens used in the assay, as discussed below, antibody prevalence in healthy European and North American adults is generally found to be 2–3 per cent (Girdwood *et al.*, 1978; de Savigny *et al.*, 1979; Glickman and Schantz, 1981; Clemett *et al.*, 1987). Most significantly, the seroprevalence may be far higher in young children — up to 26 per cent in developed countries and 60 per cent or more in developing countries (Thompson *et al.*, 1986).

VLM or other overt signs of toxocariasis occur in a much smaller proportion of the human population, generally in children less than 10 years old with a history of pica or geophagia. These symptoms include abdominal pain, hepatomegaly, anorexia, nausea, vomiting, lethargy, sleep and behavioural disturbances, headache, limb pains and cutaneous manifestations (Glickman and Schantz, 1981; Gillespie, 1987, 1988; Taylor *et al.*, 1988). It is now established that *Toxocara* larvae can be found in the human brain (Hill *et al.*, 1985) and the possibility that subclinical toxocaral infection results in significant reduction in neurological and intellectual capacity of young children remains a major concern (Marmor *et al.*, 1987).

Ocular toxocariasis occurs when a larva lodges in blood vessels at the rear of the eye, inducing a granuloma or endophthalmitis which may result in permanent eye

Table 6.1.   Seroprevalence of *T. canis*.

Positive (%)	Sample size and type	Location	Reference
0.0	70 urban children	Switzerland	Tönz *et al.* (1983)
2.0	Healthy adults	UK	Woodruff (1970)
2.2	200 blood donors	UK	Girdwood *et al.* (1978)
2.6	922 adults	UK	de Savigny *et al.* (1979)
2.8	8457 adults	USA	Glickman and Schantz (1981)
2.8	318 adults	New Zealand	Clemett *et al.* (1987)
3.6	380 children under 10 years	Puerto Rico	Berrocal (1980)
3.6	83 children under 15 years	Japan	Matsumura and Endo (1983)
3.7	530 women over 20 years	Japan	Matsumura and Endo (1983)
4.2	354 blood donors	Italy	Brunello *et al.* (1983)
4.7	41 children under 5 years	Venezuela	Lynch *et al.* (1988b)
5.4	4648 children under 15 years	USA	Marmor *et al.* (1987)
5.8	68 rural children	Switzerland	Tönz *et al.* (1983)
6.4	1369 children 1–11 years	USA	Herrmann *et al.* (1985)
7.0	660 blood donors	Australia	Nicholas *et al.* (1986)
7.1	112 children	Netherlands	van Knapen *et al.* (1983)
7.3	219 adults	Iraq	Woodruff *et al.* (1981)
7.9	114 normal adults	USA	Cypess *et al.* (1977)
11.1	261 children under 10 years	Puerto Rico	Berrocal (1980)
13.3	15 non-toxocaral eye cases	USA	Schantz *et al.* (1979)
14.1	835 individuals over 14 years	UK	Ree *et al.* (1984)
14.3	133 children 3–9 years	UK	Josephs *et al.* (1981)
14.8	168 children 6–15 years	Venezuela	Lynch *et al.* (1988b)
15.7	102 dog breeders	UK	Woodruff *et al.* (1978)
23.1	333 children 5–7 years	USA	Worley *et al.* (1984)
26.8	570 children up to 14 years	UK	Ree *et al.* (1984)
60.0	20 children 5–10 years	St Lucia	Bundy *et al.* (1987)
64.7	17 individuals with ocular pathology	USA	Schantz *et al.* (1979)
86.6	82 children under 6 years	St Lucia	Thompson *et al.* (1986)
90.2	41 individuals with ocular pathology	USA	Pollard *et al.* (1979)

damage (Schantz *et al.*, 1979; Dinning *et al.*, 1988). The presence of such a granuloma has sometimes led to the erroneous enucleation of the orbit on suspicion of a retinoblastoma. Treatment with thiabendazole, together with steroids to limit potentially harmful inflammation, usually leads to an improvement in vision, although there is rarely a complete recovery (Dinning *et al.*, 1988; Gillespie, 1987, 1988).

DIAGNOSIS AND SPECIFICITY OF *TOXOCARA* ANTIGENS

As *Toxocara* larvae do not grow or multiply in human hosts, and distribute themselves widely in the body, direct identification of parasites is rarely possible. Thus diagnosis is generally based on the detection of specific antiparasite antibody in serum, using a variety of parasite antigens which vary in specificity. Those tests employing adult worm antigens, to which the human host is never exposed, show poor sensitivity (Glickman *et al.*, 1978) and are subject to cross-reactions with antigens from other commonly found infectious nematodes. Soluble extracts from whole eggs, embryonated or unembryonated require sera to be pre-absorbed with *Ascaris* antigens to attain specificity (Glickman *et al.*, 1985), and the most satisfactory results are obtained with excretory–secretory (ES) antigens released by infective larvae cultured *in vitro* (de Savigny, 1975; de Savigny and Tizard, 1977; van Knapen *et al.*, 1982). The superiority of ES material may be in part due to the absence of the highly cross-reactive phosphorycholine specificity from parasite secretions although this determinant is well represented in whole extracts of *Toxocara* and many other nematodes (Sugane and Oshima, 1984a; Maizels *et al.*, 1987a). Currently, the ES antigens are considered sufficiently specific to be used in routine enzyme-linked immunosorbent assay (ELISA) assays to measure anti-*Toxocara* antibody titres in suspected cases of human infection (de Savigny *et al.*, 1979; Schantz *et al.*, 1979). As discussed above, positive levels vary between 0 and 90 per cent, depending on the age, geographical location and socio-economic status of the individuals examined (Table 6.1). While the highest values occur in children, in developing countries, and in individuals diagnosed with ocular pathology, the values obtained from asymptomatic populations all over the developed world suggest a high incidence of infection.

However, true prevalence rates may differ markedly from these estimates for two reasons (Smith *et al.*, 1984). Firstly, as in many other helminth infections, antibody may remain in the circulation long after living parasites have been cleared. Secondly, ES antigens contain cross-reactive specificities which could lead to false-positive reactions in the ELISA assay. For example, sera from infections with the related feline ascarid, *Toxocara cati* (Maizels *et al.*, 1987b; Kennedy *et al.*, 1987a), with *Ascaris lumbricoides* and *Ascaris suum* (Glickman *et al.*, 1985; Lynch *et al.*, 1988a; Kennedy *et al.*, 1989), *Toxocara pteropodis* (Nicholas *et al.*, 1984) and *Baylisascaris procyonis* (Boyce *et al.*, 1988) each show varying degrees of reactivity with ES antigens from *T. canis*.

Cross-reactivity between *T. canis* and the ABO blood group antigens of man has also been described. Smith and co-workers (1983) found that anti-human A and B typing sera reacted both with the surface of larvae and with ES antigens collected from cultured worms; moreover, experimental infection of rabbits with *T. canis* resulted in elevated isohaemagglutinin titres. Glickman and Schantz (1985) demonstrated that absorption of sera against AB blood cells did not decrease the titre of *T. canis* antibody, as measured in an ELISA against antigen prepared from embryonated eggs. However, at least one major carbohydrate determinant is now known not to be expressed until after the eggs hatch (Maizels *et al.*, 1987b), suggesting that egg antigen may not contain the AB cross-reactive epitopes described by Smith and co-workers.

Within this context, we decided some years ago to examine the ES antigens in greater detail to define potentially parasite-specific and ideally species-specific epitopes for diagnosis, and to explore alternative approaches to detecting infection in human beings. At the same time, we wished to investigate in what ways the production and secretion of such quantities of antigen by *Toxocara* larvae may aid the parasite in its task of establishing itself and surviving in the mammalian host, and we have therefore also considered the functional activities associated with each component of the ES material.

TOXOCARA CANIS EXCRETORY–SECRETORY ANTIGENS

The collection of ES antigens from parasites for analysis depends on the viability of the organism *in vitro*. Many parasite species, however, do not survive more than a few hours in culture and the release of material from such worms may not reflect a physiological process. In contrast, the larvae of *T. canis* survive almost indefinitely in culture (de Savigny, 1975) and there can be little doubt that the release of molecules is biologically and physiologically authentic. The longevity of *T. canis in vitro* and its production of copious amounts of material make it an ideal model with which to examine functional aspects of ES antigen immunobiology, and provides a convenient model for other tissue invasive helminths, less amenable to laboratory maintenance. In addition, the ES molecules also include protective antigens; thus, in the mouse model, immunization with *Toxocara* ES significantly reduces establishment of larvae in the tissues and in the brain (Nicholas *et al.*, 1984). Likewise, in many other nematode parasites, secreted products have been proved to induce strong levels of immune-mediated protection in both laboratory and natural hosts (reviewed by Maizels and Selkirk, 1988).

We have defined five major ES antigens (Maizels *et al.*, 1984, 1987b) of molecular weights 32, 55, 70, 120 and 400 kDa; a summary of these molecules and their properties is presented in Figure 6.3. Several of these components show considerable micro-heterogeneity: for example, both the 120 and 400 kDa products are usually seen as three bands of slightly differing molecular weight. Within each set, however, each variant shows the same immunological and biochemical properties which leads us to treat them as a single entity. Due to this heterogeneity, and the presence of several proteins of lower abundance, as many as 20 individual

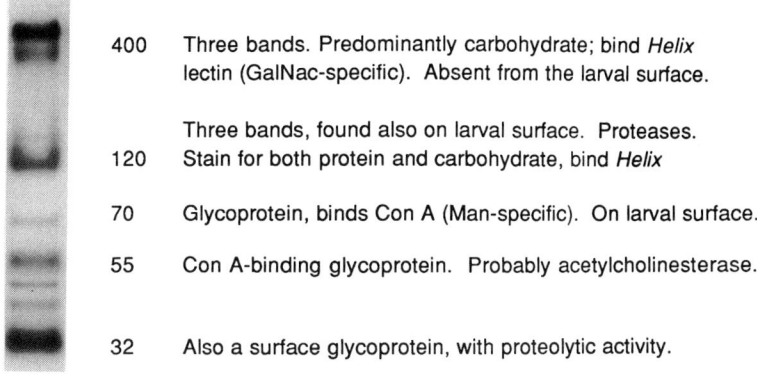

400   Three bands. Predominantly carbohydrate; bind *Helix*
lectin (GalNac-specific). Absent from the larval surface.

       Three bands, found also on larval surface. Proteases.
120   Stain for both protein and carbohydrate, bind *Helix*

70   Glycoprotein, binds Con A (Man-specific). On larval surface.

55   Con A-binding glycoprotein. Probably acetylcholinesterase.

32   Also a surface glycoprotein, with proteolytic activity.

Figure 6.3. ES antigens of *T. canis* second stage larvae

bands can be discerned by sodium dodecyl sulphate–polyacrylamide gel electrophoresis (SDS–PAGE) analysis of the ES proteins (Speiser and Gottstein, 1984; Badley *et al.*, 1987a).

Each of the major proteins of 32, 55, 70, 120 and 400 kDa show differing sensitivity to a variety of proteases and differential staining with a panel of lectins, indicating that they represent separate gene products rather than breakdown products of a common high molecular weight precursor (Meghji and Maizels, 1986). For example, while all are degraded by pronase, only the 70 and 400 kDa molecules are cleaved by the V8 protease from *Staphylococcus aureus*; in contrast, the 32, 55 and 70 kDa glycoproteins are trypsin sensitive but the larger antigens are not. Other distinctive properties of some components are unusual, particularly in the case of the high molecular weight 400 kDa molecule which cannot be stained by Coomassie Blue or even with the highly sensitive silver probe for proteins (Merrill *et al.*, 1981; Meghji and Maizels, 1986; Badley *et al.*, 1987a). Moreover, the 400 kDa consistently fails to incorporate [^{35}S]methionine in metabolic labelling experiments (Sugane *et al.*, 1985; Meghji and Maizels, 1986), further implying an unusual chemical composition for this glycoconjugate. Partly for this reason, radiolabelling with ^{125}I provides the most sensitive method of analysing *T. canis* ES antigens, labelling all the major bands (Maizels *et al.*, 1984; Meghji and Maizels, 1986) by each of the standard methods involving Iodogen, chloramine T or the Bolton–Hunter reagent.

Extrinsic iodination of larval ES antigens also permitted a direct comparison to be made with the proteins on the cuticular surface of larvae of the same stage, as iodination is a relatively surface-restricted labelling technique (Parkhouse *et al.*, 1981). This investigation showed that each of the ES proteins is also present on the larval cuticle, with the sole exception of the 400 kDa macromolecule (Maizels *et al.*, 1984). In the case of the 120 kDa glycoprotein, the same characteristic triplet was observed in both ES and surface-derived preparations. Although the physiological source of secreted antigens was not established, the same experiments did illustrate that glycoproteins labelled on the parasite surface were rapidly shed into *in vitro* culture media, as first shown by Smith *et al.* (1981). One other biological feature was revealed by this study of surface antigens, that the newly hatched larvae express relatively little surface protein and require up to 48 h of incubation to reach maximal display of the four main surface antigens of 32, 55, 70 and 120 kDa (Maizels *et al.*, 1987b).

A further striking feature of *Toxocara* ES antigens is their high degree of glycosylation. In terms of glycan to protein ratio, differential staining in SDS–PAGE gels indicates that the 400 kDa components have the highest proportion of carbohydrate (Sugane and Oshima, 1983; Meghji and Maizels, 1986; Badley *et al.*, 1987a). The individual ES antigens share at least one common oligosaccharide, defined by the studies described below with monoclonal antibodies directed to a carbohydrate epitope common to all ES antigens (Maizels *et al.*, 1987b). However, we know that there is also an element of differential glycosylation, demonstrated by the binding of ^{125}I-labelled lectins to nitrocellulose blots of ES antigen (Meghji and Maizels, 1986). Thus, the 32, 55 and 70 kDa bands bind to concanavalin A, indicating exposed mannose groups, while the 120 and 400 kDa molecules bind *Helix pomatia* agglutinin, the *N*-acetylgalactosamine and blood group A specific lectin. The presence of these and other sugars was confirmed by a gas chromatographic analysis of ES carbohydrate composition, which showed a total of 401.4 μg of sugar per mg of ES protein, with *N*-acetylgalactosamine as the predominant glycoside contributing 58 per cent of the total carbohydrate (Meghji and Maizels, 1986). Galactose, mannose and *N*-acetylglucosamine were also present in significant quantities. As will be described in the following section, these carbohydrate structures elicit a strong immune response in experimental infection, and indeed the majority of monoclonal antibodies prepared against *Toxocara* ES proved to be directed towards glycosylated epitopes (Maizels *et al.*, 1987b).

MONOCLONAL ANTIBODIES TO *TOXOCARA* ES ANTIGENS

Monoclonal antibodies raised against ES antigens have been used to elucidate the distribution and nature of epitopes carried by ES antigens (Maizels *et al.*, 1987b; Bowman *et al.*, 1987a). Of a panel of eight such monoclonal antibodies (Maizels *et al.*, 1987b), six recognized periodate sensitive epitopes (Table 6.2), reflecting the dominance of glycan specificities in the immune response. These carbohydrate epitopes are, moreover, present in a repetitive form presumably because the same oligosaccharide side-chains are coupled to protein backbones at several sites, and different monoclonal antibodies appear to bind with a hierarchy of affinities or fine specificities to related epitopes. Thus, antibody Tcn-2 will inhibit the binding of antibodies Tcn-4, -5 and -8, although these will not block Tcn-2 binding at any concentration.

One surprising result emerged from the reactivity of these monoclonals to individual ES components. Some antibodies, in particular Tcn-2 and Tcn-8, reacted with most or all of the diverse glycoproteins secreted by the parasites and could generate immunoprecipitation or immuno-blot profiles little different from

Table 6.2. Monoclonal-defined epitopes on *T. canis* larval antigens.[a]

Monoclonal antibody	Periodate sensitive epitope	Repetitive epitope	Surface exposed	Species specific	Binding to ES molecules		Binding to surface molecules
					Strong	Weak	
Tcn-2	+	+	+	+	70,120,400	55	55,120
Tcn-8	+	±	+	–	70,120,400	32,55	32,55,70,120
Tcn-4/5	+	+	–	–	70	55,120	55,120
Tcn-7	+	+	–	–	70	120	55,120
Tcn-1	+	ND	–	–	70	32,120	55,120
Tcn-6	–	ND	–	–	70,120		55,120
Tcn-3	–	±	–	–	32	70,120	32

[a] Figures indicated as molecules bound to the molecular weight of antigens recognized (in kilodaltons). For ES antigens weak binding is defined as that evident in only one of the two specificity assays employed: immunoprecipitation or Western blotting. Adapted from data presented in Maizels *et al.* (1987b).

those seen with polyclonal infection sera. Our conclusion was that the same carbohydrate side-chain occurred on each distinct ES component, and that being repetitive as well as abundant, this determinant was relatively immunogenic.

Two monoclonal antibodies bind strongly to the larval surface, confirming the finding from immunofluorescence (Smith *et al.*, 1981) and iodination (Maizels *et al.*, 1984) studies that similar molecules are represented on the surface and in the secretions. Both antibodies, Tcn-2 and Tcn-8, are directed to periodate sensitive epitopes. One monoclonal, Tcn-2 (Figure 6.4), proved to be species-specific in that it failed to react with either the surface or ES of the related feline ascarid *T. cati* (Maizels *et al.*, 1987b; Kennedy *et al.*, 1987a). A monoclonal antibody raised by Bowman *et al.* (1987a) also showed this species-specific characteristic by surface

• IgM Monoclonal recognises *Toxocara* ES antigens of 400, 120, 70 and 55 kDa; also binds the larval surface

• Directed to repetitive carbohydrate epitope shared between different secreted glycoproteins

• Target epitope is species specific

• Detects circulating *Toxocara* antigen in patients' sera by two site assay

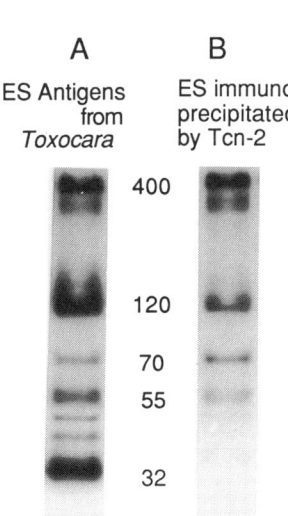

A
ES Antigens from *Toxocara*

B
ES immuno-precipitated by Tcn-2

400
120
70
55
32

Figure 6.4. Monoclonal antibody Tcn-2: specificity and applications.

fluorescence, although the target of this antibody appears to be distinct from that of Tcn-2.

These features of species-specificity and recognition of a repetitive epitope permitted us to use Tcn-2 in an effective assay for free *Toxocara* ES antigen which will also detect circulating toxocaral antigen in experimental animals or infected human patients. Based on the two-site antigen capture protocol (Figure 6.5) which has been applied widely in the diagnosis of infectious agents (for example, Burkot *et al.*, 1984; Forsyth *et al.*, 1985), the test successfully detected parasite antigen in mice immediately after infection, in rabbits over a longer term, and in a number of human patients (Figure 6.6 and Robertson *et al.*, 1988a). We are now conducting more extensive diagnostic tests with this assay on human *Toxocara* patients (Gillespie, S. H. and Maizels, R. M., unpublished). The prominence and high affinity of the Tcn-2 specificity among this panel of monoclonals make it likely to be an important contributor to the detection of circulating antigen in infected dogs with a polyclonal rabbit antiserum to ES antigens in an otherwise similar assay (Matsumura *et al.*, 1984). These workers were able to detect less than 100 ng of ES antigen with the highest levels in puppies 1 month old, declining thereafter and becoming undetectable after 6 months. Similarly, a recent report confirms that circulating ES antigens can be detected in the serum of infected mice (Bowman *et al.*, 1987b).

A further interesting point which anti-*Toxocara* monoclonals have revealed is the

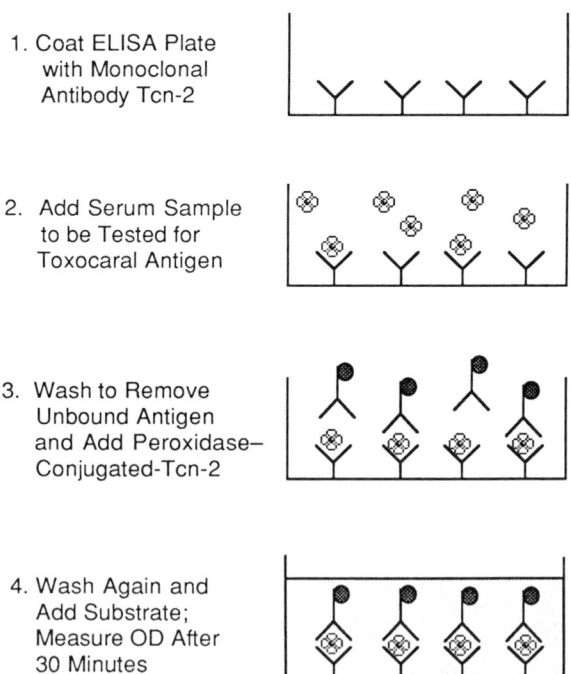

1. Coat ELISA Plate with Monoclonal Antibody Tcn-2

2. Add Serum Sample to be Tested for Toxocaral Antigen

3. Wash to Remove Unbound Antigen and Add Peroxidase–Conjugated-Tcn-2

4. Wash Again and Add Substrate; Measure OD After 30 Minutes

Figure 6.5. Two-site 'antigen capture' diagnostic test for circulating *Toxocara* antigen in serum.

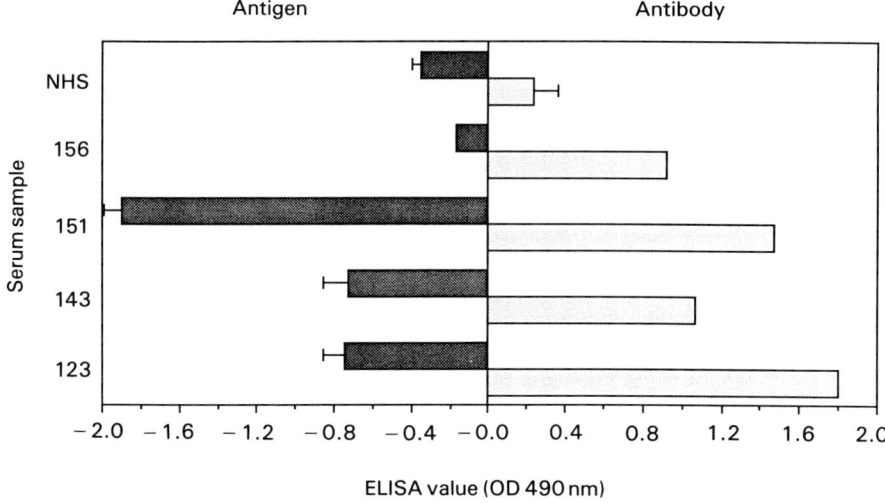

Figure 6.6. *Toxocara* antigen and antibody levels in four patients and one uninfected control case from the UK. (Reproduced from Robertson *et al.* (1988b).).

changing emphasis of the immune response over time following infection. Inhibition assays of monoclonal binding can be used to measure the quantity of a single specificity in a polyclonal infection serum (Mitchell *et al.*, 1983). In rabbits, in which a long-term infection was followed, Tcn-2-like antibody appears shortly post-infection, as measured by the ability of the rabbit serum to block Tcn-2 binding to ES antigens. However, it declines after three months and is replaced by a Tcn-3-like specificity. In part, this may reflect the maturation of the immune response and the shift from IgM anti-carbohydrate responses to IgG anti-protein antibodies, but the data do indicate the potential for differential diagnosis of new versus long-standing infections by isotype- or epitope-specific antibody assays. Some support for this notion would come from studies in dogs, in which IgG but not IgM titres increase with age, and only titres of the former correlate with the number of adult worms in the gut (Matsumura and Endo, 1982; Matsumura *et al.*, 1983).

IN VIVO ACTION OF *TOXOCARA* ES ANTIGENS

Following infection with *Toxocara* larvae, ES glycoconjugates are rapidly recognized by the immune system, and a strong antibody response ensues. There is as yet no indication that such antibody responses are protective, or if T cell mediated mechanisms are more critical in immunity to ascarid larvae; in either case, it is possible that one benefit to the parasite of its high level of ES production is diverting the immune system into synthesizing ineffective antibody. While this possibility awaits experimental verification, we should also consider a range of other possible functional roles the individual ES products may play in helping larvae to invade and survive in their mammalian hosts.

One easily envisaged role for ES antigens in parasite survival and pathogenicity is that of secreted proteases either produced by the gut dwelling stages of the worm to facilitate feeding, or released by skin-penetrating or tissue migrating helminths (Lewart and Lee, 1956), as described in invading cercariae of *Schistosoma mansoni* (McKerrow *et al.*, 1985). Adult schistosomes in the vascular system (Lindquist *et al.*, 1986; Chappell and Dresden, 1986) also produce a proteolytic enzyme capable of degrading substrates such as basement-membrane macromolecules and haemoglobin. These enzymes have been purified and characterized and include an elastinolytic metalloproteinase from cercariae (McKerrow *et al.*, 1985) and a thiol haemoglobinase from adult worms which may be involved in nutrition of the parasite (Lindquist *et al.*, 1986; Chappell and Dresden, 1986). The dog hookworm *Ancylostoma caninum* secretes an anticoagulant elastase which is presumed to aid the blood intake of the adult worm (Hotez and Cerami, 1983). This protease is antigenic, and antibodies cross-react with a product of the invasive infective larvae, suggesting that the same elastase is involved in hookworm invasion and tissue penetration (Hotez *et al.*, 1985).

To identify any similar degradative proteases in *Toxocara*, we employed three distinct assays. Firstly, we showed that living larvae could hydrolyse extracellular matrix proteins in an *in vitro* model. In this system the rat smooth muscle cell line R22 is first cultured in the presence of [³H]proline; the cells secrete a set of proteins (including fibronectin, elastin and types I and III collagen) which are radiolabelled and adhere to the plate. The R22 cells are then washed out, and parasite larvae introduced in medium (McKerrow *et al.*, 1983). Within 4 h of adding *Toxocara* larvae to these plates we could detect released radioactivity representing degraded proteins (Robertson *et al.*, 1989), and breakdown continued linearly for 54 h

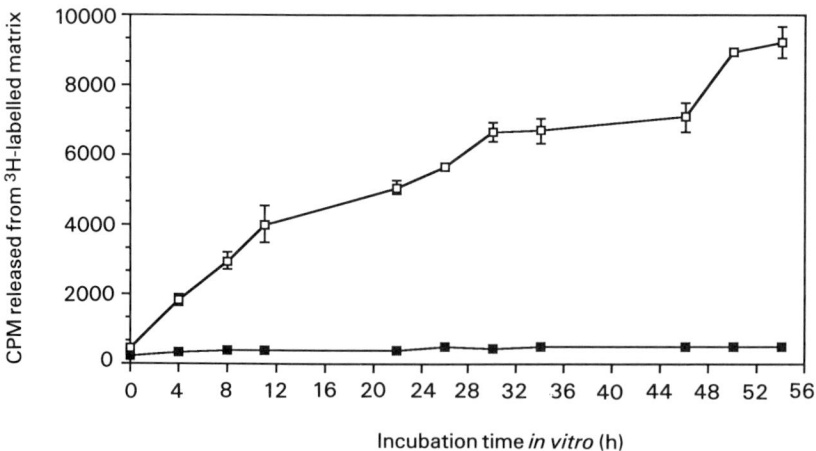

Figure 6.7. *T. canis* larvae degrade extracellular matrix proteins *in vitro* (Reproduced from Robertson *et al.* (1989).)

(Figure 6.7). Analysis of the proteins digested indicated that the secreted protease activity from *Toxocara* was probably an elastase or trypsin-like enzyme.

Further characterization of the secreted protease activity employed concentrated *Toxocara* ES in a sensitive assay measuring release of ^{125}I-labelled gelatin precoated onto plastic wells (Robertson *et al.*, 1988b). This assay indicated that more than one protease may be present: although proteolysis was optimal at pH 9, the overall pH profile indicated other optima at pH 7 and pH 5. However, the same assay facilitated an inhibitor profile, with the result that only inhibitors of serine proteases, and not inhibitors of thiol or metalloproteases blocked the activity of the nematode enzyme. Thus, if there are two or more secreted proteases, they are probably each a serine protease. Finally, substrate gel electrophoresis of ES proteins was undertaken to determine the molecular weights of the proteolytic enzymes. This showed that the major activity is the 120 kDa secreted glycoprotein, although evidence was obtained that the 32 kDa molecule may also be a protease (Robertson *et al.*, 1989).

Proteolytic enzymes may also form part of the range of immunomodulatory functions which we assume to mediate the evasion of effective immune mechanisms by *Toxocara*. Cleavage of surface bound immunoglobulin is a route taken by other helminth parasites such as *S. mansoni* (Auriault *et al.*, 1981) while *Taenia* and other cestodes produce protease inhibitors which block C5a-mediated chemotaxis (Leid *et al.*, 1987). It is equally conceivable that inhibitors prevent the physiological proteolytic cleavage of complement components during complement activation at any of the stages from C1 to C5 which require enzymatic processing. So far, it has been shown that *T. canis* larvae will bind C3, C5 and C9 (Kennedy and Kuo, 1988) but it is not known whether these components are present in an inactivated or an assembled form, and no other studies have addressed either proteolysis or protease inhibitors in this species.

In addition to direct interference by helminths in humoral immune mechanisms, some modulation at the level of immune cell activation and effector pathways may be employed. A possible influence on immunocyte activation is the acetylcholinesterase (AChE) found in the ES products of many nematodes including *Nippostrongylus brasiliensis* (Ogilvie *et al.*, 1973), the filarial worm *Brugia malayi* (Rathaur *et al.*, 1987) and also *T. canis* (Robertson *et al.*, 1987). The *Toxocara* enzyme is antigenic and like that from *B. malayi*, shows a cross-reaction with antibody raised to electric eel AChE. It has been partially purified using copper chelating and concanavalin A affinity chromatography, yielding a major component in the purified fraction as the 55 kDa glycoprotein, although this may be a contaminant co-purifying under the same conditions as the enzyme. The role of parasite derived AChE in tissue dwelling nematodes such as *B. malayi* and *T. canis* is not clear, but as acetylcholine (ACh) is known to enhance a number of immunological functions, such as the release of mediators by mast cells, the cytotoxicity of neutrophils and lymphocytes, and the production of γ-interferon by T cells, the degradation of ACh may locally down-regulate the immune response so allowing the parasite to escape immune mediated attack (Rhoads, 1984; Rathaur *et al.*, 1987).

Various effector cell populations may be directed by the host immune system to attack nematode larvae. Again, it is uncertain which cell type is most effective at this task, but some countermeasures employed by parasites have been suggested. The trematode *S. mansoni* produces an effective detoxification enzyme, glutathione-*S*-transferase (Mitchell, 1989), which may also be expressed in some nematode species. More evidence is currently available for the expression of superoxide dismutase (SOD), produced by parasites such as *Trichinella spiralis* (Rhoads, 1983), which would provide protection against one of the more potent products of the respiratory burst, the superoxide ion (O_2^-). The product of SOD, hydrogen peroxide, would be scavenged by glutathione peroxidase and catalase, and each of these activities may also be present as surface proteins or as products secreted into the immediate vicinity of the *Toxocara* worm.

One phenomenon often linked with helminth infection is eosinophilia, and this is associated with both animal (Sugane and Oshima, 1980; Kayes *et al.*, 1985) and human toxocariasis (Dent *et al.*, 1956; Beaver, 1962; Lloyd, 1987). For example, a recent study reported that 73 per cent of Irish antibody positive cases were eosinophilic (Taylor *et al.*, 1988). Significantly, eosinophilia can be induced in mice directly by administering ES antigens in the absence of larvae (Sugane and Oshima, 1984b), while the introduction of as few as five infective eggs to mice produced a significant increase in blood eosinophil levels (Kayes *et al.*, 1985). It would therefore appear that *Toxocara* larvae release substances which act as potent eosinophil-promoting factors, prompting the question of why a parasite should amplify a potentially lethal response in such a direct manner. The possibility should therefore be considered that the host eosinophilic reaction might actually be helpful to the parasite, perhaps playing a diversionary role stimulating cells which are ineffective against this species of nematode? Support for this perhaps heretical view can be found in the fact that, despite the high eosinophilia observed in experimental and human infections, the degree of eosinophilia does not influence the number of larvae recovered from infected mice (Sugane and Oshima, 1984c).

Whether or not this speculative view is correct, it is certainly true that *Toxocara* larvae prove to be resistant to direct killing by either guinea-pig (Rockey *et al.*, 1983; Badley *et al.*, 1987b) or human eosinophils (Fattah *et al.*, 1986). Eosinophils rapidly adhere to the surface of larvae, showing characteristic activation and degranulation, but larvae show little signs of damage and indeed slough off the cells together with the extracuticular material, subsequently remaining viable in culture for at least one week (Fattah *et al.*, 1986). This extracuticular layer, which appears electron dense by transmission electron microscopy (Maizels and Selkirk, 1988), may be the equivalent of the polysaccharide surface coat on the surface of *Dirofilaria immitis* microfilariae (Cherian *et al.*, 1980), and has been suggested to consist of surface antigens and antibody, the presence and subsequent shedding of which may prevent effector cells and molecules from damaging the epicuticle (Rockey *et al.*, 1983; Badley *et al.*, 1987b). Possibly, this sloughing-off is a larger scale form of the normal turnover of surface antigens from the larval cuticle (Maizels *et al.*, 1984; Smith *et al.*, 1981). Interestingly, although surface proteins can readily be released from their anchorage in the epicuticle, the outermost lipid membrane of *Toxocara* does not show the lateral mobility so characteristic of the mammalian plasma membrane (Kennedy *et al.*, 1987b). However, the relative positions of the epicuticular lipids, surface proteins and the extracuticular layer have yet to be defined, and any model for the *Toxocara* surface must take into account the rapidity with which surface proteins may be cast off.

Toxocara surface antigens may therefore play a dual role in immune avoidance; firstly, by the rate with which they are shed to join the ES they show a vertical mobility which may prevent humoral and cellular immune mechanisms from exerting any damaging effect. Once released, they appear also to perform a functional role in directly interacting with host molecules, substrates or receptors, to break down extracellular matrix and possibly complement or other soluble host mediators and to activate host cell subpopulations selectively in a manner least injurious to the parasite. With the definition of individual molecular species on the cuticular surface and in the larval ES we now aim to test *in vitro* and *in vivo* the biological and immunological activity of each component from these pathologenic larvae in order to bring a new level of precision to the study of nematode–host interactions.

CONCLUDING REMARKS

Nematodes are among the most abundant animals on earth, with numerous species both free living and parasitic. The study of these organisms provides the opportunities both to alleviate a number of widespread chronic and debilitating diseases, and to study many aspects of parasite biology. Features such as chronicity and persistence in nematodes must indicate that the immune system is being evaded or compromised in a fundamental way, and we would anticipate that a spectrum of specific molecular and physiological interactions with host cells and mediators are employed to permit these long-lived parasites to accommodate themselves. While *T. canis* is not of major importance as a human pathogen, it provides an opportunity to study a wide range of features in nematode parasite biology, such as arrested development and host specificity, antigen processing, secretion and turnover, and adaptation and survival in paratenic hosts. Each of these are applicable to other parasites of more major medical importance which share with *Toxocara* the ability to invade and establish themselves in the human body. We hope that by continuing our study of the structure, function and immune recognition of these defined glycoconjugates from *T. canis*, we will be able to provide a more fundamental understanding of nematode parasitism in general, which will in turn facilitate the range of practical control measures which are so clearly needed for the major human nematodiases.

ACKNOWLEDGEMENTS

The authors gratefully acknowledge support from the Wellcome Trust, and the provision of a research studentship from the Medical Research Council.

REFERENCES

Auriault, C., Ouaissi, M. A., Torpier, G., Eisen, H. and Capron, A., 1981, Proteolytic cleavage of IgG bound to the Fc receptor of *Schistosoma mansoni* schistosomula, *Parasite Immunology*, **3**, 33–44.

Badley, J. E., Grieve, R. B., Bowman, D. D., Glickman, L. T. and Rockey, J. H., 1987a, Analysis of *Toxocara canis* larval excretory–secretory antigens: physiochemical characterization and antibody recognition, *Journal of Parasitology*, **73**, 593–600.

Badley, J. E., Grieve, R. B., Bowman, D. D. and Glickman, L. T., 1987b, Immune-mediated adherence of eosinophils to *Toxocara canis* infective larvae: the role of excretory–secretory antigens, *Parasite Immunology*, **9**, 133–43.

Beaver, P. C., 1962, Toxocarosis (visceral larva migrans) in relation to tropical eosinophilia, *Bulletin de la Société de Pathologie Exotique*, **55**, 555–76.

Berrocal, J., 1980, Prevalence of *Toxocara canis* in babies and in adults as determined by the ELISA test, *Transactions of the American Ophthalmological Society*, **75**, 376–413.

Bisseru, B., 1969, Studies on the liver, lung, brain and blood of experimental animals infected with *Toxocara canis*, *Journal of Helminthology*, **43**, 267–72.

Borg, O. A. and Woodruff, A. W., 1973, Prevalence of infective ova of *Toxocara* species in public places, *British Medical Journal*, **4**, 470–2.

Bowman, D. D., Mika-Grieve, M. and Grieve, R. B., 1987a, *Toxocara canis*: monoclonal antibodies to larval excretory–secretory antigens that bind with genus and species specificity the cuticular surface of infective larvae, *Experimental Parasitology*, **64**, 458–65.

Bowman, D. D., Mika-Grieve, M. and Grieve, R. B., 1987b, Circulating excretory–secretory antigen levels and specific antibody responses in mice infected with *Toxocara canis*, *American Journal of Tropical Medicine and Hygiene*, **36**, 75–82.

Boyce, W. M., Branstetter, B. A. and Kazacos, K. R., 1988, Comparative analysis of larval excretory–secretory antigens of *Baylisascaris procyonis*, *Toxocara canis* and *Ascaris suum* by Western blotting and enzyme immunoassay, *International Journal for Parasitology*, **18**, 109–13.

Brunello, F., Genchi, C. and Falagiani, P., 1983, Detection of larva specific IgE in human toxocariasis, *Transactions of the Royal Society of Tropical Medicine and Hygiene*, **77**, 279–80.

Bundy, D. A. P., Thompson, D. E., Robertson, B. D. and Cooper, E. S., 1987, Age relationship of *Toxocara canis* seropositivity and geohelminth infection prevalence in two communities in St Lucia, West Indies, *Tropical Medicine and Parasitology*, **38**, 309–12.

Burke, T. M. and Roberson, E. L., 1985a, Prenatal and lactational transmission of *Toxocara canis* and *Ancylostoma caninum*: experimental infection of the bitch before pregnancy, *International Journal for Parasitology*, **15**, 71–5.

Burke, T. M. and Roberson, E. L., 1985b, Prenatal and lactational transmission of *Toxocara canis* and *Ancylostoma caninum*: experimental infection of the bitch at midpregnancy and at parturition, *International Journal for Parasitology*, **15**, 485–90.

Burkot, T. R., Zavala, F., Gwadz, R. W., Collins, F. H., Nussenzweig, R. S. and Roberts, D. R., 1984, Identification of malaria-infected mosquitoes by a two-site enzyme-linked immunosorbent assay, *American Journal of Tropical Medicine and Hygiene*, **33**, 227–31.

Chappell, C. L. and Dresden, M. H., 1986, *Schistosoma mansoni*: proteinase activity of 'hemoglobinase' from the digestive tract of adult worms, *Experimental Parasitology*, **61**, 160–7.

Cherian, P. V., Stromberg, B. E., Wiener, D. J. and Soulsby, E. J. L., 1980, Fine structure and cytochemical evidence for the presence of polysaccharide surface coat of *Dirofilaria immitis* microfilariae, *International Journal for Parasitology*, **10**, 227–33.

Clemett, R. S., Allardyce, R. A., Williamson, H. J. E., Stewart, A. C. and Hidajat, R. R., 1987, Ocular *Toxocara canis* infections: diagnosis by enzyme immunoassay, *Australian and New Zealand Journal of Ophthalmology*, **15**, 145–50.

Cypess, R. H., Karol, M. H., Zidian, J. L., Glickman, L. T. and Gitlin, D., 1977, Larva specific antibodies in patients with visceral larva migrans, *Journal of Infectious Diseases*, **135**, 633–40.

Dada, B. J. O. and Lindquist, W. D., 1979, Prevalence of *Toxocara* spp. eggs in some public grounds and highway rest areas in Kansas, *Journal of Helminthology*, **53**, 145–6.

Dent, J. H., Nichols, R. L., Beaver, P. C., Carrera, G. M. and Staggers, R. J., 1956, Visceral larva migrans, *American Journal of Pathology*, **32**, 777–803.

de Savigny, D. H., 1975, *In vitro* maintenance of *Toxocara canis* larvae and a simple method for the production of *Toxocara* ES antigen for use in serodiagnostic tests for visceral larva migrans, *Journal of Parasitology*, **61**, 781–2.

de Savigny, D. H. and Tizard, I. R., 1977, Toxocaral larva migrans: the use of larval secretory antigens in haemagglutination and soluble antigen fluorescent antibody tests, *Transactions of the Royal Society of Tropical Medicine and Hygiene*, **71**, 501–7.

de Savigny, D. H., Voller, A. and Woodruff, A. W., 1979, Toxocariasis: serological diagnosis by enzyme immunoassay, *Journal of Clinical Pathology*, **32**, 284–8.

Dinning, W. J., Gillespie, S. H., Cooling, R. J. and Maizels, R. M., 1988, *Toxocariasis*: a practical approach to management of ocular disease, *Eye*, **2**, 580–2.

Dolinsky, Z. S., Burright, R. G., Donovick, P. J., Glickman, L. T., Babish, J., Summers, B. and Cypess, R. H., 1981, Behavioral effects of lead and *Toxocara canis* in mice, *Science*, **213**, 1142–4.

Dunsmore, J. D., Thompson, R. C. A. and Bates, I. A., 1983, The accumulation of *Toxocara canis* larvae in the brains of mice, *International Journal of Parasitology*, **13**, 517–21.

Fattah, D. I., Maizels, R. M., McLaren, D. J. and Spry, C. J. F., 1986, *Toxocara canis*: interaction of human eosinophils with the infective larvae, *Experimental Parasitology*, **61**, 421–31.

Forsyth, K. P., Spark, R., Kazura, J., Brown, G. V., Peters, P., Heywood, P., Dissanayake, S. and Mitchell, G. F., 1985, A monoclonal antibody-based immunoradiometric assay for detection of circulating antigen in Bancroftian filariasis, *Journal of Immunology*, **134**, 1172–7.

Gillespie, S. H., 1987, Human toxocariasis, *Journal of Applied Bacteriology*, **63**, 473–9.

Gillespie, S. H., 1988, The epidemiology of *Toxocara canis*, *Parasitology Today*, **4**, 180–2.

Girdwood, R. W. A., Quinn, R., Smith, H. V. and Bruce, R. G., 1978, Assessment of some aspects of the potential human health hazard presented by canine toxocariasis in the Glasgow area, *Communicable Diseases, Scotland*, **12**, 7–8.

Glickman, L. T. and Schantz, P. M., 1981, Epidemiology and pathogenesis of zoonotic toxocariasis, *Epidemiological Reviews*, **3**, 23–250.

Glickman, L. T. and Schantz, P. M., 1985, Do *Toxocara canis* larval antigens used in the enzyme-linked immunosorbent assay for visceral larva migrans cross-react with AB isohaem-agglutinins and give false positive results? *Zeitschrift für Parasitenkunde*, **71**, 395–400.

Glickman, L. T., Schantz, P. M., Dombroske, R. and Cypess, R. H., 1978, Evaluation of serological tests for visceral larva migrans, *American Journal of Tropical Medicine and Hygiene*, **27**, 492–8.

Glickman, L. T., Schantz, P. M. and Cypess, R. H., 1979, Canine and human toxocariasis: review of transmission pathogenesis, and clinical disease, *Journal of the American Veterinary Medical Association*, **175**, 1265–9.

Glickman, L. T., Grieve, R. B., Lauria, S. S. and Jones, D. L., 1985, Serodiagnosis of ocular toxocariasis: a comparison of two antigens, *Journal of Clinical Pathology*, **38**, 103–7.

Griesemer, R. A., Gibson, J. P. and Elsasser, D. S., 1963, Congenital ascariasis in gnotobiotic dogs, *Journal of the American Veterinary Medical Association*, **143**, 962–4.

Herrmann, N., Glickman, L.T., Schantz, P.M., Weston, M.G. and Domanski, L.M., 1985, Seroprevalence of zoonotic toxocariasis in the United States: 1971–1973, *American Journal of Epidemiology*, **122**, 890–96.

Hill, I.R., Denham, D.A. and Scholtz, C.L., 1985, *Toxocara canis* larvae in the brain of a British child, *Transactions of the Royal Society of Tropical Medicine and Hygiene*, **79**, 351–4.

Hotez, P.J. and Cerami, A., 1983, Secretion of a proteolytic anticoagulant by *Ancylostoma* hookworms, *Journal of Experimental Medicine*, **157**, 1594–603.

Hotez, P.J., Le Trang, N., McKerrow, J.H. and Cerami, A., 1985, Isolation and characterization of a proteolytic enzyme from the adult hookworm *Ancylostoma caninum*, *Journal of Biological Chemistry*, **260**, 7343–8.

Josephs, D.S., Bhinder, P. and Thompson, A.R., 1981, The prevalence of *Toxocara* infection in a child population, *Public Health, London*, **95**, 273–5.

Kayes, S.G. and Oakes, J.A., 1976, Effect of inoculum size and length of infection on the distribution of *Toxocara canis* larvae in the mouse, *American Journal of Tropical Medicine and Hygiene*, **25**, 573–80.

Kayes, S.G., Omholt, P.E. and Grieve, R.B., 1985, Immune responses of CBA/J mice to graded infections with *Toxocara canis*, *Infection and Immunity*, **48**, 697–703.

Kennedy, M.W. and Kuo, Y.-M., 1988, The surfaces of the parasitic nematodes *Trichinella spiralis* and *Toxocara canis* differ in the binding of post-C3 components of human complement by the alternative pathway, *Parasite Immunology*, **10**, 459–63.

Kennedy, M.W., Maizels, R.M., Meghji, M., Young, L., Qureshi, F. and Smith, H.V., 1987a, Species-specific and common epitopes on the secreted and surface antigens of *Toxocara canis* infective larvae, *Parasite Immunology*, **9**, 407–20.

Kennedy, M.W., Foley, M., Kuo, Y.-M., Kusel, J.R. and Garland, P.B., 1987b, Biophysical properties of the surface lipid of parasitic nematodes, *Molecular and Biochemical Parasitology*, **22**, 233–40.

Kennedy, M.W., Qureshi, F., Fraser, E.M., Haswell-Elkins, M.R., Elkins, D.B. and Smith, H.V., 1989, Antigenic relationships between the surface-exposed, secreted, and somatic materials of the nematode parasites *Ascaris lumbricoides*, *Ascaris suum*, and *Toxocara canis*, *Clinical and Experimental Immunology*, **75**, 493–500.

Leid, R.W., Grant, R.F., and Suqet, C.M., 1987, Inhibition of equine neutrophil chemotaxis and chemokinesis by a *Taenia taeniaeformis* proteinase-inhibitor, taeniaestatin, *Parasite Immunology*, **9**, 195–204.

Lewart, R.M. and Lee, C.-L., 1956, Quantitative studies of the collagenase-like enzymes of cercariae of *Schistosoma mansoni* and the larvae of *Strongyloides ratti*, *Journal of Infectious Diseases*, **99**, 1–14.

Lindquist, R.N., Senft, A.W., Petit, M. and McKerrow, J.H., 1986, *Schistosoma mansoni*: purification and characterization of the major acidic proteinase from adult worms, *Experimental Parasitology*, **61**, 398–404.

Lloyd, S., 1987, Immunobiology of *Toxocara canis* and visceral larva migrans, in Soulsby, E.J.L. (Ed.) *Immune Responses in Parasitic Infections: Immunology, Immunopathology, and Immunoprophylaxis, Vol. I, Nematodes*, pp. 299–324, Boca Raton, Florida: CRC Press.

Lynch, N.R., Wilkes, L.K., Hodgen, A.N. and Turner, K.J., 1988a, Specificity of *Toxocara* ELISA in tropical populations, *Parasite Immunology*, **10**, 323–37.

Lynch, N.R., Eddy, K., Hodgen, A.N., Lopez, R.I. and Turner, K.J., 1988b, Seroprevalence of *Toxocara canis* infection in tropical Venezuela, *Transactions of the Royal Society of Tropical Medicine and Hygiene*, **82**, 275–81.

McKerrow, J.H., Keene, W.E., Jeong, K.H. and Werb, Z., 1983, Degradation of extracellular matrix by larvae of *Schistosoma mansoni*. I. Degradation by cercariae as a model for initial parasite invasion of host, *Laboratory Investigation*, **49**, 195–200.

McKerrow, J.H., Pino-Heiss, S., Lindquist, R. and Werb, Z., 1985, Purification and characterization of an elastinolytic proteinase secreted by cercariae of *Schistosoma mansoni*, *Journal of Biological Chemistry*, **260**, 3703–7.

Maizels, R. M. and Selkirk, M. E., 1988, Immunobiology of nematode antigens, in Englund, P. T. and Sher, F. A. (Eds) *The Biology of Parasitism*, pp. 285–308, New York: Alan R. Liss.

Maizels, R. M., de Savigny, D. and Ogilvie, B. M., 1984, Characterisation of surface and excretory–secretory antigens of *Toxocara canis* infective larvae, *Parasite Immunology*, 6, 23–37.

Maizels, R. M., Burke, J. and Denham, D. A., 1987a, Phosphorylcholine-bearing antigens in filarial nematode parasites: analysis of somatic extracts and *in vitro* secretions of *Brugia malayi* and *B. pahangi*, *Parasite Immunology*, 9, 49–66.

Maizels, R. M., Kennedy, M. W., Meghji, M., Robertson, B. D. and Smith, H. V., 1987b, Shared carbohydrate epitopes on distinct surface and secreted epitopes of the parasitic nematode *Toxocara canis*, *Journal of Immunology*, 139, 207–14.

Marmor, M., Glickman, J., Shofer, F., Faich, L. A., Rosenberg, C., Cornblatt, B. and Friedmans, S., 1987, *Toxocara canis* infection in children: epidemiologic and neuropsychologic findings, *American Journal of Public Health*, 77, 554–9.

Matsumura, K. and Endo, R., 1982, Investigation of antibodies against *Toxocara canis* in naturally infected puppies, *Zentralblatt Bakteriologie Hygiene*, 253, 139–43.

Matsumura, K. and Endo, R., 1983, Seroepidemiological study on toxocaral infection in man by enzyme-linked immunosorbent assay, *Journal of Hygiene, Cambridge*, 90, 61–5.

Matsumura, K., Kazuta, Y., Endo, R. and Tanaka, K., 1983, The IgM antibody activities in relation to the parasitologic status of *Toxocara canis* in dogs, *Zentralblatt Bakteriologie Hygiene*, 255, 402–5.

Matsumura, K., Kazuta, Y., Endo, R. and Tanaka, K., 1984, Detection of circulating toxocaral antigens in dogs by sandwich enzyme-immunoassay, *Immunology*, 51, 609–13.

Meghji, M. and Maizels, R. M., 1986, Biochemical properties of larval excretory–secretory (ES) glycoproteins of the parasitic nematode *Toxocara canis*, *Molecular and Biochemical Parasitology*, 18, 155–70.

Merrill, C. R., Goldman, D., Sedman, S. A. and Elbert, M. H., 1981, Ultrasensitive stain for proteins in polyacrylamide gels shows regional variation in cerebrospinal fluid proteins, *Science*, 211, 1437–8.

Mitchell, G. F., 1989, Glutathione-*S*-transferases — potential components of anti-schistosome vaccines? *Parasitology Today*, 5, 34–7.

Mitchell, G. F., Premier, R. R., Garcia, E. G., Hurrell, J. G. R., Chandler, H. M., Cruise, K. M., Tapales, F. P. and Tiu, W. U., 1983, Hybridoma antibody-based competitive ELISA in *Schistosoma japonicum* infection, *American Journal of Tropical Medicine and Hygiene*, 32, 114–17.

Nicholas, W. L., Stewart, A. C. and Mitchell, G. F., 1984, Antibody responses to *Toxocara canis* using sera from parasite-infected mice, and protection from toxocariasis by immunization with ES antigens, *Australian Journal of Experimental Biology and Medical Science*, 62, 619–26.

Nicholas, W. L., Stewart, A. C. and Walker, J. C., 1986, Toxocariasis: a serological survey of blood donors in the Australian Capital territory together with observations on the risks of infection, *Transactions of the Royal Society of Tropical Medicine and Hygiene*, 80, 217–21.

Ogilvie, B. M., Rothwell, T. L. W., Bremner, K. C., Schnitzerling, H. J., Nolan, J. and Keith, R. K., 1973, Acetylcholinesterase secretion by parasitic nematodes. I. Evidence for secretion by a number of species, *International Journal for Parasitology*, 3, 589–97.

Parkhouse, R. M. E., Philipp, M. and Ogilvie, B. M., 1981, Characterization of surface antigens of *Trichinella spiralis* infective larvae, *Parasite Immunology*, 3, 339–52.

Pollard, Z. F., Jarrett, W. H., Hagler, W. S., Allain, D. S. and Schantz, P. M., 1979, ELISA for diagnosis of ocular toxocariasis, *Ophthalmology*, 86, 743–9.

Quinn, R., Smith, H. V., Bruce, R. G. and Girdwood, R. W. A., 1980, Studies on the incidence of *Toxocara* and *Toxascaris* spp. ova in the environment. I. A comparison of flotation procedures for recovering *Toxocara* spp. ova from soil, *Journal of Hygiene (Cambridge)*, **84**, 83–9.

Rathaur, S., Robertson, B. D., Selkirk, M. E. and Maizels, R. M., 1987, Secretory acetycholinesterases from *Brugia malayi* adult and microfilarial parasites, *Molecular and Biochemical Parasitology*, **26**, 257–65.

Ree, G. H., Voller, A. and Rowland, H. A. K., 1984, Toxocariasis in the British Isles 1982–3, *British Medical Journal*, **288**, 628–9.

Rhoads, M. L., 1983, *Trichinella spiralis*: identification and purification of superoxide dismutase, *Experimental Parasitology*, **56**, 41–54.

Rhoads, M. L., 1984, Secretory cholinesterases of nematodes: possible functions in the host–parasite relationship, *Tropical Veterinarian*, **2**, 3–10.

Robertson, B. D., Rathaur, S. and Maizels, R. M., 1987, Antigenic and biochemical analyses of the excretory–secretory molecules of *Toxocara canis* infective larvae, in Geerts, S., Kumar, V. and Brandt, J. (Eds) *Current Topics in Veterinary Medicine and Animal Science 'Helminth Zoonoses'*, pp. 167–73, Dordrecht: Martinus Nijhoff Publishers.

Robertson, B. D., Burkot, T. R., Gillespie, S. H., Kennedy, M. W., Wambai, Z. and Maizels, R. M., 1988a, Detection of circulating parasite antigen and specific antibody in *Toxocara canis* infections, *Clinical and Experimental Immunology*, **74**, 236–41.

Robertson, B. D., Kwan-Lim, G.-E. and Maizels, R. M., 1988b, A sensitive microplate assay for the detection of proteolytic enzymes using radiolabeled gelatin, *Analytical Biochemistry*, **172**, 284–7.

Robertson, B. D., Bianco, A. E., McKerrow, J. H. and Maizels, R. M., 1989, Proteolytic enzymes secreted by larvae of the nematode *Toxocara canis*, *Experimental Parasitology*, **69**, 30–6.

Rockey, J. H., John, T., Donnelly, J. J., McKenzie, D. F., Stromberg, B. E. and Soulsby, E. J. L., 1983, *In vitro* interaction of eosinophils from Ascarid-infected eyes with *Ascaris suum* and *Toxocara canis* larvae, *Investigative Ophthalmology and Visual Science*, **24**, 1346–57.

Schantz, P. M., Meyer, D. and Glickman, L. T., 1979, Clinical, serologic and epidemiologic characteristics of ocular toxocariasis, *American Journal of Tropical Medicine and Hygiene*, **28**, 24–8.

Scothorn, M. W., Koutz, F. R. and Groves, H. F., 1965, Prenatal *Toxocara canis* infection in pups, *Journal of the American Veterinary Medical Association*, **146**, 45–8.

Smith, H. V., Quinn, R., Kusel, J. R. and Girdwood, R. W. A., 1981, The effect of temperature and antimetabolites on antibody binding to the outer surface of second stage *Toxocara canis* larvae, *Molecular and Biochemical Parasitology*, **4**, 183–93.

Smith, H. V., Kusel, J. R. and Girdwood, R. W. A., 1983, The production of human A and B blood group like substances by *in vitro* maintained second stage *Toxocara canis* larvae: their presence on the outer larval surfaces and in their excretions/secretions, *Clinical and Experimental Immunology*, **54**, 625–33.

Smith, H. V., Girdwood, R. W. A. and Kusel, J. R., 1984, Misinterpretation of toxocaral serodiagnostic tests, *British Medical Journal*, **288**, 1235.

Speiser, F. and Gottstein, B., 1984, A collaborative study on larval excretory–secretory antigens of *Toxocara canis* for the immunodiagnosis of human toxocariasis with ELISA, *Acta Tropica*, **41**, 361–72.

Sprent, J. F. A., 1958, Observations on the development of *Toxocara canis* (Werner, 1782) in the dog, *Parasitology*, **48**, 184–210.

Sugane, K. and Oshima, T., 1980, Recovery of large numbers of eosinophils from mice infected with *Toxocara canis*, *American Journal of Tropical Medicine and Hygiene*, **29**, 799–802.

Sugane, K. and Oshima, T., 1983, Purification and characterization of excretory and secretory antigen of *Toxocara canis* larvae, *Immunology*, **50**, 113–20.

Sugane, K. and Oshima, T., 1984a, Activation of complement in C-reactive protein positive sera by phosphorylcholine bearing component isolated from parasite extract, *Parasite Immunology*, **51**, 385–95.

Sugane, K. and Oshima, T., 1984b, Induction of peripheral blood eosinophilia in mice by excretory and secretory antigen of *Toxocara canis* larvae, *Journal of Helminthology*, **58**, 143–7.

Sugane, K. and Oshima, T., 1984c, Interrelationships of eosinophilia and IgE antibody production to larval ES antigen in *Toxocara canis* infected mice, *Parasite Immunology*, **6**, 409–20.

Sugane, K., Howell, M. J. and Nicholas, W. L., 1985, Biosynthetic labelling of the excretory and secretory antigen of *Toxocara canis* larvae, *Journal of Helminthology*, **59**, 147–51.

Taylor, M. R. H., Keane, C. T., O'Connor, P., Mulvihill, E. and Holland, C., 1988, The expanded spectrum of toxocaral disease, *Lancet*, **i**, 692–5.

Thompson, D. E., Bundy, D. A. P., Cooper, E. S. and Schantz, P. M., 1986, Epidemiological characteristics of *Toxocara canis* zoonotic infection of children in a Caribbean community, *Bulletin of the World Health Organization*, **64**, 283–90.

Tönz, M., Speiser, F. and Tönz, O., 1983, Toxocariasis bei Schweizer kindern, *Schweizerische Medizinische Wochenschrift*, **113**, 1500–7.

van Knapen, F., van Leusden, J. and Buys, J., 1982, Serodiagnosis of toxocaral larva migrans in monkeys by enzyme-linked immunosorbent assay (ELISA) with somatic adult and secretory larval antigens, *Journal of Parasitology*, **68**, 951–2.

van Knapen, F., van Leusden, J., Polderman, A. M. and Franchimont, J. H., 1983, Visceral larva migrans: examination by means of enzyme-linked immunosorbent assay of human sera for antibodies to excretory–secretory antigens of the second-stage larvae of *Toxocara canis*, *Zeitschrift für Parasitenkunde*, **69**, 113–18.

Woodruff, A. W., 1970, Toxocariasis, *British Medical Journal*, **3**, 663–9.

Woodruff, A. W., de Savigny, D. and Jacobs, D. E., 1978, Study of toxocaral infection in dog breeders, *British Medical Journal*, **23 Dec.**, 1747–8.

Woodruff, A. W., Watson, J., Shikara, I., Azzi, N. S. H. A., Hadithi, T. S. A. I., Adhami, S. B. H. A. I. and Woodruff, P. W. R., 1981, *Toxocara* ova in soil in the Mosul district, Iraq and their relevance to public health measures in the Middle East, *Annals of Tropical Medicine and Parasitology*, **75**, 555–7.

Worley, G., Green, J. A., Frothingham, T. E., Sturner, R. A., Walls, K. W., Pakalnis, V. A. and Ellis, G. S., 1984, *Toxocara canis* infection: clinical and epidemiological associations with seropositivity in kindergarten children, *Journal of Infectious Diseases*, **149**, 591–7.

7. Immune evasion and immunopathology in *Toxocara canis* infection

H. V. Smith

INTRODUCTION

One of the features of most host–parasite interactions is the rigidity of choice of final host. In some instances, however, humans can become infected by nematodes of animal origin — these are the nematode zoonoses. Such parasites do not usually develop to maturity, remaining instead in their larval stages, and the infections are normally destroyed by the host (e.g. cutaneous larva migrans caused by the canine hookworm *Ancylostoma caninum*). Nematode zoonotic infections of humans are best exemplified by those of the family Ascarididae, included among which are *Toxocara canis*, *Anisakis marina* and *Baylisascarish procyonis*. By far the best example of ascarid zoonosis is the first of these, *T. canis*, which can survive in humans for several years (see Figure 6.1, Chapter 6, for life cycle).

The major clinical consequences of prolonged migration of *T. canis* larvae in humans are visceral larva migrans (VLM) and ocular toxocariasis (OT). It is generally accepted that VLM and OT result from the ingestion of large or small numbers of infective eggs, respectively. The manifestation of VLM and OT are different (Table 7.1) and are primarily a consequence of the size of the infecting dose rather than the host immune response (Glickman and Schantz, 1981). Thus the generalized eosinophilia, hypergammaglobulinaemia and hepatomegaly associated with VLM (Beaver *et al.*, 1952; Beaver, 1956; Schantz, 1989) are absent in OT (Brown, 1970). OT can, however, occur concurrently with VLM in instances where large infecting doses are ingested (Glickman and Schantz, 1981). The clinical and histo-pathological features of human VLM indicate that it is a multisystem disease and patients are most likely to present during the chronic phase, when immuno-pathological events have occurred. As larvae do not multiply within the host, the severity of the disease is proportional to the infecting dose, the frequency of reinfection, and the immunoreactivity of the host. Recently Taylor *et al.* (1987,

Table 7.1. Some general symptoms/signs and findings that may accompany toxocariasis.

Visceral larva migrans	Ocular toxocariasis	Covert toxocariasis
Fever	Visual loss	Cough
Pallor	Strabismus	Abdominal pain
Malaise	Retinal granuloma	Headache
Irritability	Pars planitis	Sleep disturbance
Weight loss	Endophthalmitis	Behaviour disturbance
Skin rash	Choroidoretinitis	
Hepatomegaly	Uveitis	
Asthma	Retinal detachment	
Hypergammaglobulinaemia		
Respiratory symptoms/signs		
Nervous symptoms/signs		
Myocarditis		
Persistant eosinophilia		
Leucocytosis		
Elevated anti-A and anti-B isohaemagglutinins		

1988) expanded the classification of the syndromes accompanying human toxocariasis to include a third category of clinically inapparent infection with or without
eosinophilia which was termed covert toxocariasis (CT) (Table 7.1). Thus human
toxocariasis can be subdivided clinically into VLM + OT, VLM, OT or CT.

TOXOCARA AS AN EXPERIMENTAL MODEL

Three aspects of toxocariasis present us with an excellent opportunity to examine
the pathobiology of nematode infection in general. Firstly, the lesions elicited in

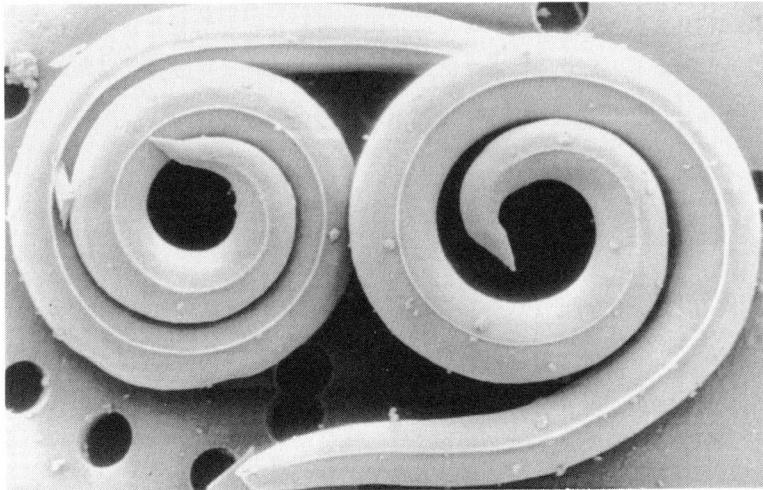

Figure 7.1. Scanning electron micrograph of two *in vitro* maintained second stage *T. canis* larvae. Note
lateral alae running down the length of the larvae. The parasites were mounted on a membrane with
pores of approximately 10 μm diameter.

Table 7.2. Some models of toxocariasis.

Syndrome[a]	Host species/strain	Reference
VLM		
Histopathology, liver	Mouse	Hoeppli *et al.* (1949)
Migratory routes	Mouse (white)	Sprent (1952, 1955)
Larval distribution	Mouse	Smith and Beaver (1953)
Diagnostic morphometrics	Mouse	Nichols (1956)
Histopathology	Pig	Done *et al.* (1960)
Anthelmintic study	Mouse (white)	Pike (1960)
Infection dynamics	Mouse (Swiss white)	Oshima (1961)
Hypersensitivity	Guinea-pig	Olson and Schultz (1963)
Maze solving	Rat	Olson and Rose (1966)
Serology	Rabbit	Hogarth-Scott (1966)
Histopathology	Mouse (OS1)	Burren (1968)
Histopathology	Mouse (white)	Bisseru (1969)
Larval distribution, CNS	Mouse (OS1)	Burren (1971)
Histopathology	Mouse, rat, guinea-pig	Mossalam *et al.* (1972)
Histopathology	Mouse (albino)	Zyngier (1974)
Ultrastructure, liver	Mouse (albino)	Zyngier and Brockbank (1974)
Serology	*Macaca sinica*	Fernando and Soulsby (1974)
P-K skin tests	Guinea-pig	Collins and Ivey (1975)
Serology	*Macaca speciosa*	Hogarth-Scott and Feery (1976)
Resistance to reinfection	Mouse	Kondo *et al.* (1976)
Serology	Rabbit	de Savigny and Tizzard (1977)
Histopathology (ultrastructure)	Mouse (HM/ICR; Swiss)	Kayes and Oaks (1978)
Serology, histopathology	Mouse (C3H)	Przyjalkowski *et al.* (1979)
T-cell function	Mouse (C57B/6J, AKR/J & F1 cross, HM/ICR)	Kayes and Oaks (1980)
Behavioural disorders	Mouse (outbred)	Dolinsky *et al.* (1981)

Serology	Rabbit	Smith *et al.* (1982)
Immunopathogenesis	Mouse (BALB/c)	Sugane and Oshima (1982)
Larval distribution, brain	Mouse (C57BL)	Dunsmore *et al.* (1983)
Serology, pathology	*Macaca fascicularis*	Glickman and Summers (1983)
Larval entrapment	Mouse (BALB/c)	Sugane and Oshima (1983a)
Larval distribution	Mouse (NIH, CD1, outbred)	Abo-Shehada and Herbert (1984a)
Anthelmintic study	Mouse (NIH, CD1)	Abo-Shehada and Herbert (1948b)
Vaccination with TES	Mouse (CBA)	Nicholas *et al.* (1984)
Immune response (graded infection)	Mouse (CBA/J)	Kayes *et al.* (1985)
Immunohistopathology	Mouse (BALB/c ByJ)	Parsons *et al.* (1986)
Immunopathology, liver	Mouse (BALB/c)	Akao *et al.* (1986)
Behavioural disorders	Mouse (outbred albino)	Hay *et al.* (1986)
Autoradiographic tracing	Mouse CD1	Wade and Georgi (1987)
Analyses of TES levels	Mouse (BALB/c ByJ)	Bowman *et al.* (1987)
Effect of irradiation on larval migration	Mouse (BALB/c ByJ)	Barriga and Myser (1987)
TES–antibody interactions	Rabbit, mouse (BALB/c, CBA/Ca)	Robertson *et al.* (1988)
Pulmonary granuloma formation, transfer of hypersensitivity	Mouse (C57BL/6J)	Kayes *et al.* (1988)
Anti-TES antibody analyses	Mouse (ICR)	Boyce *et al.* (1988)
OT (VLM + OT)		
Infection dynamics, larval distribution	Mouse (Fairfield Webster)	Olson *et al.* (1970)
Larval distribution	Mouse (Swiss Yale)	Olson (1976)
Eosinophilogenesis	Guinea-pig (Hartley)	Rockey *et al.* (1979)
Ocular serology	Guinea-pig (Hartley)	Soulsby *et al.* (1980)
Eosinophil function	Guinea-pig (Hartley)	Rockey *et al.* (1983)
Histopathology	*Macaca fascicularis*	Glickman and Summers (1983)
Pathogenesis	*Macaca fascicularis*	Watzke *et al.* (1984)
Histopathology	Mouse (C57BL)	Ghafoor *et al.* (1984)

[a] TES, *Toxocara canis* excretory–secretory products.

experimental animal models and humans are similar, their nature being largely governed by the size of infecting dose (Olson, 1976; Ghafoor *et al.*, 1984; Schantz, 1989). Secondly, the second stage (infective) larvae (Figure 7.1) are stable *in vitro* for over 1 year in defined culture media (de Savigny, 1975). Lastly, *in vitro* released antigens (excretory–secretory) have been shown to be produced by the parasite *in vivo* in both laboratory animals and infected humans (Robertson *et al.*, 1988). Direct comparisons between *in vivo* and *in vitro* findings can, therefore, be made.

Toxocara also offers an opportunity to understand the biology of tissue migratory larval nematodes in general because its parasitic larvae migrate through the soft tissues of mammalian hosts and induce similar pathology to that elicited in infections with other nematode species. In addition, the invasion of the eye in experimental animals offers a parallel for less tractable ocular nematodiases such as onchocerciasis. Toxocariasis has most often been studied in the mouse (Table 7.2), the usefulness of which resides not only in the fact that the infection's histopathology is similar to that of human beings, but also that murine genetics and immune system are well characterized. However, most effort has centred on experimental VLM with OT having received little attention.

We have developed a murine model for OT which mimics the pathology in humans, using the C57BL strain (Ghafoor *et al.*, 1984). The reason for the use of this strain is that many other inbred strains have either neonatal degenerative retinal changes causing blindness, or non-pigmented retinae and are, therefore, inappropriate. At most of the infective doses we have investigated, C57BL mice show symptoms which parallel VLM + OT in humans. However, at doses below 50 infective eggs *per os*, VLM without OT predominates (Table 7.3). We have been unable to induce OT alone but would anticipate that an infecting dose below 10 eggs *per os* would be required.

Table 7.3. Dose level and predominance of visceral or ocular involvement in murine toxocariasis.

	Infective dose of larvae					
	5000	2000	1000	500	100	50
Gross visceral pathology	+	+	+	+	+	+
Larval recoveries from carcass	+	+	+	+	+	+
Day larva first identified in eye	3	9	9	21	35	62

T. CANIS LARVAL MIGRATION PATTERNS AND THE INDUCTION OF PATHOLOGY

Immediately following penetration of the murine small intestine, larvae appear in the hepatic portal and mesenteric veins and mesenteric lymph nodes. From there they migrate to the liver and via the heart to the lungs, and exit from the lungs either *via* the pulmonary veins again into the heart and general circulation or *via* the bronchioles. The initial stages of migration whereby larvae can be found in various organs and tissues at specific times (Figure 7.2) implies a defined migratory route

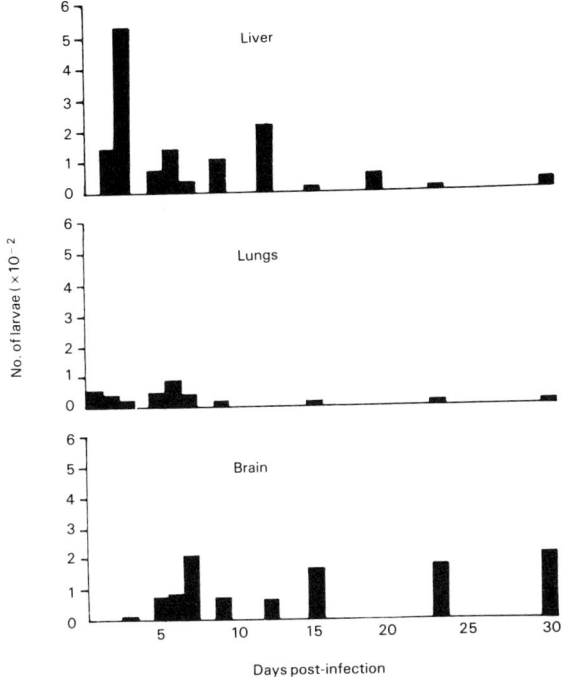

Figure 7.2. Migratory patterns of *T. canis* larvae in NIH mice. Distribution of second stage larvae over the first 21 days of a primary infection in male mice infected orally with 2000 infective *T. canis* eggs at 8 weeks of age.

(Sprent, 1952, 1955; Burren, 1968; Bisseru, 1969; Abo-Shehada and Herbert, 1984a). Larvae are disseminated throughout the soft tissues by the general circulation and reach the skeletal muscles and the central nervous system (Burren, 1971; Dunsmore *et al.*, 1983; Abo-Shehada and Herbert, 1984a; McLure, 1988) and can be found in the latter 400 days post-infection (Abo-Shehada and Herbert, 1984a). Larval invasion into the eye and the severity of disease is dose dependent (Olson *et al.*, 1970; Olson, 1976) and can be demonstrated from 3 days post-infection onwards. The early appearance of larvae in the eye can be correlated with a higher infecting dose and parallels the syndrome of VLM + OT, whereas their later appearance may well parallel the OT syndrome alone. One of the consequences of toxocaral larva migrans is the prolonged migration of larvae and the sequential development of pathology over time in various tissues and organs. This progression can be prolonged because the larvae can survive for up to 2 years in mice (Sprent, 1958). Inflammatory lesions with both acute and chronic characteristics can occur within a tissue at the same time, indicating that the genesis of pathological lesions is induced continuously during infection.

The tissue response often matures into a granuloma and granulomatous entrapment of larvae occurs, although not in the early stages of infection. According to Kayes and Oaks (1978), predominantly eosinophilic granulomata from 11-week-old infections of mice contain living larvae. Attachment of polymorphs and (primarily) eosinophils to larval surfaces can be demonstrated readily but, in general, they are not effective in killing larvae (Kayes and Oaks, 1978; Rockey *et al.*, 1983; Fattah *et al.*, 1986; Badely *et al.*, 1987a). Indeed, larvae with adherent cells recovered either from the peritoneal cavity or from the mesenteric lymph nodes of infected mice rapidly lose their adherent cells *in vitro*. Entrapment leading to larval death and the development of gross pathology have been consistent findings in the murine model, especially in the liver (Hoeppli *et al.*, 1949; Burren, 1968; Bisseru, 1969; Mossalam *et al.*, 1972; Zyngier, 1974) which may be the major site of attrition (Sugane and Oshima, 1983a; Akao, 1985). Migration of large numbers of larvae stimulates a profound generalized inflammatory response in the host, and eosinophilic and monocytic leucocytosis, in conjunction with neutropenia and lymphocytopenia, occur early in infection (Figure 7.3).

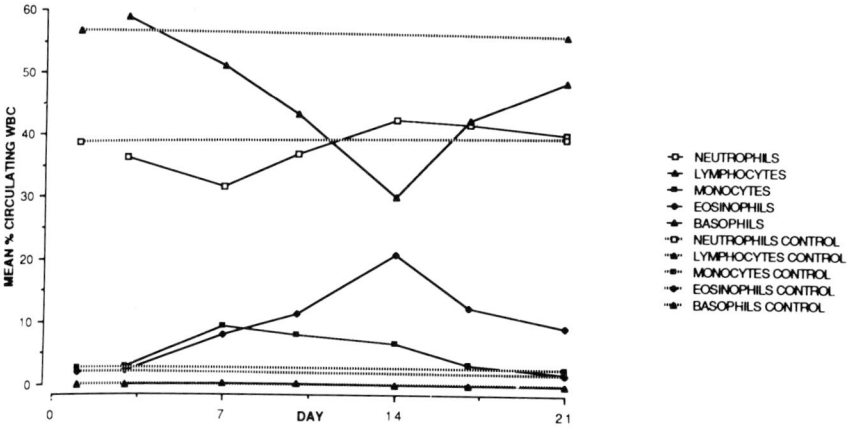

Figure 7.3. Effect of *T. canis* infection on circulating white blood cells (WBC). C57BL mice were infected with 2000 infective eggs of *T. canis* at 8 weeks of age.

Antibody to the surface of larvae or their secretions can be detected readily in experimental animals. The excretory–secretory products released by *in vitro* maintained larvae (TES) (de Savigny, 1975) are the first mosaic of antigens to which an antibody response is detectable (Smith *et al.*, 1982) and provide us with the most specific antigen preparation currently available for serodiagnosis (de Savigny and Tizzard, 1977; de Savigny *et al.*, 1979; Smith *et al.*, 1980; van Knapen *et al.*, 1983; Glickman *et al.*, 1985).

The fact that larvae can survive in immunologically responsive hosts for considerable periods of time prompted us to investigate the mechanism(s) by which second stage larvae (L2) are able to evade the host immune responses.

ANTIGEN SHEDDING AND EVASION OF THE IMMUNE RESPONSE

The first insight into the antigenicity and dynamic nature of the nematode surface was afforded by the studies on *A. caninum*, *Trichinella spiralis* and *T. canis* (Vetter and Klaver-Wesseling, 1978; Klaver-Wesseling *et al.*, 1978; Philipp *et al.*, 1980; Smith *et al.*, 1981). These studies, using fluorescent antibodies, or the extrinsic radiolabelling of surface-associated proteins, demonstrated that antigens present on nematode surfaces were secreted into the tissue culture medium *in vitro*. In our original studies (Smith *et al.*, 1981), we could only demonstrate anti-*T. canis* excretory–secretory (TES) antibody binding to toxocaral epicuticular surfaces when larvae were metabolically arrested. The two procedures which allowed us to demonstrate this were incubation of larvae at 2°C or incubation at 37°C in the presence of antimetabolites, both of which treatments caused the larvae to become coiled and immotile. The removal of these inhibitors not only resulted in a resumption of larval motility but also in the shedding of surface bound anti-TES antibody. Antibody shedding was complete in 3 h (Table 7.4) and the same surface could then be relabelled with antibody if larvae were placed under the same metabolic constraints. Further evidence that surface molecules are released by second stage larvae came from the work of Maizels *et al.* (1984) who demonstrated that about 25 per cent of radioactivity incorporated into surface proteins was released in less than 1 h of incubation *in vitro*.

From the above findings we argued that rapid shedding of epicuticular antigens constituted a mechanism for evading both antibody-dependent and antibody-mediated cellular cytotoxicity *in vivo* (Smith *et al.*, 1981; Ghafoor *et al.*, 1984). Such a mechanism would also explain the absence of larvae or larval fragments in granulomatous lesions elicited by antigen shedding by migrating larvae. In addition, we speculated that the prodigious production of TES in the host tissues could divert immune reactivity away from the larval surfaces towards TES depots, inducing greater pathology (Ghafoor *et al.*, 1984).

Table 7.4. Shedding of surface-bound antibody by *Toxocara canis* larvae.[a]

Immunofluorescent labelling		Subsequent incubation		Fluorescence after		
Temperature (°C)	Antimetabolites	Temperature (°C)	Antimetabolites	0 h	1 h	3 h
2	–	2	–	+ +	+ +	+ +
2	–	37	–	+ +	+	–
37	–	37	–	–	–	–
37	+	37	+	+ +	+ +	+ +
37	+	37	–	+ +	+	–

[a] Infective larvae of *T. canis* were incubated in rabbit antiserum to TES, then labelled with fluorescein conjugated antibody to rabbit immunoglobulin. This was performed under the temperature conditions indicated under 'immunofluorescent labelling', with or without metabolic inhibitors. The larvae were then incubated for various periods of time under the conditions indicated under 'subsequent incubation' before being examined by fluorescence microscopy. The brightness of the resultant fluorescence is indicated in arbitrary relative levels.

The rapid turnover of surface components could, at first sight, be taken to indicate an epicuticular 'membrane' possessing an unrestricted lateral mobility of lipids to allow the insertion and reexpression of surface-exposed molecules. However, direct biophysical measurements have raised the paradox that the L2 surface apparently has lipid which is highly restricted in its lateral diffusion (Kennedy *et al.*, 1987a) (see also Chapter 1).

The biochemical and immunochemical natures of the surface and TES compartments of *Toxocara* larvae have been the subject of much interest recently (Sugane and Oshima, 1983b; Maizels *et al.*, 1984, 1987; Sugane *et al.*, 1985; Meghji and Maizels, 1986; Badley *et al.*, 1987b) and are the subject of Chapter 6.

We suggest that larvae which slowly shed epicuticular antigens are more susceptible to immunological damage than those which shed their antigens rapidly. Larvae which induced chronic infections would, therefore, have to shed surface antigens continually and rapidly to avoid entrapment. As will be enlarged upon below, experiments on the dynamic nature of the external surface demonstrate that surface turnover and motility can be reduced in tandem.

LARVAL DEATH AND ANTIGEN SHEDDING

Our original studies on antigen shedding did not take into account any biological effects of serum components on the rate of surface turnover and larval motility. Antibody binding and surface antigen shedding were demonstrated with antibody raised to TES and no deleterious effect of anti-TES antibody on larvae was detected. As larval killing occurs *in vivo* (Kayes and Oaks, 1978), the possibility that sera raised in the context of infection might be qualitatively different from serum raised to purified TES was investigated.

Serum antibody raised in the context of infection of rabbits bound the surface of larvae metabolically arrested at 2°C in a similar manner to anti-TES. Unexpectedly, sera for some animals also bound the surfaces of larvae maintained at 37°C, albeit to varying degrees. Under this condition, it had been expected that bound antibody would have been rapidly shed. Instead, a gradual increase in the ability of infection serum antibody to bind surfaces of larvae maintained at 37°C was demonstrable with time after infection (Table 7.5). Thus, serum sampled early in infection exhibited a weak surface fluorescence, whereas later sera exhibited a strong fluorescence which covered the whole larval surface (Table 7.5).

This increased antibody binding at 37°C appeared in conjunction with a reduction in larval motility. L2 normally display vigorous sinusoidal movement in culture, but became coiled and motionless when co-cultured with certain infection-derived sera. Moreover, the rate of shedding of surface-bound antibody decreased with increasing time after infection (Table 7.5). These effects were most apparent between 28 and 35 days post-infection and, in sera taken at these times, 2–8 per cent of larvae died within 21 h. All sera were heat inactivated prior to use, so this effect was not due to complement, although the surface of the larvae do activate complement by the alternative pathway (Kennedy and Kuo, 1988).

The unexpected results of this study provided evidence that a 30-min exposure of

Table 7.5. Binding to outer larval surfaces of antibodies from infection sera and their effect on the surface turnover and motility of *in vitro* maintained L2 *Toxocara canis*.

Treatment[a] (Days post-infection of sera)	Temperature (°C)	Surface fluorescence after[b]			Motility
		0 h	3 h	13 h	
0	37	0	0	0	+
7		0	0	0	+
14		0	0	0	+
21		0	0	0	+
28		1	0	0	+
35		1	0	0	+
42		2	0	0	+
Control		0	0	0	+
0	2	0	0	0	−
7		1	0	0	−
14		1	0	0	−
21		3	3	3	−
28		2	2	2	−
35		2	2	2	−
42		2	2	2	−
Control		3	3	3	−
0	2→37	0	0	0	+
7		0	0	0	+
14		1	0	0	+
21		3	2	2	−
28		2	2	2	−
35		2	2	2	−
42		2	2	2	−
Control		3	0	0	+

[a] Serum from rabbits infected with *T. canis* for various periods of time were immunofluorescently labelled as per Table 7.4 and incubated for the indicated periods of time, at the temperatures given, before being viewed and scored for brightness under fluorescence microscope and motility. The control serum was antiserum raised against TES in rabbits.
[b] Intensity of fluorescence: 0, no surface fluorescence; 1, dull fluorescence; 2, moderate surface fluorescence; 3, bright fluorescence.
[c] Motility: +, motile, with sinusoidal movement; −, immotile.

viable larvae to antibody from infection sera could down-regulate both surface turnover and larval motility. Immunoglobulin isotype analyses of sequentially sampled sera revealed a predominantly immunoglobulin G (IgG) response to TES antigens by enzyme-linked immunosorbent assay (ELISA). Moreover, immunoprecipitation and sodium dodecylsulphate/polyacrylamide gel electrophoresis (SDS-PAGE) revealed no qualitative change in the TES components bound by antibody as the infection progressed.

These *in vitro* findings provide evidence for the reduction of surface turnover by antibody in the absence of cells and for the gradual increase in the larvicidal quality of serum with time. In addition, it parallels observed histological results which indicate that more larvae are entrapped in inflammatory lesions as infection progresses (Kayes and Oaks, 1978). Reduction in larval surface turnover and motility at 37°C occurred only with infection sera – it could not be demonstrated with hyperimmune anti-TES serum. These results could be taken to indicate that there are antigens or epitopes present on the larval surface which are not present in TES or that the antibody subclass differs in some critical way between the two methods of immunization.

Second stage larvae of *T. canis* possess a non-functional gut and it is possible that larvae absorb their nutrients transcuticularly or via the oral cavity alone and that antibody inhibition of nutrient acquisition could lead to reduced motility. This biological effect of antibody might operate through binding to receptors necessary for either the transcuticular transfer of metabolites or neurophysiological interactions with the external environment. It must also not be forgotten that the change of biological activity of antibody with time after infection could be due to maturation of antibody affinity, alteration in the subclass bias in IgG, or changes in epitope recognition. Evidence for the latter has come from the study of antibody produced during long-term infection of rabbits (Robertson *et al.*, 1988).

HUMAN VLM SERA HAVE DIFFERENT BIOLOGICAL ACTIVITIES FROM RABBIT SERA

In contrast to our findings with rabbit sera, we were unable to demonstrate antibody binding to larvae held at 37°C in OT sera from humans with OT (unpublished). About 3 per cent of VLM sera, typically from children with high titres for both total and immunoglobulin E (IgE) antibody, did bind. The surface fluorescence obtained, however, was weak and granular, little reduction in parasite motility occurred, and bound antibody was shed completely within 3 h.

The binding of human antibody to larval surfaces 37°C is more transient than the binding of rabbit antibody obtained following infection with large numbers of larvae (5×10^3 per kg body weight). If the induction of a protective antibody response were dose related then substantial infections of humans should also elicit antibody which not only reduces surface turnover and motility but is also larvicidal to a proportion of larvae.

Glickman and Schantz (1981) hypothesized that a threshold level of infection exists in humans below which the level of immunological activity is insufficient to impede larval migration and cause larval death. This will result in OT, whereas VLM and larval death are likely to ensue if the threshold were exceeded. The increase in larval death in VLM could, therefore, be a direct consequence of increased levels of antibody capable of binding larval surfaces at 37°C for prolonged periods.

TOXOCARA LARVAE PRODUCE HUMAN-BLOOD-GROUP-LIKE SUBSTANCES

The presence, in helminths, of blood-group-like substances resembling human A and B blood group antigens has long been known (Oliver-Gonzalez and

Torregrosa, 1944; Oliver-Gonzalez, 1946; Oliver-Gonzalez and Gonzalez, 1949). These workers demonstrated that a polysaccharide fraction from *Ascaris lumbricoides* and other helminths immunologically resembled human A and B blood group determinants. This substance also neutralized anti-A agglutinins of human serum and could be released from the parasites during culture *in vitro*.

One of the interesting laboratory findings in human toxocariasis is that elevated circulating isohaemagglutinin levels can often be demonstrated in the sera of toxocariasis patients (Huntley *et al.*, 1965, 1969). Elevated circulating anti-A and B-activities are similarly found in experimental animals infected with *T. canis*. In rabbits, for instance, antibody titres to human A and B erythrocyte antigens peaked 21 days post-infection following a primary infection (Smith *et al.*, 1983). Our investigation of the production of human-blood-group-like substances by second stage larvae demonstrated that polyclonal human anti-A and anti-B typing sera bound only to the surfaces of metabolically arrested larvae and that this antibody was shed as for anti-TES antibody (Smith *et al.*, 1983). In addition, we were able to demonstrate that TES was capable of ablating the reactivity of polyclonal anti-A with its respective erythrocyte epitope (*N*-acetylgalactosamine) and reduce the binding of polyclonal anti-B antibody to B erythrocyte antigen (galactose). Thus, these human-blood-group-like epitopes were exposed on the parasite surface and are present in TES.

We were also able to demonstrate that elevated haemagglutinins to both A and B human erythrocyte antigens could be induced experimentally in rabbits following either infection or immunization with TES. This indicated that both surface-exposed epitopes and TES were sugar rich. Second stage larvae have long been known to be rich in oligosaccharides and/or polysaccharides by periodic acid Schiff's staining (Zyngier, 1974), and this stain has been used by ourselves and others to stain larvae, larval surfaces and TES in tissue sections. The highly glycosylated nature of TES was confirmed by gas chromatographic analyses which showed a total of 401.1 μg of sugar per mg of TES protein, with *N*-acetylgalactosamine (the immunodominant sugar of human A blood group antigen) contributing 58 per cent of the total carbohydrate (Meghji and Maizels, 1986) (see also Chapter 6). Galactose, the immundominant sugar of human B blood group substance, was also detected, albeit in small amounts.

Two monoclonal antibodies (Tcn 2 and Tcn 8) which bind to larval surfaces, and recognize periodate sensitive epitopes in TES (Maizels *et al.*, 1987) also agglutinate human A group erythrocytes. Erythrocyte microagglutination analyses demonstrated the strongest reactivity of both Tcn 2 and 8 to the human A1 blood subgroup. Further analyses to identify the relevant molecules expressing blood group reactivity revealed anti-A antibody reactivity to only the 70 kDa component of TES (Figure 7.4). Also, pretreatment of living larvae with human anti-A antibody can block the binding of anti-TES. We speculate that, *in vivo*, naturally occurring isohaemagglutinins could interfere with the modulatory effect of infection-derived antibody on larval surface turnover and motility either by the direct sequestration of the pertinent epitopes, or by their masking.

No surface expression of either Lewis a (Lewis a$^+$) or Lewis b (Lewis b$^+$) antigens

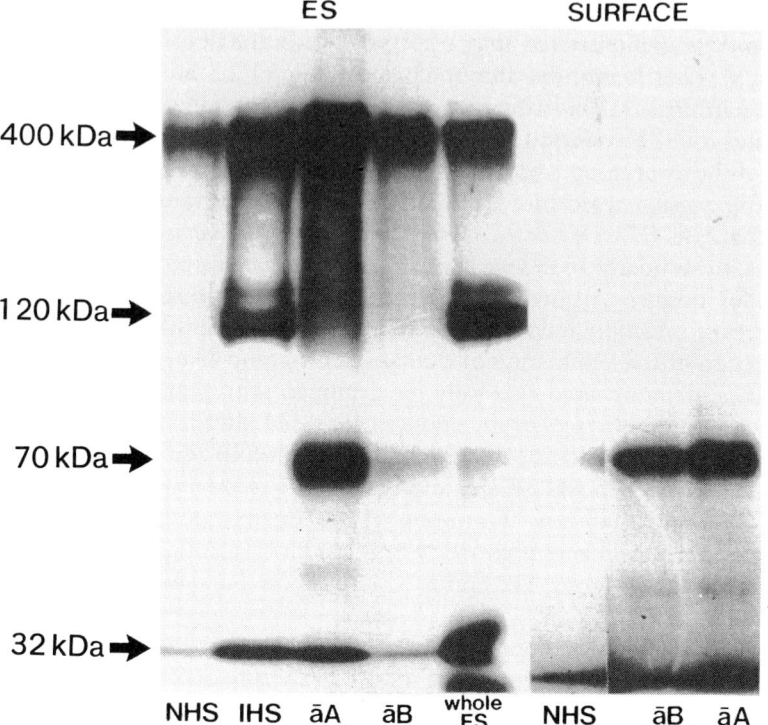

Figure 7.4. Presence of A and B blood group determinants on ES and surface glycoproteins of *T. canis* larvae. The ES (left-hand panel of five tracks) and surface (right-hand panel of three tracks) of the parasite were labelled with ^{125}I and immunoprecipitated with the following sera: pooled normal human serum (NHS), serum pooled from infected individuals (IHS), polyclonal human anti-A blood group typing serum (āA) or polyclonal human anti-B blood group typing serum (āB). The profile of ES before immunoprecipitation is also given (whole ES). Immunoprecipitates were analysed by gradient SDS-PAGE before autoradiography.

could be demonstrated on *in vitro* cultured larvae. However, larvae acquire Lewis b$^+$ antigen when incubated with erythrocytes bearing this antigen, albeit to a small degree, but do not acquire Lewis a$^+$ antigen when incubated with Lewis a$^+$ cells. This acquisition of host materials might have some relevance to the immunobiology of infection (see below) or to modification of the parasite surface (see Chapter 1).

PATHOLOGY IS INDUCED BY TES AND LARVAE

A consequence of the immune killing of migratory larvae in toxocariasis is the development of disease. The histopathogenesis of toxocaral infection has been widely studied (Burren, 1968; Bisseru, 1969; Zyngier and Brockbank, 1974; Kayes and Oaks, 1978) and, in most tissues, pathology is primarily of polymorphonuclear and lymphocytic infiltrates which change from being predominantly eosinophilic to mononuclear epitheloid infiltrates as the lesion ages (Kayes and Oaks, 1978). Early in murine infections, solitary larvae can be demonstrated in numerous tissues, in the absence of noticeable inflammatory responses. Later, however, large eosinophilic infiltrates are apparent, some containing larvae, especially in the liver.

The absence of larvae or larval fragments within many granulomata is a frequent finding, and this presents a major problem in attempting to establish a toxocaral aetiology to a lesion. Immunohistochemical analyses using either polyclonal anti-TES serum (Akao *et al.*, 1986; Parsons *et al.*, 1986) or monoclonal antibodies (Figure 7.5a) have demonstrated the presence of TES antigen depots in the absence of identifiable larvae or larval fragments. Later lesions (Figure 7.5b) may contain entrapped larvae or larval debris with or without TES depots.

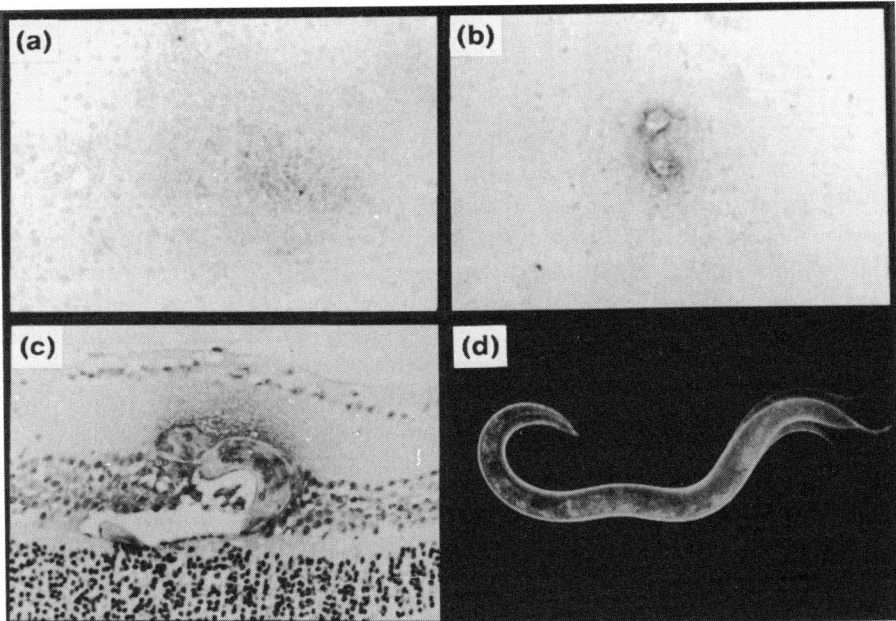

Figure 7.5. (a–c) Immunoperoxidase localization of TES in murine tissue. Formalin fixed sections of infected liver (a and b) reacted with biotinylated polyclonal anti-TES serum (Smith *et al.*, 1981) or retina (c) reacted with biotinylated monoclonal anti-TES (Tcn 2) (Maizels *et al.*, 1987; Kennedy *et al.*, 1987b). Sites of antibody binding were visualized by deposition of a brown reaction product using streptavidin peroxidase, hydrogen peroxide and diaminobenzidine. (a) Section through liver of 12-week-old male C57BL mouse infected orally with 1500 infective eggs, 14 days post-infection. Note the presence of TES deposit, but absence of larvae or larval fragments. (b) As for (a), but taken at 42 days post-infection. Note presence of TES deposit surrounding two fragments of coiled larva. (c) Section through retina of 12-week-old C57BL mouse infected orally with 1500 infective eggs, 14 days post-infection. Note the presence of a larva and its TES deposit in the retina and the physical disruption of the photoreceptor layer by the migration of the larva. (d) Living second stage larva of *T. canis* microdissected from the brain of an NIH mouse 7 days post-infection with 1500 eggs. The larva was reacted with monoclonal anti-Thy 1.2 antibody and fluorescein conjugated anti-mouse antibody. Note fluorescence covering the entire surface of the larva. The Thy 1.2 antigens is found on T cells and brain cells of the mouse.

A diversionary role for TES glycoproteins can be inferred from the results of Maizels *et al.* (1987) and Kennedy *et al.* (1987b) (see also Chapter 6), who demonstrated that anti-polypeptide monoclonal antibodies (mAbs) which bound TES epitopes were unable to bind surface-exposed epitopes of intact larvae. These mAbs, however, can be used to detect TES depots, indicating that polypeptide epitopes which are not exposed on the larval surface become accessible once shed.

Both the isotype reactivity and larval surface binding capacity of antibody from patients with CT can differ from those with other syndromes of toxocariasis (Smith *et al.*, 1988). The lack of larval surface-binding antibody in such sera is due to the lower concentrations of antibody directed against carbohydrate epitopes. Anti-polypeptide antibody, which is unable to bind to the carbohydrate rich larval surface, do bind to TES depots. The possible consequences of this imbalance are two-fold. First, larval entrapment and death following antibody-dependent cytotoxicity may be reduced, allowing a larger proportion of larvae to survive. Secondly, the consequence of the increased tissue deposition of TES could be increased pathology.

T. CANIS AND THE EYE

In OT, larvae invade the retina and choroid inducing inflammatory changes which can lead to the unilateral loss of vision (Duguid, 1961; Shields, 1984). A useful experimental model of this condition is that using C57BL mice described by Ghafoor *et al.* (1984). That this model parallels human OT was apparent when the classical lesions of human ocular disease, namely solitary retinal granulomata, diffuse chronic endophthalmitis and pars planitis, were demonstrated in the eyes of infected mice. Invasion into the retina occurs via the superficial retinal blood vessels, then larvae migrate into the deeper retinal tissues including the photoreceptor layer and, whilst there, produce TES which diffuses into the surrounding tissues. Large quantities of TES are produced locally by single larvae (Figure 7.5c) and the ensuing host response engulfs the antigen depot resulting in a large inflammatory lesion. In some instances, TES depots occur with no larvae detectable locally, presumably testifying to the mobility of their originator.

The pattern of deposition and sequestration of TES glycoconjugates in tissues could relate to the variety of lesions produced in toxocariasis. Whereas the deposition of small amounts of TES over a large area (as a result of linear movement of the parasite) is likely to produce a diffuse low grade inflammatory response, such as diffuse chronic endophthalmitis, the deposition of large amounts of TES locally (as a result of concentric movement or stasis) is likely to produce an intense inflammatory response such as a solitary granuloma. A rapid loss in visual acuity is likely to ensue when each of such responses occurs in the photoreceptor layer.

Inflammatory changes in the murine eye do not appear to be as aggressive as those in the myotropic phases of migration. The former may therefore represent an intermediate response between myotropic and neurotropic responses due to the unique character of the formative tissues of the eye (Ghafoor *et al.*, 1984). The development of the inflammatory focus in that organ is possibly slower in toxocariasis because of the continued maintenance of the integrity of the blood–retinal barrier following the penetration of larvae into the retina.

The anti-TES monoclonal antibodies (Maizels *et al.*, 1987; Kennedy *et al.*, 1987b) (see also Chapter 6) can be used to localize TES depots in the eye. Tcn 2 is the most effective mAb for this, followed by the other anti-carbohydrate mAbs (Tcn 1, 2, 4, 5, 7 and 8), but the anti-peptide mAbs Tcn 3 and 6 tend to be poor (but nevertheless useful). TES depots were readily demonstrated in early retinal lesions but not in late, mature granulomatous lesions. In early retinal lesions, where no histopathological change had occurred, TES could be detected attached onto the surfaces of, and presumably within, the cells of the plexiform layer (Figure 7.5c).

OT AND THE RELEASE OF RETINAL ANTIGENS

The effects of OT on the integrity of the murine retina can be followed by exploiting an autoimmune reaction to retinal soluble (S) antigen which, like retinal particulate (P) antigen, does not normally elicit an immune response. When released into the circulation, however, they stimulate both cellular and humoral autoimmune responses in experimental animals. For example, in our unpublished experiments, mice with only three larvae per eye have demonstrable circulating antibodies to retinal S antigen, whereas sera from mice with no ocular involvement were negative. Mice with intermediate levels of ocular involvement had lower levels of anti-S autoantibody. No histopathological evidence of retinal S antigen autoimmunity (photoreceptor destruction) was obtained from any of the eyes examined, nor were any lesions in *Toxocara* infected eyes typical of the granulomatous lesions encountered in retinal S antigen autoimmunity.

The role of auto antibody to retinal S antigen in the aetiology of toxocaral retinopathies is unknown, but it could contribute to the induction and/or development of eye lesions. Such antibody could, on the other hand, be used to follow recovery from OT after chemotherapy in humans, but it is unlikely to be specific to OT.

RESTRAINED INFLAMMATORY RESPONSES IN THE CENTRAL NERVOUS SYSTEM

Since the demonstration of solitary larvae entrapped in granulomata in the brains of children who died of other causes (Dent *et al.*, 1956; Beautyman *et al.*, 1966; Schochet, 1967, Hill *et al.*, 1985), the possibility that subclinical toxocaral infection of the brain can cause a reduction in neurological and possibly intellectual capacity has become of increasing concern (Marmor *et al.*, 1987). Episodic events such as focal epilepsy (Brain and Allan, 1964), eosinophilic meningitis (Gould *et al.*, 1985) and involvement of other parts of the central nervous system (CNS) (Russegger and Schmutzhard, 1989) are often the only presenting features of neural toxocariasis. Transient or permanent aberrations might be a consequence of larval presence or entrapment in the brain.

Larval migration into the CNS of mice produces behavioural disturbances (Dolinsky *et al.*, 1981; Hay *et al.*, 1986). Those which evade granulomatous entrapment in the viscera accumulate in the grey and white matter of the brain, although the majority occur in the grey matter during the early stages (Dunsmore *et al.*, 1983). This is presumably because of the greater blood supply to grey matter (Le Gros Clark, 1975), although a proportion can also be found in the spinal cord (Burren, 1971; McLure, 1988) and as many as 17 per cent of the larval burden can be found in both grey and white matters two weeks after infection with 2000 infective eggs *per os* (McLure, 1988).

In contrast to inflammatory responses induced by *Toxocara* in other parts of the body, little reaction to larvae is apparent in the CNS (Burren, 1968; Glickman and Summers, 1983; Dunsmore *et al.*, 1983). It has been suggested, therefore, that the accumulation of larvae in the brain may represent a population surviving in an immunologically privileged site.

In our mouse model, a low grade astrocytosis, microglial enlargement and a diffuse accumulation of macrophages is demonstrable in white matter early in infection. No cellular infiltrate could be demonstrated in proximity to a parasite, but immunohistochemical studies revealed both immunoglobulin and complement (C3) around apparently intact larvae. We had concluded that the immune response in neural tissue was less aggressive than in other tissue (Ghafoor *et al.*, 1984), but the granuloma described by Hill *et al.* (1985) indicates that this might not always be the case.

The fact that living larvae do not on the whole stimulate an intense cellular response in the brain raises the possibility that they may be in some way protected from the cellular and humoral mechanisms effective in other organs. This could either be due to limits placed upon the immune response in the brain in order to prevent neurological damage, or that larvae are in some way less provocative when in CNS. Living larvae microdissected from murine brains can be shown to have acquired Thy 1.2 antigen (Figure 7.5d) and this antigen is lost within 3 hours *in vitro* by the shedding process detailed earlier, and not re-expressed. This presumably passive acquisition of Thy 1.2-like brain antigens may protect larval surfaces from immune mediated attack and damage whilst resident in the brain.

Larval antigen might, however, reach astrocytes (the antigen presenting cells of the brain) which remove antigens from their deposition site and concentrate them at astrocyte end feet which form a continuous layer around blood vessels, thereby focusing immune responses on perivascular areas (Wekerle *et al.*, 1987; Pentreath, 1989). This increase in astrocyte and macrophage numbers might inhibit the induction of focal inflammatory responses in brain parenchyma.

The acquisition of Thy 1.2-like brain antigens could serve a dual role. First, it could protect larval surfaces from immune-mediated damage in the brain by masking parasite epitopes. Secondly, and possibly more importantly, it could mask the potentially pathogenic epitopes in TES from being recognized in the parenchyma, until deposited perivascularly, thereby avoiding the likelihood of neural disorganization elicited by the immune recognition of TES produced by larvae in the brain. When larval recognition does occur in the CNS, a granulomatous response can ensue (Hill *et al.*, 1985).

GRANULOMATOUS ENTRAPMENT IN SKELETAL MUSCLE

Larvae which enter the skeletal muscles can reside there for long periods of time and are surrounded by an inflammatory infiltrate. The degree of inflammatory responsiveness ranges from the development of a mild inflammatory response with diffuse myositis with larvae rarely present (Zyngier, 1974; Dunsmore *et al.*, 1983) to the development of large white exudative lesions which are then replaced by granulomata containing living larvae (Kayes and Oaks, 1978). This granulomatous response is T lymphocyte dependent (Sugane and Oshima, 1982). In our murine model, the degree of inflammatory cell infiltration into the musculature of infected mice is moderate and a generalized myositis with few entrapped larvae is evident.

The musculature is the site from which reactivated arrested larvae are thought to arise during pregnancy in the dog and from which definitive hosts contract larval infection by carnivores. Dunsmore *et al.* (1983) suggested that the brain might also serve this function.

TOXOCARA LARVAE AS INTRACELLULAR PARASITES

Larvae resident in musculature are, according to Kayes and Oaks (1978), able to survive early granulomatous encapsulation. There is no doubt that larval attrition can occur in musculature but they can also survive for long periods in this site. One of the possible reasons for prolonged survival in muscle is for larvae to invade muscle cells, and hence be isolated from immune effectors.

In our histological examination of infected psoas muscle of mice, individual larvae were found within muscle cells. No inflammatory changes or cellular infiltrates were present in the vicinity of these intracellular larvae. Classical inflammatory infiltrates comprising of eosinophils, macrophages and lymphocytes were present in other areas of the psoas and in the liver. How long individual larvae can remain intracellular is not known.

Intracellular nematode parasites are an uncommon finding in mammalian hosts, and to the author's knowledge, this phenomenon has not been reported with any nematode other than *T. spiralis* and *T. pseudospiralis*.

CONCLUDING REMARKS

T. canis infections of humans are not usually thought to be of major importance worldwide. Whether this is true or not in both developed and developing countries remains a matter for debate, but in the present context its importance lies in what it can tell us about nematode immunobiology. The features described herein, surface turnover, the expression of host-like antigens, the development of an intracellular niche and the induction of larvicidal antibody, could shed considerable light on host–parasite interactions and add to our understanding of the fundamental processes underlying the induction of morbidity and mortality by nematode parasites.

ACKNOWLEDGEMENTS

I wish to acknowledge the continued support of the Medical Research Council for the provision of four consecutive project awards, and the support of the W. H. Ross Foundation (Scotland) for the study of the Prevention of Blindness. I am also indebted to Mae McCulloch for help with the typing.

REFERENCES

Abo-Shehada, M. N. and Herbert, I. V., 1984a, The migration of larval *Toxocara canis* in mice. II. Post-intestinal migration in primary infections, *Veterinary Parasitology*, **17**, 75–83.

Abo-Shehada, M. N. and Herbert, I. V., 1984b, Anthelmintic effect of levamisole, ivermectin, albendazole and fenbendazole on larval *Toxocara canis* infection in mice, *Research in Veterinary Science*, **36**, 87–91.

Akao, N., 1985, Immune responses to excretory–secretory products of second stage larvae of *Toxocara canis*: humoral immune response relating to larval trapping in the liver of reinfected mice, *Japanese Journal of Parasitology*, **34**, 293–300.

Akao, N., Kondo, K., Saki, H. and Yoshimura, H., 1986, An immunopathological study of the liver of the mice infected with *Toxocara canis*, *Japanese Journal of Parasitology*, **35**, 135–40.

Badley, J. E., Grieve, R. B., Bowman, D. D. and Glickman, L. T., 1987a, Immune-mediated adherence of eosinophils to *Toxocara canis* infective larvae: the role of excretory–secretory antigens, *Parasite Immunology*, **9**, 133–43.

Badley, J. E., Grieve, R. B., Bowman, D. D., Glickman, L. T. and Rockey, J. H., 1987b, Analysis of *Toxocara canis* larval excretory–secretory antigens: physicochemical characterization and antibody recognition, *Journal of Parasitology*, **73**, 593–600.

Barriga, O. O. and Myser, W. C., 1987, Effects of irradiation on the biology of the infective larvae of *Toxocara canis* in the mouse, *Journal of Parasitology*, **73**, 89–94.

Beautyman, W., Beaver, P. C., Buckley, J. J. C. and Woolf, A. L., 1966, Review of a case previously showing an ascarid larva in the brain, *Journal of Pathology and Bacteriology*, **91**, 271–3.

Beaver, P. C., 1956, Larva migrans, *Experimental Parasitology*, **5**, 587–621.

Beaver, P. C., Snyder, M. D., Carrera, G. M., Dent, J. H. and Lafferty, J. W., 1952, Chronic eosinophilia due to visceral larva migrans, *Pediatrics*, **9**, 7–19.

Bisseru, B., 1969, Studies on the liver, lung, brain and blood of experimental animals infected with *Toxocara canis*, *Journal of Helminthology*, **43**, 267–72.

Bowman, D. D., Mika-Grieve, M. and Grieve, R. B., 1987, Circulating excretory–secretory antigen levels and specific antibody responses in mice infected with *Toxocara canis*, *American Journal of Tropical Medicine and Hygiene*, **36**, 75–82.

Boyce, W. M., Brandstetter, B. A. and Kazacos, K. A., 1988, Comparative analysis of larval excretory–secretory antigens of *Baylisascaris procyonis*, *Toxocara canis* and *Ascaris suum* by western blotting and ensyme immunoassay, *International Journal of Parasitology*, **18**, 109–13.

Brain, L. and Allen, B., 1964, Encephalitis due to infection with *Toxocara canis*. Report of a suspected case, *Lancet*, i, 1355–7.

Brown, D. H., 1970, Ocular *Toxocara canis*, *Journal of Pediatric Ophthalmology*, **7**, 182–92.

Burren, C. H., 1968, Experimental toxocariasis. 1. Some observations on the histopathology of the migration of *Toxocara canis* larvae in the mouse, *Zeitschrift für Parasitenkunde*, **30**, 152–61.

Burren, C. H., 1971, The distribution of *Toxocara* larvae in the central nervous system of the mouse, *Transactions of the Royal Society of Tropical Medicine and Hygiene*, 65, 450–3.

Collins, R. F., and Ivey, M. H. 1975, Specificity and sensitivity of skin test reactions to extracts of *Toxocara canis* and *Ascaris suum*. 1. Skin tests done on infected guinea pigs, *American Journal of Tropical Medicine and Hygiene*, 24, 455–9.

Dent, J. H., Nichols, R. L., Beaver, P. C., Carrera, G. M. and Staggers, R. J., 1956, Visceral larva migrans with a case report, *American Journal of Pathology*, 32, 777–803.

de Savigny, D. H., 1975, *In vitro* maintenance of *Toxocara canis* larvae and a simple method for the production of Toxocara ES antigen for use in serodiagnostic tests for visceral larva migrans, *Journal of Parasitology*, 61, 781–2.

de Savigny, D. H. and Tizzard, I. R., 1977, *Toxocara* larva migrans: the use of larval secretory antigens in haemagglutination and soluble antigen fluorescent tests, *Transactions of the Royal Society of Tropical Medicine and Hygiene*, 71, 501–7.

de Savigny, D. H., Voller, A. and Woodruff, A. W., 1979, Toxocariasis: serological diagnosis by enzyme immunoassay, *Journal of Clinical Pathology*, 32, 284–8.

Dolinsky, Z. S., Burright, R. G., Donovick, P. J., Glickman, L. T., Babish, J., Summers, B. and Cypess, R. H., 1981, Behavioral aspects of lead and *Toxocara canis* in mice, *Science*, 213, 1142–4.

Done, J. T., Richardson, M. D. and Gibson, T. E., 1960, Experimental visceral larva migrans in the pig, *Research in Veterinary Science*, 1, 133–51.

Duguid, I. M., 1961, Features of ocular infestation by *Toxocara*, *British Journal of Ophthalmology*, 45, 789–96.

Dunsmore, J. D., Thompson, R. C. A. and Bates, I. A., 1983, The accumulation of *Toxocara canis* larvae in the brains of mice, *International Journal for Parasitology*, 13, 517–21.

Fattah, D. I., Maizels, R. M., McLaren, D. J. and Spry, C. J., 1986, *Toxocara canis*: interaction of human blood eosinophils with the infective larvae, *Experimental Parasitology*, 61, 421–31.

Fernando, S. T. and Soulsby, E. J. L., 1974, Immunoglobulin class of antibodies in monkeys infected with *Toxocara canis*, *Journal of Comparative Pathology*, 84, 569–76.

Ghafoor, S. Y., Smith, H. V., Lee, W. R., Quinn, R. and Girdwood, R. W. A., 1984, Experimental ocular toxocariasis: a mouse model, *British Journal of Ophthalmology*, 68, 89–96.

Glickman, L. T. and Schantz, P. M., 1981, Epidemiology and pathogenesis of zoonotic toxocariasis, *Epidemiological Reviews*, 3, 230–50.

Glickman, L. T. and Summers, B. A., 1983, Experimental *Toxocara canis* infection in cynomolgus macaques (*Macaca fascicularis*), *American Journal of Veterinary Research*, 44, 2347–54.

Glickman, L. T., Grieve, R. B., Lauria, S. S. and Jones, D. L., 1985, Serodiagnosis of ocular toxocariasis: a comparison of two antigens, *Journal of Comparative Pathology*, 38, 103–7.

Gould, I. M., Newell, S., Green, S. H. and George, R. H., 1985, Toxocariasis and eosinophilic meningitis, *British Medical Journal*, 291, 1239–40.

Hay, J., Arnott, M. A., Aitken, P. P. and Kendall, A. T., 1986, Experimental toxocariasis and hyperactivity in mice, *Zeitschrift für Parasitenkunde*, 72, 115–20.

Hill, I. R., Denham, D. A. and Scholtz, C. L., 1985, *Toxocara canis* larva in the brain of a British child, *Transactions of the Royal Society of Tropical Medicine and Hygiene*, 79, 351–4.

Hoeppli, R., Feng, L. C. and Li, F., 1949, Histological reactions in the liver of mice due to larvae of different *Ascaris* species, *Peking Natural History Bulletin*, 18, 119–32.

Hogarth-Scott, R. S., 1966, Visceral larva migrans — an immunofluorescent examination of rabbit and human sera for antibodies to the ES antigens of the second stage larvae of *Toxocara canis*, *Toxocara cati* and *Toxascaris leonina* (Nematoda), *Immunology*, 10, 217–23.

Hogarth-Scott, R. S. and Feery, B. J., 1976, The specificity of nematode allergens in the diagnosis of human visceral larva migrans, *Australian Journal of Experimental Biology and Medical Science*, **54**, 317–27.

Huntley, C. C., Costas, M. C. and Lyerly, A., 1965, Visceral larva migrans syndrome: clinical characteristics and immunological studies in 51 patients, *Paediatrics*, **36**, 523–36.

Huntley, C. C., Lyerly, A. and Patterson, M. V., 1969, Isohaemagglutinins in parasitic infections, *Journal of the American Medical Association*, **208**, 1145–8.

Kayes, S. G. and Oaks, J. A., 1978, Development of the granulomatous response in murine toxocariasis. Initial events, *American Journal of Pathology*, **93**, 277–94.

Kayes, S. G. and Oaks, J. A., 1980, *Toxocara canis*: T lymphocyte function in murine visceral larva migrans and eosinophilia onset, *Experimental Parasitology*, **49**, 47–55.

Kayes, S. G., Omholt, P. E. and Grieve, R. B., 1985, Immune responses of CBA/J mice to graded infections with *Toxocara canis*, *Infection and Immunity*, **48**, 697–703.

Kayes, S. G., Jones, R. E. and Omholt, P. E., 1988, Pulmonary granuloma formation in murine toxocariasis; transfer of granulomatous hypersensitivity using bronchoalveolar lavage cells, *Journal of Parasitology*, **74**, 950–6.

Kennedy, M. W. and Kuo, Y.-M., 1988, The surfaces of the parasitic nemodes *Trichinella spiralis* and *Toxocara canis* differ in the binding of post-C3 components of human complement by the alternate pathway, *Parasite Immunology*, **10**, 459–63.

Kennedy, M. W., Foley, M., Kuo, Y-M., Kusel, J. R. and Garland, P. B., 1987a, Biophysical properties of the surface lipid of parasitic nematodes, *Molecular and Biochemical Parasitology*, **22**, 233–40.

Kennedy, M. W., Maizels, R. M., Meghji, M., Young, L., Qureshi, F. and Smith, H. V., 1987b, Shared and species-specific epitopes on the secreted and surface antigens of *Toxocara cati* and *Toxocara canis* infective larvae, *Parasite Immunology*, **9**, 407–20.

Klaver-Wesseling, J. C. M., Vetter, J. C. M. and Visser, W. K., 1978, A comparative *in vitro* study of antibody binding to different stages of the hookworm *Ancylostoma caninum*, *Zeitschrift für Parasitenkunde*, **56**, 147–59.

Le Gros Clark, W. E., 1975, *The Tissues of the Body*, Oxford: Clarendon Press.

Kondo, K., Shimada, Y., Kurimoto, H. and Oda, K., 1976, Experimental studies on 'Visceral larva migrans'. 2. On resistance of mice receiving sensitisation and challenge with *Toxocara canis* eggs, *Japanese Journal of Parasitology*, **25**, 371–6.

Maizels, R. M. and Robertson, B. D., 1990, *Toxocara canis*: secreted glycoconjugate antigens in immunobiology and immunodiagnosis, in Kennedy, M. W. (Ed.) *Parasitic Nematodes — Antigens, Membranes and Genes*, London: Taylor and Francis.

Maizels, R. M., de Savigny, D. and Ogilvie, B. M., 1984, Characterization of surface and excretory–secretory antigens of *Toxocara canis* infective larvae, *Parasite Immunology*, **6**, 23–37.

Maizels, R. M., Kennedy, M. W., Meghji, M., Robertson, B. D. and Smith, H. V., 1987, Shared carbohydrate epitopes on disinct surface and secreted antigens of the parasitic nematode *Toxocara canis*, *Journal of Immunology*, **139**, 207–14.

Marmor, M., Glickman, J., Shofer, F., Faich, L. A., Rosenberg, C., Cornblatt, B. and Friedmans, S., 1987, *Toxocara canis* infection in children: epidemiologic and neuro-psychologic findings, *American Journal of Public Health*, **77**, 554–9.

McLure, J. M., 1988, 'The pathology of an early *Toxocara canis* infection. A model in Balb/c and NIH mice investigated by haematology, serology, parasitology and histo-pathology', FIMLS Thesis, Institute of Medical Laboratory Sciences, London.

Meghji, M. and Maizels, R. M., 1986, Biochemical properties of larval excretory–secretory glycoproteins of the parasitic nematode *Toxocara canis*, *Molecular and Biochemical Parasitology*, **18**, 155–70.

Mossalam, I., Attallah, O. A. and Hosney, Z., 1972, Experimental studies on visceral larva migrans of *Toxocara canis* in laboratory animals, *Acta Veterinaria Academiae Scientiarum Hungaricae*, **22**, 71–80.

Nichols, R. L., 1956, The etiology of visceral larva migrans. 1. The diagnostic morphology of infective second-stage *Toxocara* larvae, *Journal of Parasitology*, **42**, 349–62.

Nicholas, W. L., Stewart, A. C. and Mitchel, G. F., 1984, Antibody responses to *Toxocara canis* using sera from parasite-infected mice and protection from toxocariasis by immunisation with ES antigens, *Australian Journal of Experimental Biology and Medical Science*, **62**, 619–26.

Oliver-Gonzalez, J., 1946, Functional antigens in helminths, *Journal of Infectious Diseases*, **78**, 232–7.

Oliver-Gonzalez, J. and Gonzalez, L., 1949, Release of the A_2 isoagglutinogen-like substance of infectious organisms into human blood, *Journal of Infectious Diseases*, **85**, 66–71.

Oliver-Gonzalez, J. and Torregrosa, M. V., 1944, A substance in animal parasites related to the human isoagglutinogens, *Journal of Infectious Diseases*, **74**, 173–7.

Olson, L. J., 1976, Ocular toxocariasis in mice: distribution of larvae and lesions, *International Journal of Parasitology*, **6**, 247–51.

Olson, L. J. and Rose, J. E., 1966, Effect of *Toxocara canis* infection on the ability of white rats to solve maze problems, *Experimental Parasitology*, **19**, 77–84.

Olson, L. J. and Schultz, C. W., 1963, Some biochemical and immunological aspects of host–parasite relationships. Nematode-induced hyper-sensitivity reactions in guinea pigs: onset of eosinophilia and positive Schultz–Dale reactions following graded infections with *Toxocara canis*, *Annals of the New York Academy of Sciences*, **133**, 440–55.

Olson, L. J., Izzat, N. N., Petteway, M. B. and Reinhart, J. A., 1970, Ocular toxocariasis in mice, *American Journal of Tropical Medicine and Hygiene*, **19**, 238–43.

Oshima. T., 1961, Standardization of techniques for infecting mice with *Toxocara canis* and observations on the normal migration routes of the larvae, *Journal of Parasitology*, **47**, 652–6.

Parsons, J. C., Bowman, D. D. and Grieve, R. B., 1986, Tissue localisation of excretory–secretory antigens of larval *Toxocara canis* in acute and chronic murine toxocariasis, *American Journal of Tropical Medicine and Hygiene*, **35**, 974–81.

Pentreath, V. W., 1989, Neurobiology of sleeping sickness, *Parasitology Today*, **5**, 215–18.

Philipp, M., Parkhouse, R. M. E. and Ogilvie, B. M., 1980, Changing proteins on the surface of a parasitic nematode, *Nature*, **287**, 538–40.

Pike, E. H., 1960, Effect of diethylcabamazine, oxophenarsine hydrochloride and piperazine citrate on *Toxocara canis* larvae in mice, *Experimental Parasitology*, **9**, 223–32.

Przyjalkowski, Z., Zapart, W. and Starzynski, S., 1979, Investigation of intravital diagnosis of *Toxocara canis* larva migrans in experimentally infected mice, *Bulletin de l'Academie Polonaise des Sciences*, **XXVL**, 875–80.

Robertson, B. D., Burkot, T. R., Gillespie, S. H., Kennedy, M. W., Wambai, Z. and Maizels, R. M., 1988, Detection of circulating parasite antigen and specific antibody in *Toxocara canis* infections, *Clinical and Experimental Immunology*, **74**, 236–41.

Rockey, J. H., Donnelly, J. J., Stromberg, B. E. and Soulsby, E. J. L., 1979, Immunopathology of *Toxocara canis* and *Ascaris suum* infections of the eye: the role of the eosinophil, *Investigative Ophthalmology and Visual Science*, **18**, 1172–84.

Rockey, J. H., John, T., Donnelly, J. J., McKenzie, D. F., Stromberg, B. E. and Soulsby, E. J., 1983, *In vitro* interaction of eosinophils from ascarid-infected eyes with *Ascaris suum* and *Toxocara canis* larvae, *Investigative Ophthalmology and Visual Science*, **24**, 1346–57.

Russegger, L. and Schmutzhard, E., 1989, Spinal toxocaral abcess, *Lancet*, ii, 398.

Schantz, P. M., 1989, *Toxocara* larva migrans now, *American Journal of Tropical Medicine and Hygiene*, **41**, 21–34.

Schochet, S. S., 1967, Human *Toxocara canis* encephalopathy in a case of visceral larva migrans, *Neurology*, **17**, 227–9.

Shields, J. A., 1984, Ocular toxocariasis. A review, *Survey of Ophthalmology*, **28**, 361–81.

Smith, M. H. D. and Beaver, P. C., 1953, Persistence and distribution of *Toxocara* larvae in the tissues of children and mice, *Pediatrics*, **12**, 491–7.

Smith, H. V., Quinn, R., Bruce, R. G. and Girdwood, R. W. A., 1980, A paper radio-immunosorbent test (PRIST) for the detection of larva-specific antibodies to *Toxocara canis* in human sera, *Journal of Immunological Methods*, **37**, 47–55.

Smith, H. V., Quinn, R., Kusel, J. R. and Girdwood, R. W. A., 1981, The effect of temperature and antimetabolites on antibody binding to the outer surface of second stage *Toxocara canis* larvae, *Molecular and Biochemical Parasitology*, **4**, 183–93.

Smith, H. V., Quinn, R., Bruce, R. G. and Girdwood, R. W. A., 1982, Development of the serological response in rabbits infected with *Toxocara canis* and *Toxascaris leonina*. *Transactions of the Royal Society of Tropical Medicine and Hygiene*, **76**, 89–94.

Smith, H. V., Kusel, J. R. and Girdwood, R. W. A., 1983, The production of human A and B blood group like substances by *in vivo* maintained second stage *Toxocara canis* larvae; their presence on the outer larval surfaces and in their excretions/secretions, *Clinical and Experimental Immunology*, **54**, 625–33.

Smith, H. V., Hinson, A., Girdwood, R. W. A. and Taylor, M. R. H., 1988, Variation in antibody isotype responses in clinically covert toxocariasis, *Lancet*, ii, 167.

Soulsby, E. J. L., Stromberg, B. E., Donnelly, J. J. and Rockey, J. H., 1980, Intraocular immunoglobulin E induced by intravitreal infections with *Ascaris* and *Toxocara* spp. larvae, *Ophthalmic Research*, **12**, 45–53.

Sprent, J. F. A., 1952, On the migratory behaviour of the larvae of various *Ascaris* species in white mice. 1. Distribution of larvae in tissues, *Journal of Infectious Diseases*, **90**, 165–76.

Sprent, J. F. A., 1955, On the invasion of the central nervous system by nematodes. 1. The incidence and pathological significance of nematodes in the central nervous system, *Parasitology*, **45**, 31–40.

Sprent, J. F. A., 1958, Observations on the development of *Toxocara canis* (Werner, 1782) in the dog, *Parasitology*, **48**, 184–210.

Sugane, K. and Oshima, T., 1982, Eosinophilia, granuloma formation and migratory behaviour of larvae in the congenitally athymic mouse infected with *Toxocara canis*, *Parasite Immunology*, **4**, 307–18.

Sugane, K. and Oshima, T., 1983a, Trapping of large numbers of larvae in the livers of *Toxocara canis*-reinfected mice. *Journal of Helminthology*, **57**, 95–9.

Sugane, K. and Oshima, T., 1983b, Purification and characterization of excretory and secretory antigen of *Toxocara canis* larvae, *Immunology*, **50**, 113–20.

Sugane, K., Howell, M. J. and Nicholas, W. L., 1985, Biosynthetic labelling of the excretory and secretory antigens of *Toxocara canis* larvae, *Journal of Helminthology*, **59**, 147–51.

Taylor, M. R. H., Keane, C. T., O'Connor, P., Girdwood, R. W. A. and Smith, H. V., 1987, Clinical features of covert toxocariasis, *Scandinavian Journal of Infectious Diseases*, **19**, 693–6.

Taylor, M. R. H., Keane, C. T., O'Connor, P., Mulvihill, E. and Holland, C., 1988, The expanded spectrum of toxocaral disease, *Lancet*, i, 692–5.

van Knapen, F., van Leusden, J., Polderman, A. M. and Franchimont, J. H., 1983, Visceral larva migrans: examination by means of enzyme-linked immunosorbent assay of human sera for antibodies to excretory–secretory antigens of the second stage larvae of *Toxocara canis*, *Zeitschrift für Parasitenkunde*, **69**, 113–18.

Vetter, J. C. M. and Klaver-Wesseling, J. C. M., 1978, IgG antibody binding to the outer surface of infective larvae of *Ancylostoma caninum*, *Zeitschrift für Parasitenkunde*, **58**, 91–6.

Wade, S. E. and Georgi, J. R., 1987, Radiolabelling and autoradiographic tracing of *Toxocara larvae* in male mice, *Journal of Parasitology*, **73**, 116–20.

Watzke, R. C., Oaks, J. A. and Folk, J. C., 1984, *Toxocara canis* infection of the eye. Correlation of clinical observations with developing pathology in the primate model, *Archives of Ophthamology*, **102**, 282–91.

Wekerle, H., Sun, D., Oropeza-Wekerle, R. L. and Meyermann, R., 1987, Immune reactivity in the nervous system: modulation of T lymphocyte activation by glial cells, *Journal of Experimental Biology*, **132**, 43–57.

Zyngier, F. R., 1974, Histopathology of experimental toxocariasis in mice, *Annals of Tropical Medicine and Parasitology*, **68**, 225–8.

Zyngier, F. R. and Brockbank, A., 1974, Electron microscopy of the lung in experimental *Toxocara canis* infection, *Annals of Tropical Medicine and Parasitology*, **68**, 229–33.

8. Immunology, biochemistry and molecular biology of hookworm antigens

D. I. Pritchard, P. G. McKean, P. J. Tighe and R. J. Quinnell

INTRODUCTION

The successful application of recent, and perhaps prematurely optimistic, proposals to control hookworm infection and disease (Warren, 1988; Keymer and Bundy, 1989; Crompton, 1989) will depend in part on an increase in our knowledge and understanding of the basic biology of hookworms. It is also becoming apparent that the identification of hookworm antigens for vaccination, or for use in diagnosis and monitoring hookworm epidemiology, will ultimately depend on a close association between laboratory science and field work. This approach, although initially laborious and not immediately rewarding in terms of rapid publication, will eventually be more valuable than that conducted in understandably limited laboratory animal model systems.

Meanwhile, our current knowledge of hookworms, and of the immune response against them, indicates that a number of molecules are worthy of further study. These include antigens present in hookworm excretions or secretions (ES antigens), and those expressed in or on the cuticle (Pritchard, 1987). However, before giving a description of this biochemically based research, it is important to raise some points about the basic immunobiology of hookworms. This is because the parasites reside in contrasting environments during their passage from the soil to their site of reproduction, environments to which the worms have adapted biochemically and morphologically during their evolution. This obviously has a bearing on the design of strategies for identifying hookworm molecules for the purposes described above.

THE HOOKWORM LIFE CYCLE

Life in the outside world

Hookworms infect their hosts following a brief existence in the external environment. Ova passed in the faeces develop and hatch within the faecal matter between 24 and 32°C (Figure 8.1). The rhabditiform first and second larval stages (L1 and L2) feed on faecal microflora, developing within 10 days into the filariform, ensheathed, infective third stage larva (L3). The retained sheath (the L2 cuticle) protects this stage from environmental stress. Although studies on human hookworms have yet to be completed, morphological comparisons between rhabditiform and filariform stages in a rodent parasite which adopts a similar strategy for infection (*Nippostrongylus brasiliensis* (Bonner, 1979a,b)) have demonstrated that the switch is accompanied by gross morphological and, by inference, biochemical changes, particularly in the gastrointestinal tract of the worm. The L3 stage does not feed, and relies on endogenous glycogen stores for energy. Its intestine has probably regressed, in a manner similar to that described for *N. brasiliensis*, from the well-developed absorptive structure of the rhabditiform stage.

Further evidence for the relative biochemical quiescence of filariform L3 stages again comes from work on *N. brasiliensis* L3 larvae (Bonner, 1979b). These larvae fail to incorporate significant levels of radioactive uridine into their RNA, a trend

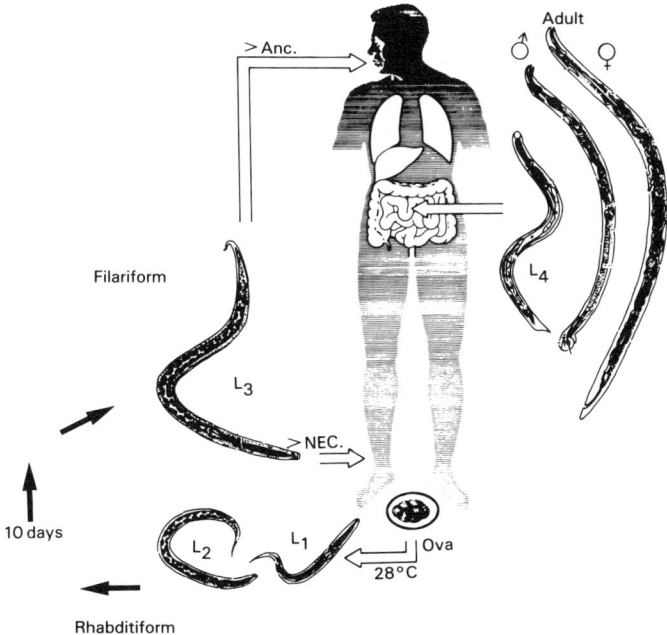

Figure 8.1. A generalized life cycle diagram for hookworms. Adapted from Schad and Banwell (1984).

which is completely reversed when the correct stimuli (from a potential host?) are applied. The environmental stimuli required to initiate this important biochemical switch seem to be multiple, and include temperature, physical contact, oxygen and carbon dioxide. Therefore, although hookworm larvae do not venture far from their point of deposition in the soil, they are sufficiently sensitive to detect subtle environmental changes; these stimuli then set into motion a chain of biochemical events which, presumably, equip the larvae for the infection process.

These stimuli promote: (a) the activation of an endocrine system established in a previous stage, which results in the release of exsheathing or hatching fluid, allowing the next stage to emerge; and (b) the initiation of DNA transcription in a hitherto dormant gene set of the next parasitic stage (Petronijevic *et al.*, 1986).

This process is of great importance to the stability and eventual outcome of the host–parasite relationship. Potential antigens present in exsheathing fluid, and in the ultimate products of the newly initiated DNA transcription, are probably the first with which the immune system of the host comes into contact.

Entry into the host

Although *Ancylostoma duodenale* is considered to be adapted for oral infection (Kendrick, 1934; Yoshida *et al.*, 1958), skin penetration is probably the most common route of entry into the host (Figure 8.2a). However, it would appear that different species of hookworm can employ contrasting mechanisms. Whilst *Necator americanus* actively secretes enzymes during penetration, some *Ancylostoma* species appear to penetrate by purely mechanical means (Matthews, 1977, 1982). *Necator* exsheaths following stimulation with skin lipids, and penetration is accompanied by a morphological change in the parasites' excretory oesophageal glands (Smith, 1976). This change is not seen in *A. ceylanicum*. *Necator* can only infect experimental hosts a finite number of times; this possibly reflects an exhaustion of, and inability to resynthesize, their supply of penetration enzymes. In contrast, *A. tubaeforme* penetrates repeatedly without exsheathment (even from the dermal aspect), and does not appear to secrete enzymes. These observations led Matthews to conclude that some *Ancylostoma* species at least do not require enzyme secretion in order to penetrate the host. However, it is worth bearing in mind that *A. ceylanicum* and *A. caninum* exsheath before or during penetration (Matthews, 1977), and that *A. caninum* larvae appear to possess a metalloprotease (Hotez *et al.*, 1985), although the latter observation was not based on a functional assay.

Figure 8.2. Opposite: The illustrated life cycle of *Necator americanus*. (a) L3 below the keratinized layer (KL) of the skin of a Balb/c mouse 48 h following a challenge infection. Stain: Giemsa. (b) L3 in the bronchial lumen (BL) of a Balb/c mouse 3 days following primary infection. Stain: haematoxylin and eosin. (c) Adult in the ileum of a hamster 35 days following a primary infection. Stain: haematoxylin and eosin. CD, cell debris; CP, cutting plate; CUT, cuticle; GL, gut lumen; GM, gut mucosa; OG, oesophageal gland. (Scale bar: 0.1 mm.)

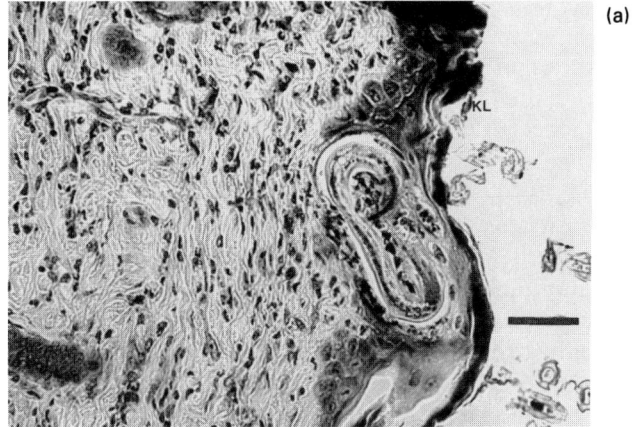

(a)

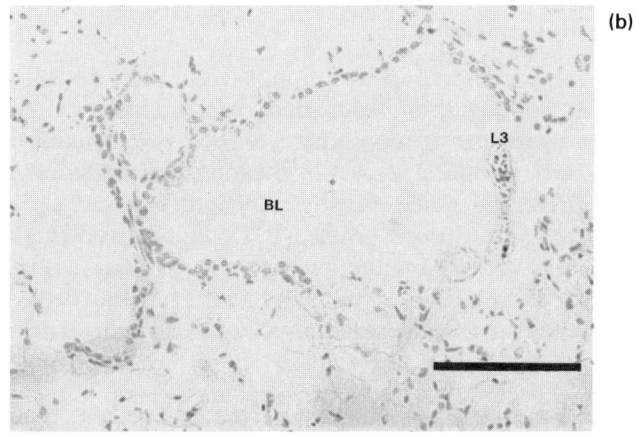

(b)

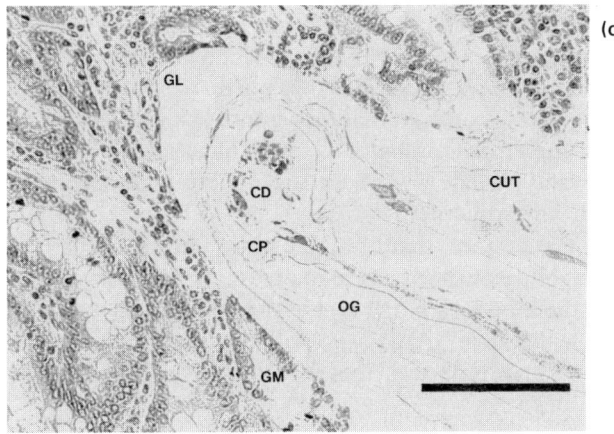

(c)

Migratory stages

Whilst little is known at present about the biochemistry of hookworm stages (see Figure 8.2b) during their migration to the gut, some work has been done using *N. brasiliensis* as a model (Croll and Ma, 1978), and the ability to passage hookworms in laboratory animals at least provides a basis for future investigations (Sen and Seth, 1967; Schad, 1979; Wells, 1988).

For example, investigations into the mechanisms by which larvae migrate systemically and whether or not proteolytic enzymes are essential to this migration might lead to immunological or pharmacological inhibition of this phase. In addition, the larvae of *A. duodenale*, faced with adverse conditions within the host, have the ability to enter period of relative quiescence (arrested development). However, the stimuli necessary for triggering this event and the methods for detecting these stimuli remain unresearched (Schad *et al.*, 1973; Nawalinski and Schad, 1974).

Life in the gastrointestinal tract

The hookworm, having entered its host from the soil, now finds itself essentially back in the outside world, within the lumen of the small intestine. Here, the worm becomes attached by its cutting mouth parts to the well vascularized ileal mucosa (Figure 8.2c). Intrinsic pumping activity by the musculature surrounding the oesophageal bulb facilitates the intake of tissue in the form of a buccal plug, and causes bleeding at the site of attachment (Wells, 1931). Blood loss occurs in two peaks, associated temporarily with the maturation of the fourth larval (L4) and adult stages (Clark *et al.*, 1961), and bleeding is promoted by the foraging behaviour of adults (Kalkofen, 1970, 1974) and by the secretion of anti-coagulants (Loeb and Fleisher, 1910; Thorson, 1956a; Hotez and Cerami, 1983) and, possibly, haemolysins (Whipple, 1909; Schwartz, 1920). Blood loss inevitably leads to the iron deficiency anaemia often associated with hookworm infection (Foy and Nelson, 1963).

Human plasma actively promotes worm feeding by stimulating oesophageal pumping (Roche and Martinez Torres, 1960). *In vitro* survival is also promoted in the presence of serum (Komiya *et al.*, 1956) and homologous sera appear to be the most effective (Yasuraoka *et al.*, 1960). At a biochemical level, Fernando and Wong (1964) demonstrated that a non-diffusable portion of dog serum enhanced the incorporation of glucose from external media into the glycogen of *A. caninum*. In addition, serum supplements enhance secretion of acetylcholinesterase by adult *N. americanus* (Burt and Ogilvie, 1975), an enzyme often described as a biochemical hold-fast for gastrointestinal nematodes.

It would appear, therefore, that human plasma actively promotes worm feeding and enhances the ability of the worm to maintain its station against intestinal peristalsis. However, the blood-sucking habit of the hookworm could transpire to be its ultimate weakness, because the parasite will be directly exposed to serum-borne immune effector mechanisms and, thereby, to effective vaccination.

THE DESIGN OF STRATEGIES TO INTERRUPT THE LIFE CYCLE

If the processes already known to be used by hookworms to perpetuate their life cycles are carefully examined, it becomes possible to design strategies to interrupt these cycles. A common feature of these strategies (Table 8.1) is the proposal to inhibit the activity, or prevent the secretion, of enzymes necessary for skin penetration and tissue migration by larval stages, and enzymes used by adults for feeding and maintaining their station in the gastrointestinal tract.

To provide some background to the development of these strategies, it should be noted that enzyme activity was first reported in hookworm secretions by Loeb and Smith in 1904. The concept of secretory enzymes as a target for immune attack was then suggested in a series of papers by Chandler (1932, 1953) and was taken up with respect to hookworms by Thorson (1956a–c). He demonstrated that sera from dogs refractory to infection with hookworm (*A. caninum*) inhibited the proteolytic activity of oesophageal extracts of adult worms and, subsequently, went on to show that these extracts could be used to vaccinate against infection. Those observations formally demonstrated that the proteolytic secretions of adult worms, later termed 'feeding enzymes', were 'functional antigens', in that their inhibition by the immune system is related to a refractory state. This pioneering work provided the impetus for further investigations into the proteolytic activity of hookworm secretions, with a view to using purified enzymes as vaccines against infection (Oya and Noguchi, 1977; Hotez and Cerami, 1983; Hotez *et al.*, 1986).

The rationale behind this approach is that prevention of feeding by immunological or pharmacological means will affect parasite fecundity and survival. Similarly, a sufficiently specific and avid immune response against the worm's secretory acetylcholinesterase (AChE), or its inhibition by drugs, would add to the demise of the adult worm.

However, an attack on the adult parasite in the gastrointestinal tract could be viewed as a rearguard action, equally achievable by anthelmintics. Chemotherapy does not, however, have the immune system's advantage of recall and is increasingly subject to drug resistance. A more worthwhile approach might be to

Table 8.1. Proposed strategies for interrupting the life cycle of hookworms.

L3 (Figure 8.2a)
(1) Pharmacological inhibition of penetration enzymes or their release: e.g. protease inhibitors
(2) Immunological neutralization of penetration enzymes
(3) Stimulation of intradermal CMI against cuticular, excretory–secretory, antigens

Migratory Larvae (Figure 8.2b)
(1) As above
(2) As above
(3) Stimulation of bronchial-associated lymphoid tissue (specific CMI and IgA)

Adult worms (Figure 8.2c)
(1) Pharmacological inhibition of feeding enzymes, anti-coagulants, AChE
(2) Immunological neutralization of enzymes
(3) Stimulation of gut-associated lymphoid tissue (CMI, specific IgA)

CMI, cell-mediated immunity; IgA, immunoglobulin A; AChE, acetylcholinesterase.

immobilize the L3 stage in the skin by blocking its penetration enzymes. It is therefore surprising that the study of hookworm larval proteases has been extremely limited to date, particularly as radiation-attenuated larvae of *A. caninum* have been used successfully to protect dogs against challenge infection (Miller, 1971). The arrest of larvae in somatic migratory locations was probably responsible for the stimulation of the highest and most uniform resistance in dogs confirming the immunogenic potential of infective larvae and their undefined products. The recent upsurge in interest in the field of skin immunology (reviewed by Bos and Kapsenberg, 1986), and the belief that the intradermal route of immunological challenge is a very efficient one with regard to the stimulation of immunity (e.g. *Schistostoma mansoni* (James, 1987) and hepatitis B (Irvine *et al.*, 1986, 1987)), adds further impetus to the development of this area of hookworm biology.

As well as developing strategies based on the neutralization of functional antigens, to combat infection and reduce pathology, there is also a requirement to identify and provide markers for the intensity and species of infection, and for the degree of exposure. Therefore, in the following sections we deal specifically with recent research into hookworm antigens.

POTENTIAL DIAGNOSTIC ANTIGENS

For diagnostic purposes, it is likely that species-specific antigens would be required, particularly when the contrasting biology of the two major hookworm species infecting man is considered (Table 8.2). Adopting the criterion of species specificity, a number of candidate molecules can be shown to exist.

Table 8.2. Some biological differences between the hookworm species *Necator americanus* and *Ancylostoma duodenale*. Adapted from Banwell and Schad (1978) and Hoagland and Schad (1978).

	N. americanus	*A. duodenale*
Eggs		
Dimensions (μm)	64–76 × 36–40	56–60 × 36–40
Larvae		
Length of body (excluding sheath) (μm)	500–600	600–700
Length of tail (anus to tip) (μm)	<73	>73
Intestine at oesophageal — intestinal junction	As wide as oesophageal bulb	Narrower than oesophageal bulb
Transverse striations	Present on sheath in tail region	Inconspicuous in tail region
Adults		
Size	Smaller and more slender	Larger and thicker
Anterior end	Bends in opposite direction to body curvature	Bends in same direction as body curvature
Buccal capsule	Four chitinous plates	Six hook-like teeth
Copulatory bursa	Dorsal ray split from base. Total number of rays 14. Two spicules fused at tip	Dorsal ray single. Total number of rays 13. Two spicules separate
Posterior end	Spine absent	Spine present
Vulval opening	Anterior	Posterior

Table 8.2. Continued.

	N. americanus	A. duodenale
Character of nematodes		
Mean female body weight (g)	0.9×10^{-3}	2.2×10^{-3}
Egg output/female/day	5000–10 000	10 000–25 000
Egg output as percentage of body weight	121–150	140–166
Typical natural life span (years)	3–5	1
Developmental arrest in humans and seasonal egg output	No	Yes
Blood loss/worm/day (ml)	0.03	0.15–0.23
Maximum survival time of larvae outside host	Less	Greater
Temperature at which 90 per cent of eggs hatch (°C)	20–35	15–35
Temperature above which eggs do not hatch (°C)	40	45
Resistance of eggs to low O_2, low temperature, desiccation, chemicals and death in faeces	Less	Greater
Resistance of larvae to chemicals and desiccation	Less	Greater
Larvae penetrating skin (laboratory infections)	77	32
Hookworm dermatitis in host	Greater	Less
Oral transmission	No	Yes
In utero or transmammary transmission	No	Possibly
Mean numbers of worms taken from one host (light infection)	29.5	11.5
Mean numbers of worms taken from one host (heavy infection)	488	193
Maximum number of worms taken from one host	2300	529
AChE activity (Δ pH g^{-1})	97.84	12.38
Sensitivity to the anthelmintic Ivermectin	Resistant	Susceptible

For example, the 28–33 kDa ES antigens (also present in homogenates) of adult *N. americanus* are recognized by sera from apparently 'Necator-only' geographical areas, and these sera fail to recognize antigens from other hookworm species (Figure 8.3a). This observation has since been extended using sera from a field study done in Papua New Guinea (Pritchard *et al.*, 1990c), where *N. americanus* and *Ascaris lumbricoides* were shown to co-exist. Recognition of the *Necator* 33 kDa antigen was retained following extensive absorption of the sera with *A. lumbricoides* pseudocoelomic fluid (the absorption removed all anti-*Ascaris* activity), indicating the apparent diagnostic potential of this ES antigen. However, it is becoming increasingly clear that recognition may not be indicative of current infection, as apparently worm-free individuals reacted with these antigens (this is perhaps evidence of immunity caused by a previous infection).

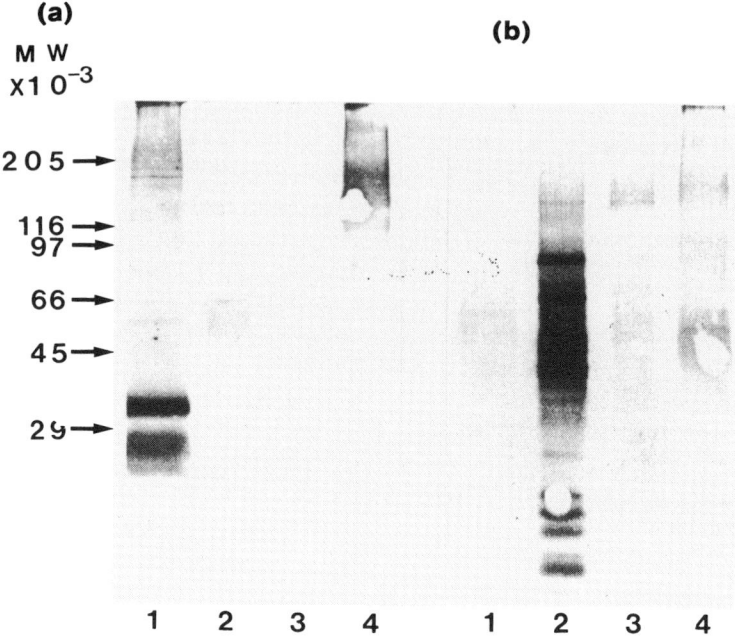

Figure 8.3. Immunoblotting analysis using human post-infection sera. Homogenate preparations (50 μg per lane) of four different hookworm species were probed with either (a) human post — *Necator americanus* infection sera, or (b) human post-*Ancylostoma duodenale* infection sera (1:1000 dilution). Blots were developed using horseradish peroxidase conjugated protein A (1:1000 dilution) and the substrate 4-chloro-1-napthol. Lanes : (1) *N. americanus* adult homogenate; (2) *A. duodenale* adult homogenate; (3) *A ceylanicum* adult homogenate; (4) *A. caninum* adult homogenate. Molecular weight standards (in kilodaltons) are shown on the left-hand side of the figure.

Additional research has indicated that a 17 kDa major accumulated protein of *N. americanus* could be useful as a diagnostic reagent. Antisera raised against this protein react only against homologous antigens on Western blotting, and completely fail to react in enzyme-linked immunosorbent assay (ELISA) against a range of parasite preparations (Pritchard *et al.*, 1990a). The protein appears to be accumulated during development, is associated with the cuticle and oesophageal-gland granules, and is both secreted and immunogenic to the infected host. It therefore has the necessary credentials for a diagnostic reagent; it is immunogenic, species specific and relatively abundant. Experiments are now in progress, in combination with a field study in Papua New Guinea, to assess the applicability of this protein to diagnosis.

Parallel studies using *A. duodenale* would also indicate the presence of species-specific antigens (Pritchard *et al.*, 1990b), as sera obtained from an individual undergoing 'biotherapy' for polycythaemia vera (by the deliberate introduction of *A. duodenale* adults into the gastrointestinal tract) (Walterspiel *et al.*, 1985)) recognized *A. duodenale* preferentially on Western blotting (Figure 8.3b).

Antigens which are species *and* stage specific probably offer the best opportunity for monitoring epidemiological studies by immunological means. For example, the distinction between the level of established infection, and the degree of exposure to infection, could be achieved using adult-derived and larval-derived species-specific antigens, respectively (see below).

Although progress on the identification of adult antigens has been relatively rapid, the identification of specific larval antigens has been hampered by a shortage of biological material for analysis. Recent studies (Gamble *et al.*, 1989) would suggest that numbers of infective nematode larvae in the region of 5×10^6 would be required for detailed biochemical and immunological experiments. Obtaining this quantity of fresh, infective larvae from hamsters infected with *N. americanus* poses a significant logistical problem for researchers. Nevertheless, a candidate molecule for the assessment of exposure has been identified (Pritchard, 1990) using labelled lectins and small quantities of living larvae. The molecule in question is a 42 kDa glycoprotein associated with refractile ring formation during exsheathment (Figure 8.4). Binding of fluorescein isothiocyanate conjugated wheatgerm agglutinin (FITC-WGA) or horseradish peroxidase conjugated wheatgerm agglutinin (HRP-WGA) to larvae or antigen on Western blots is completely inhibited by triacetyl-chitotriose but not by *N*-acetylglucosamine. This differential inhibition has been suggested to indicate the presence of chitin (Peters and Latka, 1986; Brydon *et al.*, 1987). Although this molecule is unlikely to be useful for the generation of a protective immune response because of its specific location in cast sheaths (Figure 8.4), it is potentially useful as a marker for exposure provided that (a) it is immunogenic to the infected host, and (b) the immune response against it is quantitatively related to the number and severity of infections experienced.

Stage- and species-specific *cuticular antigens* also offer opportunities for use in exposure studies (see below). Recent experiments (Pritchard, 1990; McKean, 1989) have indicated that the surface protein profiles of the L2 sheath and L3 cuticle are distinct, but further work is required to determine their immunogenicity and species specificity.

Another recent development in the immunobiology of the larval stage has been the demonstration by Sanjeev Kumar, in our laboratory (unpublished), that sera from humans infected with *N. americanus* (Pritchard *et al.*, 1990c) react exclusively against the sheath of the L3 stage in immunofluorescence assays. Little reactivity was seen against the L3 cuticle. The results of these immunofluorescence experiments were confirmed quantitatively in ELISA using detergent cetyltrimethyl ammonium bromide (CTAB) stripped antigens from the sheath. Reactivity in ELISA was also recorded against exsheathing fluid, indicating that the ensheathed larvae traverses the skin to come into contact with the host's immune system. In fact, ensheathed larval penetration was directly visualized by applying fluores-ceinated ensheathed L3 larvae to intact skin and following their progress using frozen skin sections. In these experiments, fluorescent larvae and their sheaths were positively identified deep within the dermal tissues. This formal demonstration of an immune response against the cast sheath, and antigens in exsheathing fluid, obviously has some bearing on the survival of the invading parasite. The generation of an immune response against these exoantigens could be viewed as a diversionary tactic, and might explain why apparently continuous infection with hookworm

(a)

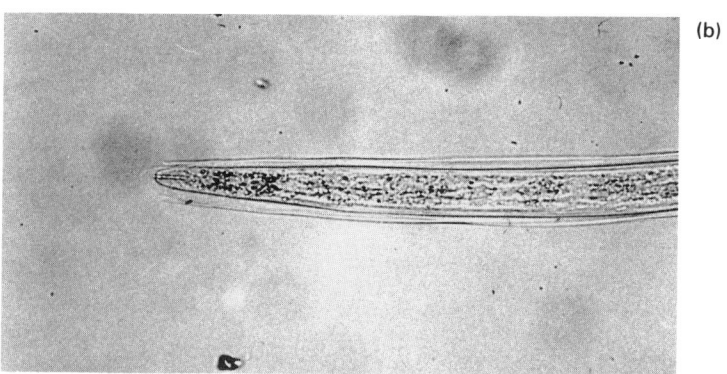

(b)

Figure 8.4. The binding of WGA to a 42 kDa molecule on exsheathing L3 *Necator americanus*. (a) FITC-WGA binding to an exsheathing larva during refractile ring formation and preceding the loss of the apical cap. (b) The same field photographed in white light. (c) FITC-WGA eventually binds to the whole of the cast sheath following exsheathment. (d) The binding HRP-labelled WGA to somatic preparations of L3, L4 and adult stages following Western blotting. The 42 kDa *N*-acetylglucosamine containing protein appears to be restricted to L3 and L4 stages, and is possibly involved in the moulting process.

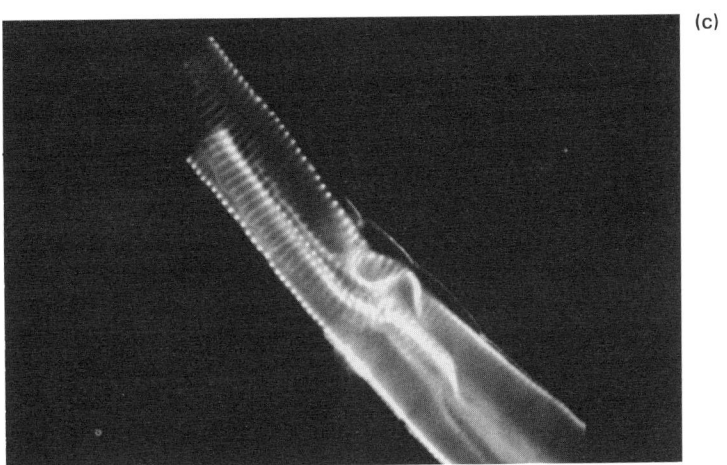

(c)

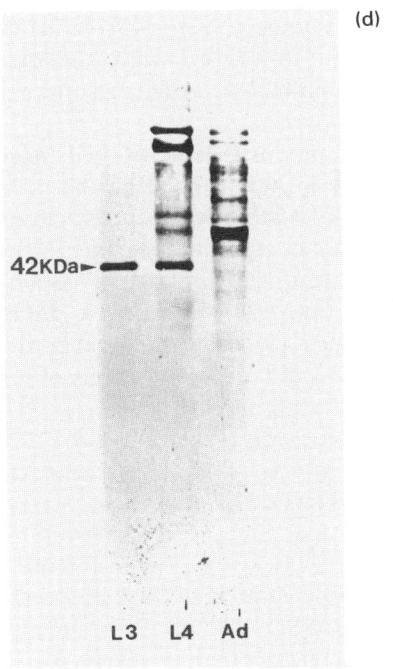

(d)

42KDa▶

L3   L4   Ad

larvae occurs in human populations in spite of a pronounced response against larval antigens (see below).

Finally, hookworm larval secretory proteinases could be used as species- and stage-specific markers of infection status. However, for the reasons outlined above, progress in this area is likely to be slow.

POTENTIAL PROTECTIVE ANTIGENS

Strategies for the interruption of the parasite life cycle, based on the stimulation of immunological effector mechanisms, are described above, where it was also suggested that secretory enzymes represent logical targets for attack. Of these, secreted proteinases and AChE are particularly attractive candidate molecules for further study, and recent and predicted progress in research into each of these types of molecule is described in this section.

Proteinases

Using an assay based on the release of radioactivity from ^{125}I-labelled fibrinogen, proteinase activity was detectable in ES products from *N. americanus* L3 larvae (Pritchard, 1990; McKean, 1989). Activity was maximal at pH 7.5, confirming the work of Matthews (1982), and the proteinase profile of larval ES products in gelatin substrate gels was distinct from that seen with ES products from adult worms. Larval proteinases resolved at 195, 166, 137, 72 and 62 kDa, whilst adult ES proteinases were relatively homogeneous, with bands of proteolysis at 158 and 64 kDA. Interestingly, and in context with the data presented above, L3 exsheathing fluid exhibited proteolytic activity against gelatin, with a single band of proteolysis at 116 kDa.

Adult *N. americanus* ES products demonstrated 'haemoglobinase' proteinase activity over a range of pH optima, suggesting the presence of several different species of proteinase. Activity is inhibited by pepstatin A and *N*-ethylmaleimide, but enhanced by the presence of aspartic acid (Julie Burleigh, personal communication) and cysteine, indicating the presence of aspartye and cysteinyl (thiol) proteinases (Bond and Butler, 1987). As serine proteinase activity has been reported in larval ES products (Salafsky, personal communication), a switch in activity during development could reflect the variety of substrates and biochemical environments encountered during the life cycle.

In a separate series of experiments (Pritchard *et al.*, 1990b), a comparison of the proteinolytic activities in extracts of adult *N. americanus*, *A. duodenale*, *A. caninum* and *A. ceylanicum* revealed a number of interesting differences between the species (Figures 8.5 and 8.6). The *Ancylostoma* species appeared to demonstrate more potent activity against casein (Figure 8.5) and, less distinctly, haemoglobin (Figure 8.6) substrates at pH 8.5 in sodium dodecylsulphate poly-acrylamide gel electrophoresis (SDS-PAGE) (Pritchard, 1990; McKean, 1989), and the molecular sizes of the proteinases in these species were similar, but distinct from those seen in *N. americanus* extracts (the exact molecular weights are published in Pritchard (1990) and McKean (1989)). This biochemical difference is probably a further example of the evolutionary divergence of the Ancylostominae and Necatorineae subfamilies, and could explain the difference in pathogenesis described in an earlier section (Table 8.2).

Continuing on the theme of proteinase activity, it would appear that *N. americanus* ES products contain exclusive proteinolytic activity for immunoglobulin A (IgA), the human secretory immunoglobulin isotype (McKean, 1989; Pritchard, 1990). At first glance this would appear to be a perfect adaptation to life in the gastrointestinal tract, but it should be remembered that hookworms are in intimate contact with whole blood and the defence mechanisms which it bears.

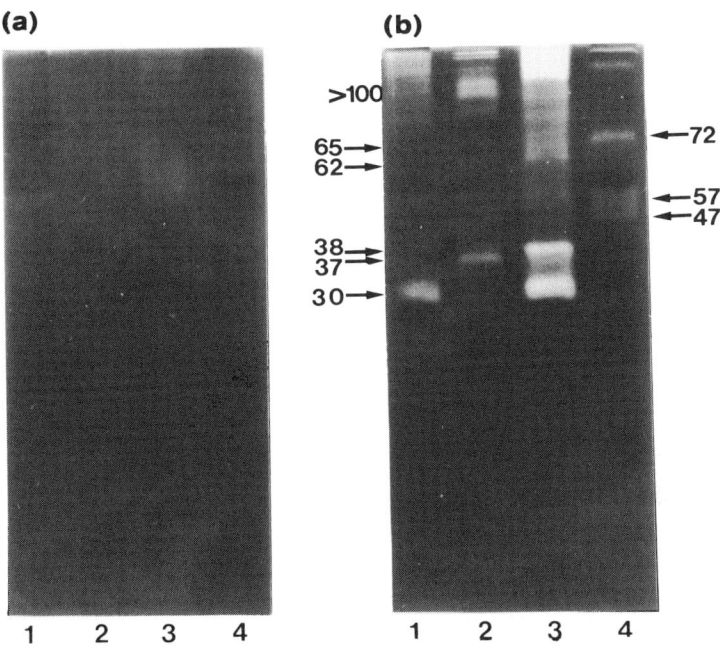

Figure 8.5. Demonstration of proteolytic activity in hookworm homogenates by *in situ* digestion of proteolytic substrates in SDS-PAGE gels and negative staining. Substrate gels were run using non-reduced samples with 75 μg of Triton X-100 extracted adult homogenate per lane. In order to detect general parasite proteases the proteolytic substrate casein ('Carnation' evaporated milk) was incorporated into the resolving gel at the time of pouring at a final concentration of 1 per cent. Gels (10 per cent) were run overnight at 4°C (40 V) then washed in 2.5 per cent Triton X-100 in distilled water for 1 h (three changes of buffer) to remove SDS and thereby renature any enzymes present in the adult homogenates.
Gels were then incubated at 37°C for 24 h in either (a) 0.1 M sodium citrate–citric acid buffer, pH 4.2 or (b) 0.1 M glycine NaOH buffer, pH 8.5, to allow digestion to take place prior to staining with Coomassie brilliant blue. Following staining, gels were destained using ethanol (50 per cent), acetic acid (10 per cent) and water (40 per cent), thereby allowing direct visualization of enzymatic polypeptides as clear areas of proteolysis. Lanes: (1) *A. caninum* adult homogenate; (2) *A. ceylanicum* adult homogenate; (3) *A duodenale* adult homogenate; (4) *N. americanus* adult homogenate.
Molecular weights (in kDa) of the major proteolytic bands are indicated on the figures. Full details of the molecular weights have been published in the review by Pritchard (1990).

(a)   **(b)**

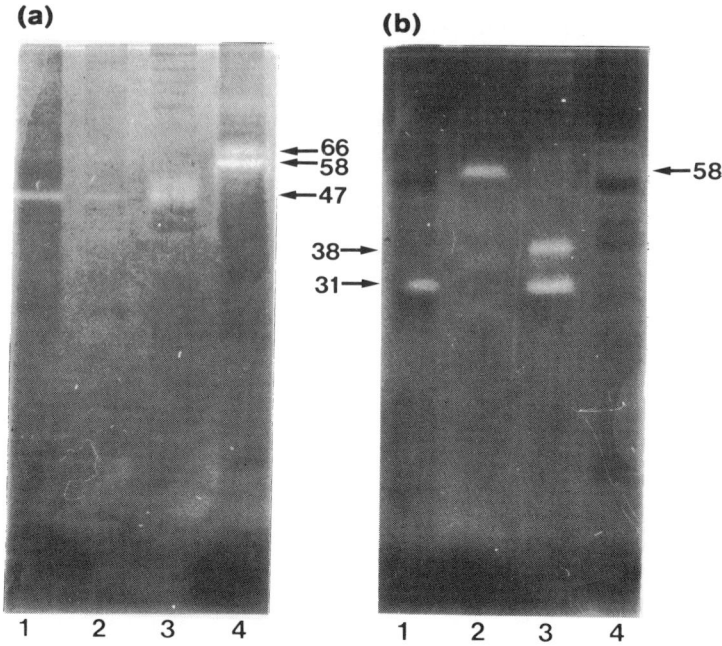

Figure 8.6. Demonstration of proteolytic activity in hookworm homogenates by *in situ* digestion of proteolytic substrates in SDS-PAGE gels and negative staining. Substrate gels were run using non-reduced samples with 75 μg of Triton X-100 extracted adult homogenate per lane. In order to detect haemoglobin digesting enzymes, human haemoglobin was incorporated at a final concentration of 0.1 per cent. Other conditions ((a) and (b)) and lanes (1–4) are as described in the legend to Figure 8.5.

Acetylcholinesterase (AChE)

Interest in hookworm AChE was generated in 1978 by the demonstration of its immunogenicity to the infected host (Ogilvie *et al.*, 1978). Recently, techniques have been developed and utilized to purify secreted AChE from collagen-supplemented cultures of *N. americanus*. In comparison, *A. ceylanicum* failed to secrete AChE under identical culture conditions (Pritchard *et al.*, 1991), an observation also recorded for *A. duodenale* (Wang Feng-lin *et al.*, 1979). The AChE secreted was shown by Sephacryl S.300 gel filtration to have a native molecular weight of 400 000, and a pI of 3.55. Using an affinity matrix of edrophonium chloride immobilized onto CNBr activated sepharose, 81 per cent of the AChE activity of ES products was retained by the matrix, representing 4.3 per cent of the total loaded protein. This AChE rich fraction was eluted with free edrophonium and used to raise specific antisera in rabbits. Immunoglobulin G (IgG) preparations of these sera had the following properties:

1. localized AChE to the oesophageal glands of the adult worm;
2. recognized subunits of 32, 60, 80, 140 and 220 kDa on electroblotting of adult ES;
3. inhibited AChE activity in ES products; and

4. inhibited AChE secretion by adult worms in culture.

It would therefore appear that a relatively specific antibody had been manufactured against a functional parasite ES product. The availability of this probe will now allow definitive experiments to be designed to ascertain the role of AChE in the host–parasite relationship.

RECENT STUDIES ON CUTICULAR ANTIGENS

From ^{125}I surface labelling studies it has become apparent that the human hookworm, *N. americanus*, expresses a limited number of proteins at the cuticle surface. To date, functional roles have not been ascribed to any of these surface proteins, and they appear to be transiently associated with the surface of the cuticle before being shed into the external environment (Pritchard *et al.*, 1985). Although this high degree of surface turnover ensures that a prominent immune response is generated against the nematode surface, such as immune responses do not appear to be effective because hookworm parasites can establish long-lasting, chronic infections.

This has led to the suggestion by some workers that nematode surface proteins possibly act in a parasite-protective manner, masking underlying molecules of structural relevance (Pritchard *et al.*, 1988a, b). In support of this notion, it has been observed that, unless the nematode epicuticle is damaged or functionally impaired in some way (for example, by the stripping action of the cationic detergent CTAB), it is not possible to detect significant labelling of the collagenous, structural components of the nematode cuticle (Pritchard *et al.*, 1988a, b). This inability to insert radioactive iodine into the deeper lying layers of the nematode cuticle suggests that epicuticular proteins can act as a protective barrier which, in conjunction with surface turnover, could prevent serious immune damage to the cuticle. Consequently, the relevance of immune responses directed against nematode surface antigens remains a matter of some debate.

The mechanism(s) by which surface proteins are anchored into the epicuticle is also unclear at present, although recent studies would suggest that a charge interaction of some nature is almost certainly involved. Incubation of adult worms in either conditions of high salt or high pH is found to enhance the release of surface labelled material, findings which are consistent with surface proteins being held in superficial attachment by electrostatic or hydrogen-bond interactions. Moreover, experiments using the non-ionic detergent TX-114 (Figure 8.7a) have demonstrated that surface proteins are predominantly hydrophilic in nature and, consequently, would not appear to be integral membrane proteins anchored into the nematode epicuticle by virtue of hydrophobic domains (McKean, 1989).

Since surface turnover is temperature dependent and anti-metabolite sensitive (Vetter and Klaver-Wesseling, 1978), the expression of surface proteins is possibly an active process. Furthermore, since surface turnover is increased in the presence of antibodies and immune effector cells (Philipp *et al.*, 1980), nematode parasites may have the capacity to detect changes in their environment and respond by

(a)

MW
x10^{-3}

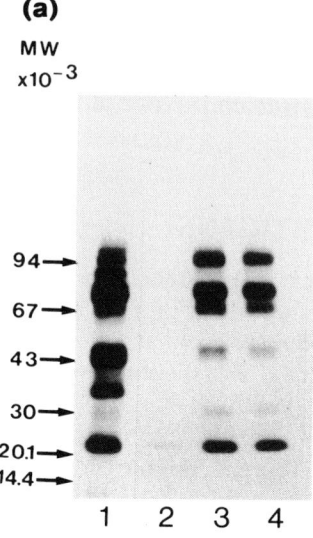

94→
67→
43→
30→
20.1→
14.4→

1   2   3   4

(b)

MW
x10^{-3}

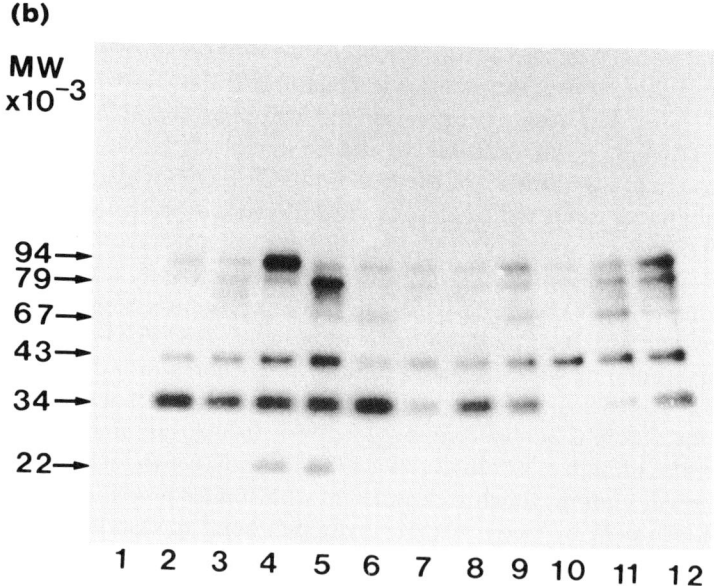

94→
79→
67→
43→
34→
22→

1   2   3   4   5   6   7   8   9   10   11   12

Figure 8.7. (a) Demonstration of the hydrophilic nature of *Necator americanus* surface proteins. Adult worms surface labelled with ^{125}I were extracted using the detergent Triton X-114 (1 per cent solution) and then phase separated into hydrophobic (detergent phase) and hydrophilic proteins (aqueous phase) by increasing the temperature of solution to above the cloud point for Triton X-114 (Bordier, 1981). Pelleted material left after Triton X-114 extraction was then further extracted in SDS-PAGE sample buffer to remove any remaining surface labelled proteins. Lanes: (1) BME extracted pellet from Triton X-114 extracted worms; (2) detergent phase from Triton X-114 worms; (3) aqueous phase from Triton X-114 extracted worms; (4) sucrose plug from Triton X-114 extracted worms. Note equal volumes rather than equal counts per minute were analysed.

(b) Demonstration of the antigenic nature of *Necator americanus* surface proteins. 10 000 TCA precipitable counts per minute of ^{125}I labelled, SDS extracted surface proteins were immunoprecipitated using various human serum samples and then resolved by SDS-PAGE (5–20 per cent gradient gels). Lanes: (1) normal human sera; (2–9) sera from hookworm positive (*Necator americanus*) patients in Kenya (courtesy of Dr P. Craig); (10) sera from a patient harbouring only *Ancylostoma duodenale* hookworms (courtesy of Professor G.A. Schad); (11) sera from a caucasian male following repeated self-infection with *Necator americanus* (courtesy Dr P. A. J. Ball); (12) pool of sera made from serum samples taken from hookworm positive (*Necator americanus*) Australian aborigines.

increasing the rate of synthesis of epicuticular proteins in order to minimize immune mediated damage. A more detailed understanding of the biochemical nature and mechanism of expression of surface proteins may, therefore, enable the formulation of drugs to prevent surface turnover, facilitating immune damage of these superficial antigen-containing layers and underlying structural elements of the cuticle (see Chapter 2). In support of this notion, it has been demonstrated that *Schistosoma mansoni* adult worms damaged by schistomocidal drugs, such as praziquantel, are more vulnerable to immune attack (Brindley *et al.*, 1989). The role of the immune system in this action has been demonstrated conclusively by a reduced efficacy of chemotherapy in T-cell deprived mice (Sabah *et al.*, 1985).

An alternative immunological strategy, however, proposed by Pritchard *et al.* (1988a, b), is that a vaccine containing both surface proteins and deeper lying structural collagens would be able to stimulate an immune response capable of causing significant damage to the adult worm. The rationale of such an approach is that an elevated response against surface proteins may be capable of overcoming surface turnover, allowing antibodies against hitherto hidden collagens to cause structural damage to the cuticle. Such an approach would be particularly effective if the vaccine contained epitopes capable of eliciting immunoglobulin E (IgE) responses to release mast cell proteases, which have been demonstrated to have some specificity for nematode cuticular collagens (McKean and Pritchard, 1989).

Immunoprecipitation analyses have revealed that surface antigens of hookworms do not exhibit the extensive cross-reactivity that is evident between parasites of the genus *Brugia*, indicating that hookworm surface proteins could be important in immunodiagnosis. Surface antigens from the infective stage larvae and adult worms of *Brugia malayi*, *B. timori* and *B. pahangi* contain cross-reacting epitopes, and antisera raised against the surface antigens from one stage of parasite react in immunoprecipitation experiments against surface antigens from other stages and species (Maizels *et al.*, 1983). In contrast, human post *A. duodenale* hookworm infection sera recognizes only one surface antigen (molecular weight 43 kDa) from *N. americanus* adults (Figure 8.7b). This suggests that even though these two hookworm species are closely related taxonomically, and occupy similar sites in the human host, surface-exposed proteins do not contain significant regions of structural similarity.

Further immunoprecipitation studies (McKean, 1989) using monoclonal antibodies raised against adult *N. americanus* ES products, have demonstrated that, while the surface antigens of *N. americanus* adults may be species specific, four of these surface molecules (molecular weights 94, 79, 67 and 32 kDa) can be coprecipitated by the same monoclonal antibody. This implies that regions of homology exist on these *N. americanus* surface molecules. Whether this homology exists at the protein sequence level, or is a consequence of shared carbohydrate determinants, remains to be investigated.

THE MOLECULAR BIOLOGY OF HOOKWORM ANTIGENS

The exquisite host-specificity of human hookworms has a significant bearing on medical research into these host–parasite relationships. Hookworms are passaged in the laboratory with difficulty, and maintenance often involves the use of immuno-logically incompetent neonatal or steroid-treated adult hosts. As a result, the supply of biological material for conventional biochemical analysis is often restricted, as is the availability of relevant post-infection sera for immunological experiments. The former problem has been partly solved by the acquisition of powerful techniques from the field of molecular genetics. This now makes it possible for parasitologists to isolate and characterize parasite antigens accurately from a small quantity of starting material. The latter problem will only be overcome by obtaining sera from the well-designed field studies alluded to earlier (Pritchard *et al.*, 1990c).

Despite these problems and reservations, significant inroads are being made into our understanding of the molecular nature of hookworm antigens.

The widespread use of cloning systems, employing the bacteriophage vectors λgt11, λgt22 and λZAP, or the plasmid vectors pRIT5, the pUC or the pPR series, allowing expression of cloned complementary DNA (cDNA) or genomic sequences in *Escherichia coli*, has opened many avenues of study to the parasitologist. By using sera from an infected individual or a laboratory host, which have a known reactivity to parasite protein antigens, one can rapidly isolate relevant cloned sequences and fusion proteins resulting from their expression.

With *Necator* a recurring problem has been the availability of sufficient mRNA for cloning, since the numerous steps involved in mRNA purification tend to pre-dispose the preparation to nuclease attack. This can be avoided using total RNA as the template for cDNA synthesis, since the use of specific oligo-(dT) based primers, and a directional expression vector such as λgt22 reduces the frequency of ribosomal based clones appearing in the library to less than 0.1 per cent (Lu and Werner, 1988), an acceptable trade-off when precious RNA is at stake. The recent con-struction of a cDNA library from the L3 of *Necator* by the above means yielded over 99.97 per cent recombinants, thereby proving the suitability of this procedure (E. Walsh, personal communication).

In addition, the recently available polymerase chain reaction (PCR), with its ability to amplify DNA tracts bordered by specific sequences up to 2^{30}-fold, should also help the molecular parasitologist to isolate cDNAs from previously limiting quantities of sample (Belyavsky *et al.*, 1989). PCR used with well-devised cloning strategies also allows the production of subtractive cDNA libraries (Palazzolo and

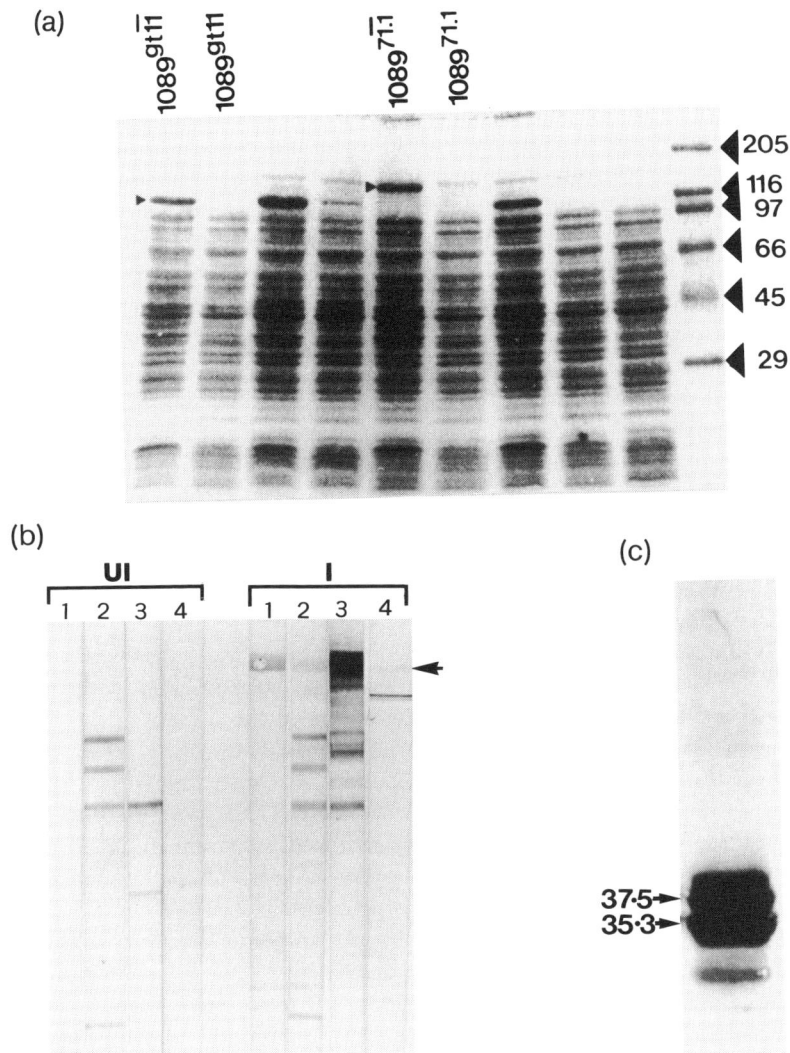

Figure 8.8. Demonstration of the cloning of hookworm antigens. (a) Coomassie blue stained 5–20 per cent SDS-polyacrylamide gel of lysed *E. coli* (Y 1089) lysogens carrying various λgt11 prophage: Y1089^{gt11}, lysogen containing the parent vector λgt11; Y1089$^{71.1}$, lysogen containing the recombinant bacteriophage λgt11.71.1; I, lysogen after induction of lytic phage growth and β-galactosidase expression with 5 mM β-D-thiolgalactopyranoside (IPTG). Arrows indicate the position of the induced protein. Molecular weights are 116 kDa for the native β-galactosidase and 140 kDa for the fusion protein from clone 71.1. (b) Western blots of Y1089$^{71.1}$ lysogen before (UI) and after (I) induction of β-galactosidase expression. The 141 kDa fusion protein is indicated by the arrow. The primary antisera used were: (1) rabbit anti-*Necator* homogenate; (2) human post-infection serum (Kenya) No. 666; (3) rabbit anti-β-galactosidase; and (4) human post-infection serum (Kenya) No. 273. The sera were not absorbed with *E. coli* lysates prior to use hence the *E. coli* specific bands visible in both UI and I blots. (c) Western blot of *E. coli* lysate (NM522) carrying a plasmid expression vector pUC9.2, with an in-frame insertion of the cDNA insert from clone 71.1. The blot was probed with *E. coli* absorbed post-infection human serum (Kenya). The major fusion protein bands are indicated.

Meyrowitz, 1987) representing only the differences (or similarities) in expressed genes between different stages of the life cycle.

Using a λgt11 based cDNA library, containing cDNAs of adult *N. americanus* (male and female worm) mRNAs, a series of clones has been isolated, expressing polypeptides which are strongly recognized by human post-*Necator* infection sera. The first of these to be studied, clone KA71.1, yielded a 613 bp sequence which was expressed as a soluble 140 kDa β-galactosidase fusion protein following induction (see Figure 8.8). The absorption of human post-infection sera onto the fusion protein resulted in the affinity purification of antibodies which recognized the 33 kDa region shown in Figure 3a. This cDNA has since been subcloned into pUC9.2 for enhanced expression of the fusion protein, where a doublet of 35.3 and 37.5 kDa proteins, and expression levels of up to 10 per cent of the soluble cell protein, were seen following Western blotting and Coomassie blue staining.

Sequence analysis of the cDNA revealed an open reading frame of open structure with a hydrophilic nature, which is consistent with the high solubility of the expressed fusion protein. However, comparison of the cDNA sequence and the predicted protein sequence with numerous sequence data bases has as yet now failed to show any significant homology with published sequences. Further cDNAs identified by screening the library with radiolabelled KA71.1 cDNA are being analysed to obtain the complete coding sequence.

The coding sequences for a number of the antigenic molecules which make up the 28–33 kDa region on Western blots have also been isolated, since sequence comparison between individual clones has indicated the presence of at least four unique sequences. Fusion proteins from these clones, and clone KA71.1, are being purified for testing in ELISA assays as potential surrogate antigens for diagnostic purposes. However, preliminary results with the KA71.1 protein (from pUC9.2) indicate that a high level of purity may be required before meaningful results are obtained. In the meantime, epitope mapping analysis of the sequence will be conducted in an attempt to overcome this problem.

The same library has also provided clones with antigenic similarity to the 17 kDa antigen and cuticular collagen previously described.

The isolation of cloned sequences corresponding to secreted acetylcholinesterase and proteolytic feeding enzymes of *Necator*, mentioned above, requires the production of specific antisera or monoclonals against peptide epitopes, an undertaking which is progressing. However, it has recently been shown that numerous proteinase sequences can be isolated using degenerate oligonucleotides as primers in PCR, which results in the amplification of partial proteinase sequences (Eakin *et al.*, 1990). Whilst this approach could be seen as useful for the serine proteinases, which have highly conserved regions, it is only recently that cysteinyl proteinases, with less conserved active-site sequences, have been isolated (Sakanari *et al.*, 1989). If aspartyl proteinase sequences (see section on hookworm proteinases) are amenable to the same approach, then the study of parasite proteinases will have been taken a major step forward, considering the tiny quantities of native proteinases available for study.

RECENT PROGRESS IN THE IMMUNOEPIDEMIOLOGY OF HOOKWORM INFECTION

In close collaboration with colleagues at Oxford University and the Christensen Research Institute in Madang, a field study site was recently established in Madang Province, Papua New Guinea, to investigate the immunoepidemiology of hookworm infection. The first visit to the study site was made in July–August 1988. Faecal egg counts were collected from 202 people, all of whom were treated with pyrantel pamoate (Combantrin–Pfizer plc), and full (48 h) worm counts were obtained from 123 people. The epidemiology of hookworm in the village was typical of that reported for hookworm infections elsewhere (Anderson, 1982). The prevalence was 90 per cent and the mean (± standard error) intensity of infection was 25.3 (± 4.0) worms/host. Age vs. prevalence and age vs. intensity profiles of infection were both monotonic, with the greatest worm burden in the oldest age class. Hookworm burdens had a highly overdispersed frequency distribution, with an aggregation parameter, k, of 0.370 (Pritchard *et al.*, 1990c).

A second field visit was made in August–September 1989. Of the 202 people who participated in 1988, faecal and blood samples were obtained from 170 people in 1989. As expected, there had been significant reinfection with hookworm, although both the prevalence and intensity of hookworm infection were lower than in 1988. The prevalence of hookworm in 1989 was 63 per cent and the mean intensity 138 eggs/g, 45 per cent of the 1988 level. Faecal egg counts in 1989 were largely independent of age (Figure 8.9), indicating a fairly uniform reinfection with

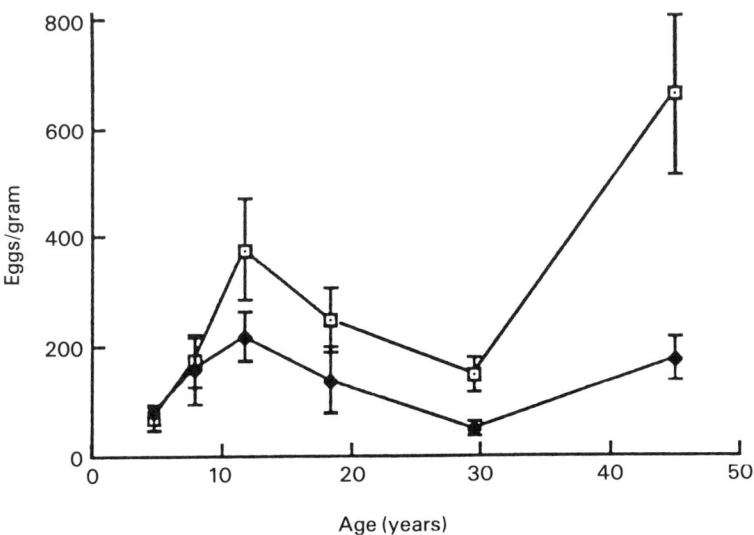

Figure 8.9. The age vs. intensity profile of hookworm infection in August 1989, after 12 months reinfection (lower line), compared with the pretreatment profile in July 1988 (upper line). Intensity is shown as the mean ± standard error eggs/g count in different age classes.

age; the percentage reinfection fell from over 100 per cent in the younger age classes to 25 per cent in adults.

There was significant predisposition to hookworm infection, with a positive correlation between egg counts in 1988 and 1989. This correlation was strong in males, but weaker and not significant in females, and did not vary consistently with host age (R. Quinnell, unpublished).

Initial immunological analysis of the study population concentrated on the combined IgG, IgA and IgM response to two antigen preparations: adult ES antigens and collagen antigens. Levels of anti-ES antibodies reflected worm burden, while those of anti-collagen antibodies were correlated with host age (Pritchard *et al.*, 1990c).

Further work has investigated the isotype-specific antibody response to both adult and larval hookworm antigens. Data have been collected on the IgG, IgA, IgM, IgD and IgE response in 1988 to adult ES antigens (after absorption with *Ascaris* antigen), to homogenized larvae, and to five larval antigen preparations: the L3 sheath-surface proteins, exsheathing fluid, L3 ES, ensheathed somatic and exsheathed somatic antigens. Total IgG, IgA, IgE and IgM levels are also known, and sera collected following reinfection (1989) have also been analysed.

These data will be compared with the epidemiological data to look for correlations with initial or reinfection intensity (worm burden or eggs/g) and host age. As an example, the IgG response to adult ES and larval homogenate, plotted vs. host age, is shown in Figure 8.10. IgG responses were strongly correlated between 1988 and 1989, as were responses against adult and larval antigens. IgG responses were positively correlated with host age, and were generally not correlated with current hookworm intensity.

With regard to the dynamics of the immune response, it can be seen from Figure 8.10 that specific IgG levels against adult ES products dropped significantly following anthelmintic treatment, whereas IgG levels against larval antigens were unchanged. Analyses of other isotype-specific responses (L. Walsh, unpublished) have revealed that specific IgM levels against adult ES products followed the pattern for IgG levels, specific IgA and IgE levels were unchanged and specific IgD levels actually increased significantly following worm removal. All specific responses against larval antigens were depressed significantly following treatment, with the exception of IgG and IgD.

A tentative explanation of these immunological phenomena is possible, but is obviously subject to the results of further investigation. For example, it could be argued that IgG and IgM responses against adult ES are largely systemic responses to antigenic stimulation in the gastrointestinal tract. Consequently, a reduction in worm burden by anthelmintic treatment reduces these responses, whilst sufficient antigen is still presented locally within the gastrointestinal tract (despite a reduction in worm burden) to maintain specific circulating levels of IgA and IgE.

With regard to the antigen- and isotype-specific responses against larval antigens, the maintenance of the immune response in the IgG isotype could reflect the possibility that a general decrease in exposure has occurred (all villagers were informed of the mode of transmission of hookworm infection), and that immunological memory for the IgG isotype is superior to that for the other isotypes.

(a)

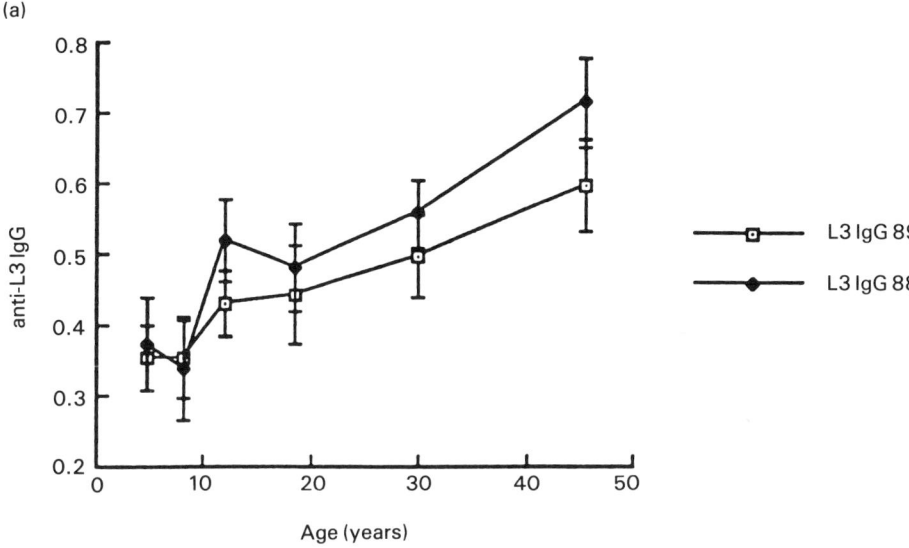

(b)

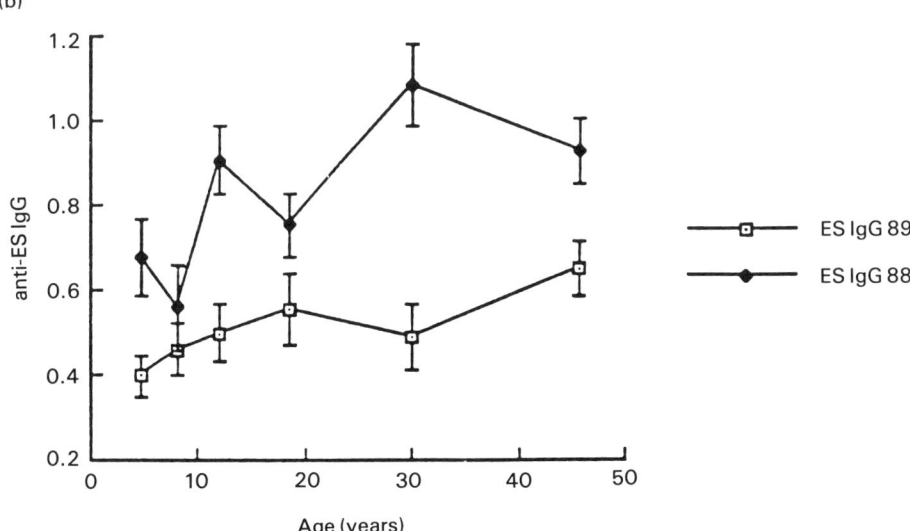

Figure 8.10. The relationship between IgG antibodies against hookworm antigens in 1988 (pretreatment) and 1989 (after 12 months reinfection) and host age. Antigen preparations were (a) larval homogenate and (b) adult ES material. Antibody values expressed as optical density (OD) at 420 nm (mean ± standard error). (a) ☐, L3 IgG 89; ◆ , L3 IgG 88. (b) ☐, ES IgG 89; ◆ , ES IgG 88.

The increase in the IgD response against adult ES could reflect a decrease in chronic antigen stimulation, as a result of anthelmintic treatment, with an increase in the number of circulating undifferentiated B cells.

The relationship between hookworm intensity and the iron status of the host was also investigated in 1988. There was no correlation between hookworm intensity and either haemoglobin concentration or packed red cell volume. There was, however, a negative correlation between hookworm intensity and serum ferritin. Thus, hookworm infection at levels below those previously reported to be pathogenic appeared to be associated with early stages of iron deficiency (Pritchard *et al.*, 1991).

Finally, hookworms recovered after anthelminthic treatment in 1988 were preserved in formalin and analysed at Oxford, where their wet and dry weight and uterine egg count was assessed. Worm weight increased with host age; there was no negative correlation with hookworm burden, suggesting that density-dependent effects on worm growth and fecundity may not be important (R. Quinnell, unpublished).

There was, however, evidence that worm weight or fecundity are affected by components of the host immune response. The significant negative correlations between female worm weight or egg count and anti-ES IgG, IgA and IgM antibodies, and between male worm weight and eosinophil count observed in our study represent the first demonstration of the possibly protective effect of an immune response against hookworms in humans.

CONCLUDING REMARKS

It was stated in a recent review on hookworm control (Warren, 1988) that

> the techniques of molecular biology and genetic engineering ... are now allowing macromolecular dissection of the parasite surfaces, structures and enzymes, and the control mechanisms that govern the different stages of the life cycle. These developments offer renewed hope of more specific and even quantitative diagnostic tests and of drugs and vaccines that will provide long-term control of infection.

Other reviewers have called for species-specific diagnostic tests, and for immunological markers of infection intensity (Keymer and Bundy, 1989; Crompton, 1989).

In the present review, therefore, we have attempted to put some of these statements and requests into context by describing recent (and projected) developments in hookworm biology.

With regard to vaccination, it is felt by the present reviewers that biologically active proteins, secreted intradermally by the exsheathing L3 larvae of *N. americanus*, offer the best candidate antigens for the induction of a protective immune response. However, the sheer paucity of antigenic material available for study could be problematic. Molecular cloning techniques offer a solution, and experiments are being conducted to ascertain the optimal conditions for obtaining nucleic acids from the infective larvae. It will also be necessary to compare the antigenic profiles of the infective larvae of *N. americanus* and *A. duodenale*, particularly as the latter adopts a different infection strategy and essentially stimulates a different compartment of the immune system. However, given time, effort and support, larval products will eventually be tested for their efficacy in partial animal model systems. In the meantime, strenuous efforts are being made to assess the immunological reactivity of human populations against larval and adult-derived hookworm antigens separately, in an attempt to provide some clues as to the relevance of particular stage- and species-specific antigens.

The testing of vaccines against establishing or established adult stages presents its own unique problems, because suitable animal models do not, and are unlikely to, exist. However, the passive transfer of antisera raised against candidate vaccines, and their effect on worm survival/fecundity, offers an alternative strategy for assessing the efficacy of selected molecules.

Immunological monitoring of the intensity of exposure to infection, and of the intensity of established infection should be possible, provided 'epidemiologically appropriate' species- and stage-specific larval and adult antigens can be discovered. Candidate antigens have been described above, and their use is under active investigation at present. However, the nature of the laboratory work involved, and the involvement of human populations, probably means that progress will be comparatively slow in this area of research.

In conclusion, although our knowledge of hookworm immunology and biochemistry is steadily increasing, and stage- and species-specific antigens are being discovered, it is felt that the recognition of these molecules by the human immune system will have to be put under strict epidemiological scrutiny before any breakthrough can be made in the areas of vaccination and differential diagnosis.

ACKNOWLEDGEMENTS

The authors would like to thank the Wellcome Trust and NATO for their support of this research area, and the SERC for providing a studentship for P. G. McKean. Table 2 was compiled from a number of sources, notably Banwell and Schad (1978) and Hoagland and Schad (1978). Thanks are also extended to Linda Timothy for providing the excellent photomicrographs for Figures 2a and 2b, to Anthony Pitt for photographic assistance and to Mrs Barbara Hill for typing this manuscript.

REFERENCES

Anderson, R. M., 1982, The population dynamics and control of hookworm and roundworm infections, in Anderson, R. M. (Ed.) *Population Dynamics of Infectious Disease*, pp. 67–106.

Banwell, J. G. and Schad, G. A., 1978, 'Hookworm' in *Clinics in Gastroenterology*, 7, 129–156.

Belyavsky, A., Vinogradova, T. and Rakewsky, K., 1989, PCR based cDNA library construction: general cDNA libraries at the level of a few cells, *Nucleic Acids Research*, 17, 2919–32.

Bond, J. S. and Butler, P. E., 1987, Intracellular proteases, *Annual Review of Biochemistry*, 56, 333–64.

Bonner, T. P., 1979a, Changes in the structure of *Nippostrongylus brasiliensis* intestinal cells during development from the free-living to the parasitic stages, *Journal of Parasitology*, 65, 745–50.

Bonner, T. P., 1979b, Initiation of development 'in vitro' of third stage *Nippostrongylus brasiliensis*, *Journal of Parasitology*, 65, 74–8.

Bordier, C., 1981, Phase separation of integral membrane proteins in Triton X-114 solution, *Journal of Biological Chemistry*, 256, 1604–7.

Bos, J. D. and Kapsenberg, M. L., 1986, The skin immune system: its cellular constituents and their interactions, *Immunology Today*, 7, 235–40.

Brindley, P. J., Shand, M., Norde, A. P. and Sher, A., 1989, Role of host antibody in the chemotherapeutic action of praziquantel against S. *mansoni*, *Molecular and Biochemical Parasitology*, 34, 99–108.

Brydon, L. J., Gooday, G. W., Chappell, L. H. and King, T. P., 1987, Chitin in eggshells of *Onchocerca gibsoni* and *Onchocerca volvulvus*, *Molecular Biochemistry and Parasitology*, 25, 267–72.

Burt, J. S. and Ogilvie, B. M., 1975, *In vitro* maintenance of nematode parasites assayed by acetylcholinesterase and allergen secretion, *Experimental Parasitology*, 38, 75–82.

Chandler, A. C., 1932, Susceptibility and resistance to helminthic infections, *Journal of Parasitology*, 18, 135–52.

Chandler, A. C., 1953, Immunology in parasite diseases, *Journal of the Egyptian Medical Association*, 36, 811–34.

Clark, C. H., Kling, J. M., Woodley, C. H. and Sharp, N., 1961, A quantitative measurement of the blood loss caused by Ancylostomiasis in dogs, *American Journal of Veterinary Medicine*, 22, 370–3.

Croll, N. A. and Ma, K., 1978, The location of parasites within their hosts: the passage of *Nippostrongylus brasiliensis* through the lungs of the laboratory rat, *International Journal of Parasitology*, 8, 289–95.

Crompton, D. W. T., 1989, Hookworm disease: current status and new directions, *Parasitology Today*, 5, 1–2.

Eakin, A. E., Bouvier, J., Sakanari, J. A., Craik, C. S. and McKerrow, J. H., 1990, Amplification and sequencing of genomic DNA fragments encoding cysteine proteases from protozoan parasites, *Molecular and Biochemical Parasitology*, 39, 1–8.

Fernando, M. A. and Wong, H. A., 1964, Metabolism of hookworms. II. Glucose metabolism and glycogen synthesis in adult female *Ancylostoma caninum*, *Experimental Parasitology*, 15, 284–292.

Foy, H. and Nelson, G. S., 1963, Helminths in the etiology of anaemia in the tropics with special reference to hookworms and schistosomes, *Experimental Parasitology*, 14, 240–62.

Gamble, H. R., Purcell, J. P. and Fetterer, R. H., 1989, Purification of a 44 kDa protease which mediates the ecdysis of infective *Haemonchus contortus* larvae, *Molecular Biochemistry and Parasitology*, 33, 49–58.

Hoagland, K. E. and Schad, G. A., 1978, *Necator americanus* and *Ancylostoma duodenale*: life history parameters and epidemiological implications of two sympatric hookworms of humans, *Experimental Parasitology*, 44, 36–49.

Hotez, P. J. and Cerami, A., 1983, Secretion of a proteolytic anticoagulant by *Ancylostoma* hookworms, *Journal of Experimental Medicine*, 157, 1594–603.

Hotez, P. J., Le Trang, N., McKerrow, J. H. and Cerami, A., 1985, Isolation and characterisation of a proteolytic enzyme from the adult hookworm *Ancylostoma caninum*, *Journal of Biological Chemistry*, 260, 7343–8.

Hotez, P.J., Newport, G., Le Trang, N., Agabian, N. and Cerami. A., 1986, Isolation, cloning and expression of a protease from *Ancylostoma* hookworms, *Journal of Cellular Biochemistry*, **10A**, 132 (Abs. C48).

Irvine, N.I., Alder, M., Kurtz, J.B. and Juel-Jensen, B., 1986, Intradermal vaccination against hepatitis B, *Lancet*, **6 Dec**, 1340.

Irvine, W.L., Parsons, A.J., Kurtz, J.B. and Juel-Jensen, B., 1987, Intradermal hepatitis B vaccine, *Lancet*, **5 Sep**, 561.

James, S.L., 1987, *Schistosoma* spp: progress towards a defined vaccine, *Experimental Parasitology*, **63**, 247–52.

Kalkofen, U.P., 1970, Attachment and feeding behaviour of *Ancylostoma caninum*, *American Journal of Tropical Medicine and Hygiene*, **23**, 1046–53.

Kalkofen, U.P., 1974, Intestinal trauma resulting from feeding activities of *Ancylostoma caninum*, *American Journal of Tropical Medicine and Hygiene*, **23**, 1046–53.

Kendrick, J.F., 1934, The length of life and the rate of loss of the hookworms *Ancylostoma duodenale* and *Necator americanus*, *American Journal of Tropical Medicine and Hygiene*, **14**, 363–79.

Keymer, A.E. and Bundy, D., 1989, Seventy five years of solicitude, *Nature*, **337**, 114.

Komiya, Y., Yasara, Oka, K. and Sato, A., 1956, Survival of *Ancylostoma caninum in vitro*, *Japanese Journal of Medical Science and Biology*, **9**, 283–92.

Loeb, L. and Smith, A.J., 1904, The presence of a substance inhibiting the coagulation of the blood in Anchilostomiasis. *Proceedings of the Pathological Society of Philadelphia*, New Series, **7**, 173–8.

Loeb, L. and Fleiser, M.S., 1910, The influence of extracts of *Anchylostoma caninum* on the coagulation of the blood and on haemolysis. *Journal of Infectious Diseases*, **7**, 625–31.

Lu, X. and Werner, D., 1988, Construction and quality of cDNA libraries prepared from cytoplasmic RNA not enriched in poly (A) + RNA, *Gene*, **71**, 157–64.

Maizels, R.M., Partono, F., Oemijati, D., Denham, D.A. and Ogilvie, B.M., 1983, Cross reactive surface antigens on three stages of *Brugia malayi*, *B. pahangi* and *B. timori*, *Parasitology*, **87**, 249–62.

Matthews, B.E., 1977, The passage of larval helminths through tissue barriers, *Symposium of the British Society for Parasitology*, **15**, 93–119.

Matthews, B.E., 1982, Skin penetration by *Necator americanus* larvae, *Zeitschrift für Parasitenkunde*, **68**, 81–6.

McKean, P.G., 1989, unpublished Ph D thesis, University of Nottingham.

McKean, P.G. and Pritchard, D.I., 1989, The action of a mast cell protease on the cuticular collagen of *Necator americanus*, *Parasite Immunology*, **11**, 293–7.

Miller, T.A., 1971, Vaccination against the canine hookworm diseases, *Advances in Parasitology*, **9**, 153–83.

Nawalinski, T.A. and Schad, E.A., 1974, Arrested development in *Ancylostoma duodenale* cause of self-induced infection in man, *American Journal of Tropical Medicine and Hygiene*, **23**, 895–8.

Ogilvie, B.M., Bartlett, A., Godfrey, R.C., Turton, J.A., Worms, M.J. and Yeats, R.A., 1978, Antibody responses in self-infections with *Necator americanus*, *Transactions of the Royal Society of Tropical Medicine and Hygiene*, **72**, 66–71.

Oya, Y. and Noguchi, I., 1977, Some properties of hemaglobin protease from *Ancylostoma caninum*, *Japanese Journal of Parasitology*, **26**, 307–13.

Palazzolo, M.J. and Meyerowitz, E.M., 1987, A family of lambda phage cDNA cloning vectors, SWAI allowing the amplification of RNA sequences, *Gene*, **52**, 197–206.

Peters, W. and Latka, I., 1986, Electron microscopic localisation of chitin using colloidal gold labelled with wheat-germ agglutinin, *Histochemistry*, **84**, 155–60.

Petronijevic, T., Rogers, W.P. and Sommerville, R.I., 1986, Organic and inorganic acids as the stimulus for exsheathment of infective juveniles of nematodes, *Journal of Parasitology*, **16**, 163–8.

Philipp, M., Parkhouse, R. M. E. and Ogilvie, B. M., 1980, Changing proteins on the surface of a parasitic nematode, *Nature (London)*, **287**, 538–40.

Pritchard, D. I., 1987, The molecular biology of gastrointestinal nematodes, *Baillière's Clinical Tropical Medicine and Communicable Diseases*, **2**, 511–34.

Pritchard, D. I., 1990, in Schad, G. A. and Warren, K. S. (Eds) *Hookworm Disease: Current Status and New Directions*, Basingstoke: Taylor and Francis.

Pritchard, D. I., Crawford, C. R., Duce, I. R. and Nehnke, J. M., 1985, Antigen stripping from the nematode epicuticle using the cationic detergent cetyltrimethyl ammonium bromide (CTAB), *Parasite Immunology*, **7**, 575–85.

Pritchard, D. I., McKean, P. G. and Rogan, M. T., 1988a, Cuticle preparations from *Necator americanus* and their immunogenicity in the infected host, *Molecular Biochemistry and Parasitology*, **28**, 275–84.

Pritchard, D. I., McKean, P. G. and Rogan, M. T., 1988b, Cuticular collagens — a concealed target for immune attack in hookworms, *Parasitology Today*, **4**, 239–41.

Pritchard, D. I., McKean, P. G., Rogan, M. T. and Schad, G. A., 1990a, The identification of a species-specific antigen from *Necator americanus*, *Parasite Immunology*, **12**, 259–67.

Pritchard, D. I., McKean, P. G. and Schad, G. A., 1990b, An immunological and biochemical comparison of hookworm species, *Parasitology Today*, **6**, 154–6.

Pritchard, D. I., Quinnell, R. J., Slater, A. F. G., McKean, P. G., Dale, D. D. S., Raiko, A. and Keymer, A. E., 1990c, Epidemiology and immunology of *Necator americanus* infection in a community in Papua New Guinea: humoral responses to excretory–secretory and cuticular collagen antigens, *Parasitology*, **100**, 317–26.

Pritchard, D. I., Quinnell, R. J., Moustafa, M., McKean, P. G., Slater, A. F. G., Raiko, A., Dale, D. D. S. and Keymer, A. E., 1991, Hookworm *Necator americanus* infection and storage iron depletion, *Transactions of the Royal Society of Tropical Medicine and Hygiene*, in press.

Roche, M. and Martinez-Torres, C., 1960, A method for *in vitro* study of hookworm activity, *Experimental Parasitology*, **9**, 250–6.

Sabah, A. A., Fletcher, C., Webbe, G. and Doenhoff, M. J., 1985, *Schistosoma mansoni*: reduced efficacy of chemotherapy in infected T-cell deprived mice, *Experimental Parasitology*, **60**, 348–54.

Sakanari, J. A., Staunton, C. E., Eakin, A. E., Craik, C. S. and McKerrow, J. H., 1989, Serine proteases from nematode and protozoan parasites: isolation of sequence homologs using generic molecular probes, *Proceedings of the National Academy of Sciences (USA)*, **86**, 4863–7.

Schad, G. A., 1979, *Ancylostoma duodenale*: maintenance through six generations in helminth naive pups, *Experimental Parasitology*, **47**, 246–53.

Schad, G. A. and Banwell, J. G., 1984, Hookworms, in Warren, K. S. and Mahmoud, A. A. F. (Eds) *Tropical and Geographical Medicine*, Chap. 42, New York: McGraw Hill.

Schad, G. A., Chowdhury, A. B., Dean, C. G., Cochar, V. K., Nawalinski, T. A., Thomas, J. and Tonascia, J. A., 1973, Arrested development in human hookworm infections: an adaptation to a seasonally unfavourable external environment, *Science*, **180**, 502–4.

Schwartz, B., 1920, Hemolysins from parasitic worms, *Archives of the Medical Institute of Chicago*, **26**, 431–5.

Sen, H. G. and Seth, D., 1967, Complete development of the human hookworm. *Necator americanus* in golden hamsters, '*Mesocricetus auratus*', *Nature*, **214**, 609–10.

Smith, J. M., 1976, Comparative ultrastructure of the oesophageal glands of third stage larval hookworms, *International Journal of Parasitology*, **6**, 9–13.

Thorson, R. E., 1956a, The effect of extracts of amphidial glands, excretory glands and esophagus of adults of *A. caninum* on coagulation of dog's blood, *Journal of Parasitology*, **42**, 26–30.

Thorson, R. E., 1956b, Proteolytic activity in extracts of the esophagus of adults of *Ancylostoma caninum*, and the effect of immune serum on this activity, *Journal of Parasitology*, **42**, 21–5.

Thorson, R. E., 1956c, The stimulation of acquired immunity in dogs by injections of extracts of the esophagus of adult hookworms, *Journal of Parasitology*, **42**, 501–4.

Vetter, J. C. M. and Klaver-Wesseling, J. C. M., 1978, IgG antibody binding to the outer surface of the infective larvae of *Ancylostoma caninum*, *Zeitschrift für Parasitenkunde*, **58**, 91–6.

Walterspiel, J. N., Buchanan, G. R., Schad, G. A. and Carpentieri, U., 1985, Erythropoietin-induced congenital erythrocytosis: treatment with myelosuppressive agents and hookworm infestation, *Journal of Paediatrics*, **107**, 575–7.

Wang Feng-Lin, Ning Kai-bi, Wang Xiu-zheng, Yang Guang-min and Wang Juying, 1979, Average values of *Ancylostoma duodenale* and *Necator americanus* cholinesterase activity in humans, *Chinese Medical Journal*, **96**, 60–2.

Warren, K. S., 1988, Hookworm control, *Lancet*, **15 Oct**, 897–8.

Wells, C., 1988, unpublished PhD thesis, University of Nottingham.

Wells, H. S., 1931, Observations on the blood sucking activities of the hookworm *Ancylostoma caninum*, *Journal of Parasitology*, **4**, 167–82.

Whipple, G. H., 1909, The presence of a weak haemolysin in the hookworm and its relationship to the anemia of uncinariasis, *Journal of Experimental Medicine*, **11**, 331–43.

Yasuraoka, K., Hosaka, Y. and Ogawa, K., 1960, Survival of *Ancylostoma duodenale in vitro*, *Japanese Journal of Medical Science and Biology*, **13**, 207–12.

Yoshida, Y., Nakanishi, Y. and Mitani, N., 1958, Experimental studies on the infection modes of *Ancylostoma duodenale* and *Necator americanus* to the definitive host, *Japanese Journal of Parasitology*, **7**, 704–14.

9. Parasite enzymes in the diagnosis and control of ruminant nematodiasis

D. P. Knox and D. G. Jones

INTRODUCTION

The potential importance of enzyme systems for the establishment and maintenance of parasites within their hosts has been recognized for several decades and can broadly be classified into three functional areas:

1. host invasion;
2. parasite feeding; and
3. provision of a stable environment including immunological avoidance.

The hypothesis that host antibodies could act as specific anti-enzymes which prevent the worms from digesting and assimilating host protein was first proposed over half a century ago by Chandler (1932, 1936). Later, Thorson (1953) suggested that antibodies against enzymes in excretions/secretions of *Nippostrongylus muris* inhibited larval migration in the rat by interfering with their nutrition and development. The author hypothesized that invading larvae would be more susceptible to encapsulation in host tissues and that the ability of worms to establish normal infections in the intestine would be impaired. Further support for the enzyme–antienzyme hypothesis was provided by demonstrating that proteolytic activity in oesophageal extracts from adult *Ancylostoma caninum* was inhibited by serum from dogs immune to reinfection but not by non-immune sera (Thorson, 1956). He concluded that enzyme–antienzyme relationships would be highly complex in helminth infections and that maximal protective immunity may depend on a mosaic of reactions.

In recent decades, attention has been increasingly focused on parasite enzyme systems. However, it is beyond the scope of this contribution to review all areas of interest and in this chapter we concentrate on the enzymes of selected nematode species, those of ruminant gastrointestinal nematodiasis.

RUMINANT NEMATODIASIS — THE PROBLEMS

The nematode helminths constitute the most widespread and economically important group of gastrointestinal parasites in domestic ruminants. Infections are common and affect several regions of the alimentary tract including the abomasum, small intestine and large intestine. Pathogenicity has been attributed to numerous species whose relative importance varies both geographically and in accordance with the prevailing climate. Production loss is primarily due to impaired weight gain, wool growth or milk production and poor reproductive performance, although mortality can occur in severe cases (Holmes, 1985). Control currently depends on anthelmintic therapy and/or clean or rotational grazing. Whilst these strategies have been applied successfully in countries with efficient farming industries, they are too costly and impracticable elsewhere (Miller, 1984). Moreover, anthelmintic resistance has emerged as the most important problem in the successful control of nematode parasites in grazing animals (Waller, 1990). Anthelmintic dosing strategies must be continually monitored and modified to minimize the selection pressure that creates anthelmintic-resistant nematode populations. Further definition of nematode biochemistry offers possibilities of identifying novel enzymes or functional pathways through which specific inhibition of the parasite might be achieved (Wang, 1984). Finally, there is a need to define the molecular nature of host–parasite relationships with the long-term goal of developing vaccination strategies (Soulsby, 1985).

ENZYMES IN THE DIAGNOSIS AND CONTROL OF NEMATODIASIS

Considerable research effort is now being directed towards developing novel strategies for serodiagnosis and control of ruminant gastrointestinal nematodiasis based on enzymes of putative excretory–secretory origin.

A difficulty in applying the detection of parasite antigens to serodiagnosis of infection is that they provoke antibody responses and are, therefore, rapidly cleared from the circulation of the host (Parkhouse, 1987). However, functional activity of parasite-derived enzymes may remain unimpaired in antigen–antibody complexes in the host circulation and could provide the basis for the quantitative serodiagnosis of parasite burdens. A well characterized enzyme to which this would apply is acetylcholinesterase (AChE) (reviewed by Rhoads, 1984).

A more immediate application of parasite biochemistry lies in monitoring the viability of the parasite. Modifications in absolute enzyme activities, isoenzyme distribution or enzyme release can define the response of parasites to chemotherapy (Rapson *et al.*, 1985) or immunological attack (Edwards *et al.*, 1971; Jones and Ogilvie, 1972).

IS DIFFERENTIAL SERODIAGNOSIS IMPORTANT?

Among the general epidemiological principles of ruminant helminthiasis are basic assumptions that every animal is exposed to infection and that parasitic contamination of the environment is continuous. However, the nature and extent of infection shows considerable seasonal variation which depends both on the life cycle of individual species and prevailing climatic conditions. The availability of differential serodiagnostic tests, capable of defining the type, site and extent of an infection, would have several applications in the control of gastrointestinal nematodiasis. Firstly, the economic impact of subclinical infestations could be accurately assessed and the extent of infection correlated to production loss. Secondly, early detection of anthelmintic resistance would be facilitated and alternative optimized strategies could be introduced. This flexibility would be an important benefit for the conservation of existing anthelmintics.

SERODIAGNOSIS — THE PRESENT

At present, the only widely applied serodiagnostic marker of ruminant gastro-intestinal parasitism is the elevation of blood pepsinogen, a specific gastric pro-enzyme, indicative of abomasal parasitism (Thomas and Waller, 1975; Coop *et al.*, 1977). An enzyme-linked immunoabsorbent essay (ELISA) has been reported for ovine pepsinogen (Turner and Shanks, 1982), although problems have been encountered in developing a similar assay for the bovine proenzyme (Eckersall, 1988).

In the search for organ-specific correlates of intestinal damage, Blackmore and Palmer (1977), using L-phenylalanine sensitivity to distinguish intestinal alkaline phosphatase (AP) (Fishman *et al.*, 1963), reported elevated serum activity in 54 out of 67 cases of unspecified intestinal parasitism in the horse. Initial attempts to apply this approach to the serodiagnosis of chronic ovine *Trichostrongylus* infections proved unsuccessful due, at least in part, to the broad range of serum AP isoenzyme activities found to occur in sheep (Jones and Knox, 1984).

SERODIAGNOSIS — THE FUTURE

Definition of intestinal parasitism depends on the positive identification of the causative nematode(s), quantitative assessment of infection and the correlation of the severity of infection with host production efficiency. Such diagnosis must involve either the detection of specific products of the invading organism or of the host immune response against the parasite. Detection of specific, anti-parasite antibodies in serum is not necessarily definitive of current infection. Moreover, grazing ruminants are continually exposed to a variety of parasites evoking a continuous and variable immunoglobulin turnover. Detection of specific parasite products can also be restricted by host antibody responses (Parkhouse, 1987). However, measurement of parasite-specific enzymes, identified on the basis of functional activity, could provide a new approach to serodiagnosis.

ACETYLCHOLINESTERASE

Acetylcholinesterase and nematode infection

The presence of AChE in parasitic helminths and the potential roles of this enzyme in the host–parasite relationship are well recognized and have been reviewed elsewhere (Rhoads, 1984; Rathaur *et al.*, 1987). High activities of AChE have been detected in the secretory glands of the intestinal nematodes *Nippostrongylus brasiliensis* (Lee, 1970) and *Trichostrongylus colubriformis* (Ogilvie *et al.*, 1973) and release of the enzyme by both parasites has been demonstrated *in vitro* (Ogilvie *et al.*, 1973; Rothwell *et al.*, 1973). Antibodies against nematode AChE have been detected in the sera of infected hosts (Jones and Ogilvie, 1972; Rothwell *et al.*, 1973) and there is evidence that the enzyme has antigenic specificity at the genus but not the species level (Rothwell *et al.*, 1976). Intestinal mucosal activity of AChE was significantly elevated during chronic subclinical infection with *Trichostrongylus vitrinus* (Jones, 1982) and *T. colubriformis* (Jones, 1983) and in the latter study the level of AChE was directly related to worm burden. In the same studies, no significant differences were observed when total host circulatory AChE activities were compared in worm-free and infected sheep, but differences in specific AChE isoenzymes could not be precluded. Although infected hosts produced antibody to *T. colubriformis* AChE, interaction between antibody and antigen did not inhibit enzyme activity (Rothwell and Merritt, 1974). The hypothesis that parasite AChE or antibody-AChE complexes might be detected in the host circulation, by fractionation and specific staining of isoenzyme activity, has been examined (Jones and Knox, 1988, 1990) and these experiments are summarized below.

AChE isoenzymes present in the infected host

AChE isoenzymes were fractionated by a combination of gel filtration (Sephadex G200) and polyacrylamide gel electrophoresis (PAGE), and enzyme activity detected by specific histochemical staining (Karnowsky and Roots, 1964). AChE isoenzymes, present in plasma and intestinal mucosal homogenates, were compared in lambs chronically infected with *T. colubriformis* and worm-free controls. Typical profiles obtained in plasma and mucosal homogenates, following densitometric scanning, are illustrated in Figure 9.1(a) and 9.1(b) respectively.

AChE in plasma from worm-free lambs was resolved into four peaks electrophoretically (Figure 9.1(a)). Similar peaks were present in plasma extracts from chronically parasitized lambs but, in the latter, a further three AChE bands (arrowed in Figure 9.1(a)) were detected. These isoenzymes corresponded, in terms of electrophoretic mobility, to peaks found in whole worm homogenates of adult *T. colubriformis*. The latter contained two further isoenzymes, at least one of which had an identical mobility to an enzyme present in all sheep plasmas analysed.

Extracts from intestines of chronically infected lambs contained two AChE isoenzymes not present in equivalent preparations from worm-free animals (Figure 9.1(b)). Again these isoenzymes had similar electrophoretic mobilities to

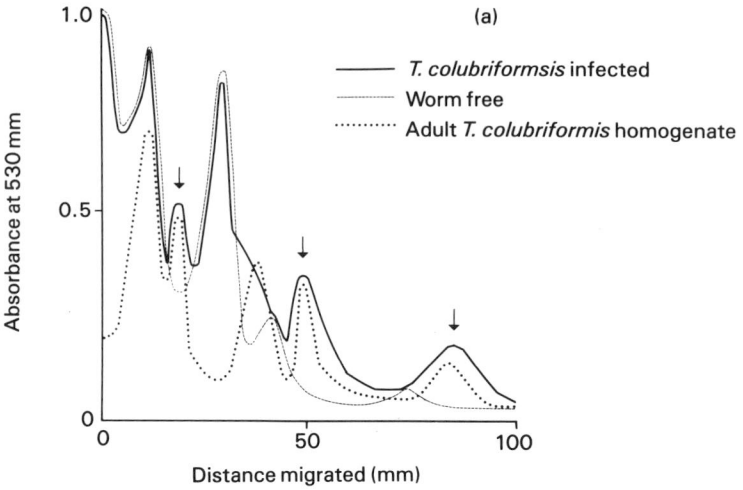

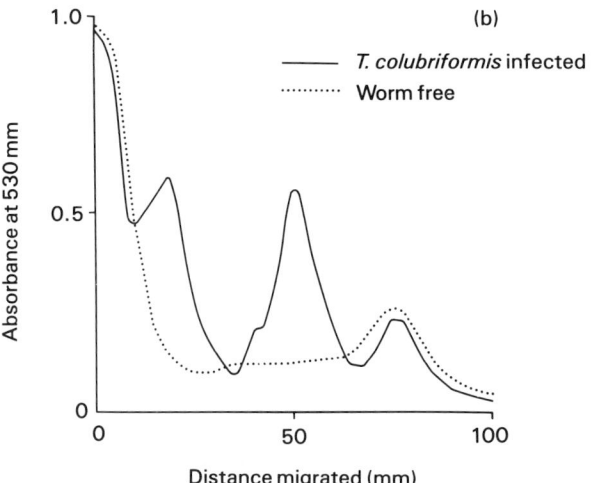

Figure 9.1. Electrophoretic profiles of AChE isoenzymes present in plasma extracts (a) and mucosal homogenates (b) from sheep infected with *T.colubriformis* and worm-free controls. Polyacrylamide gels (7 per cent; 80 mm × 4 mm internal diameter rods) in Tris/borate buffer pH 8.35 were run at 20 mA (80V) for 5 h at room temperature and stained, specifically, for AChE activity. (Reproduced from Jones and Knox (1989))

isoenzymes evident in adult worm homogenates and neither peak was observed in worm-free controls. We can speculate that these isoenzymes were of nematode origin and may have contributed to the elevated total AChE activity reported previously and correlated to worm burden, in sheep chronically infected with *T. colubriformis* (Jones, 1983).

Diagnostic potential of AChE

The coincident electrophoretic mobilities of AChE in plasma and parasite suggested that some or all of the changes noted could be due to direct detection of parasite AChE in the circulation of infected hosts. These results, in association with the correlation of mucosal AChE activity with worm burden in lambs chronically infected with *T. colubriformis* (Jones, 1983), indicated that parasite-derived AChE could fulfil the requirements for a serodiagnostic marker of ovine *Trichostrongylosis*. Further studies are, however, required to confirm the parasite origin of the candidate AChE isoenzymes. Immunological probes could then be used to develop a specific diagnostic immunoassay(s) for parasitic AChE. There remains a need to define the time of onset of isoenzyme changes and their quantitative relationship to worm burden. In recent experiments, significant alterations in plasma AChE isoenzyme profiles were demonstrated in guinea-pigs between 8 and 12 days after a single oral challenge with 3000 third larval stage (L3) *T. colubriformis* (D. P. Knox and M. Taylor, unpublished). However, several different AChE isoenzymes were detected in control animals, and some overlap with parasite AChE, emphasizing the complexity of the problem and the need for parasite-specific AChE identification.

PROTEINASES

Nematode proteinases — functional aspects

The release of proteolytic enzymes (proteinases and proteases) by parasitic nematodes is well documented (Von Brand, 1973) and they have been ascribed numerous roles in the aetiology of parasite disease including penetration of host tissue barriers (Matthews, 1977), anticoagulation (Hotez and Cerami, 1983) and proteolytic cleavage of surface bound immunoglobulin (Auriault *et al.*, 1981a). It has also been suggested that parasite proteinases may inactivate complement and generally facilitate evasion of host responses to the parasite (Leid, 1987). Evidence supporting the 'evasion' hypothesis has been provided in studies with the trematodes *Schistosoma mansoni* (Auriault *et al.*, 1981a) and *Fasciola hepatica* (Chapman and Mitchell, 1982) in which a trypsin like neutral serine proteinase and a metalloproteinase, released by *S. mansoni* schistosomula, can hydrolyse surface-bound immunoglobulin and free immunoglobulin G (IgG) (Auriault *et al.*, 1981a). Moreover, the peptides produced by enzymic hydrolysis of IgG inhibit macrophage stimulation, enzyme release and immunoglobulin E (IgE)-mediated cytotoxicity against schistosomula *in vitro*. Macrophage-mediated cytotoxicity was, however, unimpaired by the proteinases the products of IgE hydrolysis (Auriault *et al.*, 1980, 1981b). Proteolytic digestion of surface-bound IgG could also facilitate removal of Ig–Fc fragments from the parasite surface and, hence, inhibit complement fixation and attachment of Fc-receptor bearing phagocytes. Indeed, immature *F. hepatica* release a 'fabulating' enzyme capable of hydrolysing mouse, rat, rabbit and sheep immunoglobulin (Chapman and Mitchell, 1982).

Host responses to parasite proteinases

It is also pertinent to consider the response of the host to parasite proteinases, which could involve inactivation by specific antibody or by non-specific, endogenous proteinase inhibitors, such as α-2-macroglobulin or α-1-antitrypsin. Antibody from rabbits infected with *Ascaris suum* inhibited proteinases released *in vitro* by the parasitic larval stages of the parasite (Knox and Kennedy, 1988). Similarly, proteolytic activity in extracts of the oesophagus of adult *Ancylostoma caninum* was inhibited by sera from dogs immune to reinfection but not by normal sera (Thorson, 1956). In contrast, antibodies to filarial collagenase have been detected in sera of patients infected with *Onchocerca volvulus* or *Brugia malayi* (Petralanda *et al.*, 1986): the antibodies immunoprecipitated filarial collagenase but did not inhibit enzyme activity. The binding of antibody to proteinase, although not necessarily inhibiting activity against an unnatural substrate, may nevertheless prevent proteolytic cleavage of the native protein substrate by steric hindrance. Whether or not a given parasite proteinase is inhibited by a host may depend on the site of action of antibody and whether or not endogenous proteinase inhibitors have access to it. Indeed, it has been proposed that endogenous mammalian proteinase inhibitors are under selective evolutionary pressure from extrinsic proteinases, probably of parasite origin (Hill and Hastie, 1987).

Biochemical characterization of proteinases

When examining proteinases present in and released by parasitic nematodes, two important points need to be considered. Firstly, the proteinase content of whole worm homogenates may be underestimated due to the presence of proteinase inhibitors endogenous to the parasite such as those reported in ascarids and tapeworms (Von Brand, 1973) and in smaller nematodes such as *N. brasiliensis* (Juhasz and Kassai, 1981). Endogenous inhibitors are usually of low molecular weight and may protect parasites against the lytic action of host proteinases (Von Brand, 1973) but their effect on parasite proteinases has yet to be defined. Secondly, it is difficult quantitatively to relate *in vitro* release of parasite proteinases to the *in vivo* situation since culture conditions may not be optimal and not reflect the changing environments encountered within the host.

Initial characterization of proteolytic activity usually depends on examining effects on non-specific protein substrates such as casein, gelatin, haemoglobin or azocoll, at a variety of pH values. Where a specific functional role is suspected proteolysis of candidate host proteins can be examined, activity being detected by release of α-amino nitrogen (Matthews, 1984), of a radioactive label (Hotez and Cerami, 1983; Petralanda *et al.*, 1986; Robertson *et al.*, 1989), or by viscosity determinations (Dresden and Asch, 1972). When the release of more than one proteinase is suspected, or different stages of the same parasite are being compared, useful preliminary information can be obtained by determining pH optima of activities with low molecular weight peptide substrates. Using this approach activity can be differentiated into, for example: chymotrypsin-, trypsin- or elastase-like activity with esterolytic or amidolytic specificity. Finally, by use of a variety of inhibitors with specificity for different proteinase classes, proteinase activity can be defined on the basis of functional groups or metal ions essential for catalytic activity (Hartley, 1960). Some of the inhibitors used and the class of proteinase activity sensitive to their action are summarized in Table 9.1.

Alternatively, mixtures of proteinases can be fractionated and visualized directly by the incorporation or diffusion of protein substrates into polyacrylamide gels in which the proteinases have been separated (Andary and Dabich, 1974). Electrophoresis is usually performed under non-reducing conditions and, after an appropriate incubation period, zones of proteolysis appear as clear bands against a blue background following Coomassie blue staining (see also Chapters 6 and 8). In addition, activity can now be detected using a range of fluorogenic substrates (North *et al.*, 1990) and a combination of these methods can provide extensive data on peptide sequences that can be hydrolysed by individual enzymes. The molecular weight of resolved proteinases can then be estimated and, by the incorporation of specific inhibitors in the incubation buffer, their catalytic dependencies can be established. This technique has now received wide application in nematode biology (e.g. Robertson *et al.* (1989); see also Chapters 6 and 8) and has been used to provide evidence for the inhibition of parasite proteinases by antibody derived from infected hosts (Knox and Kennedy, 1988).

Table 9.1. Classification of proteinase activity on the basis of inhibitor sensitivity.

Inhibitor	Abbreviation	Working conc. (mM)	Proteinase class indicated
Phenlymethanesulphonylfluoride	Pms-F	1.0	Serine
4-Hydroxymercuribenzoate	4HMB	0.1	Thiol (cysteine)
N-Ethylmaleimide	NEM	1.0	Thiol
L-Transepoxysuccinylleucylamido-4(guanido)butane	E64	0.2	Thiol
Ethylenediaminotetraacetic acid	EDTA	2.0	Metallo
1,10-Phenanthroline	1,10 Phe	1.0	Metallo
Pepstatin	—	2.0 μm	Carboxyl (aspartate)
Dithiothreitol[a]	DTT	2.0	Thiol Metallo

[a] Reduction of activity, in the presence of inhibitor, is indicative of the classes listed with the exception of DTT which protects 'thiol' proteinase activity. This protection is usually observed as an apparent stimulation of enzyme activity in comparison to a control preparation incubated in the absence of DTT. This compound (DTT) inhibits metallo proteinase activity. Bracketed amino acids indicate alternative terms used to define proteinase class on the basis of amino acids essential to catalytic activity.

Proteinase release by gastrointestinal nematodes

The release of proteolytic enzymes by gastrointestinal nematodes is well documented. Proteinases secreted by *Anisakis simplex* larvae have been related to penetration of the gut wall and stomach mucosa of both intermediate and final hosts (Matthews, 1984). Larval and adult *Ancylostoma caninum* release a 37 kDa proteinase which facilitates larval penetration and acts as an anticoagulant facilitating adult feeding (Hotez and Cerami, 1983; Hotez *et al.*, 1985). Moreover, a recent report (Cox *et al.*, 1990) described a 35 kDa cysteine (thiol) proteinase with fibrinolytic properties present in extracts of adult *Haemonchus contortus*. There is histological evidence for the degradation of skin collagen by *N. brasiliensis* infective larvae (Lee, 1972), both adults and larvae of which contain collagenolytic activity (Harper and Bloch, 1974). Release of proteolytic activity from the excretory pores of adult *Necator americanus* and *T. colubriformis* has also been demonstrated (McLaren *et al.*, 1974), the former activity being related to skin penetration by invading larvae (Matthews, 1982). In addition, proteinases from adult *N. americanus* and *A. caninum* show antigenic cross-reactivity (Pritchard, 1986; and Chapter 8). ES antigens, harvested following *in vitro* maintenance of the tissue invasive larval stages of *Toxocara canis* (Robertson *et al.*, 1989) and *Ascaris suum* (Knox and Kennedy, 1988) contain multiple proteolytic activities. In a recent report (Knox and Jones, 1990) we have provided evidence for the release of proteinases by a variety of gastrointestinal nematodes of ruminants.

Characterization of proteinases released by *A. suum*

The general approach outlined above is exemplified by a series of experiments which examined proteinase release by the parasitic larval stages of *A. suum in vitro* (Knox and Kennedy, 1988).

pH optima

Examination of the pH optima of proteolytic activity, using a range of low and high molecular weight substrates (Figure 9.2a and 9.2b), showed that the *in vitro* released (IVR) products of the tissue invasive infective (L2) and lung stage (L3/L4) larvae of this parasite contained proteolytic activity over a broad range.

Although L2 and L3/L4 IVR products had similar alkaline pH optima with azocoll, a collagen-based general protein substrate, differences were observed between and within the two IVR products when specific peptide substrates were used. In general, proteinase activity released by both developmental stages had major pH optima at pH 6 and 9, although clear differences in distribution were apparent between stages (Figures 9.2(a) and 9.2(b)).

Substrate specificity

Studies on substrate specificity, using a range of peptide and protein substrates at

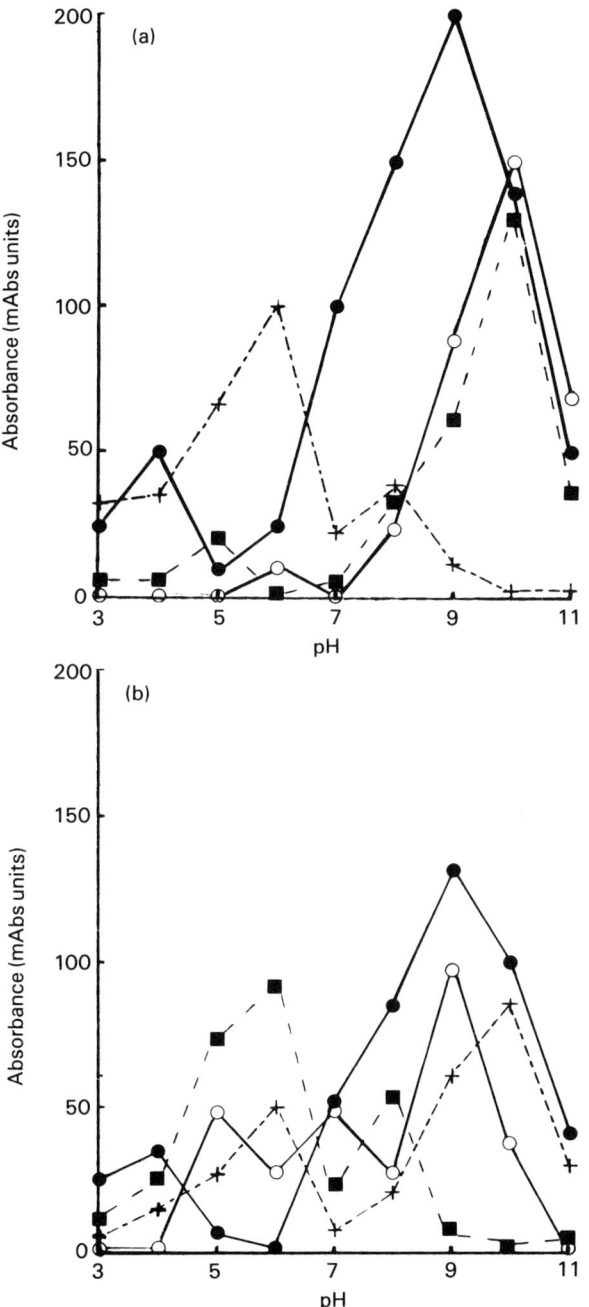

Figure 9.2. The effect of pH on proteinase activity of L2 (a) and L3 / L4 (b) IVR of *A. suum*. Substrates used: (●) azocoll, absorbance read at 520 nm; (○) CBZ-L-Ala-4NPE, (■) CBZ-L-Lys-4NPE, and (+) CBZ-L-Try-4NPE all read at 405 nm.
CBZ, carbobenzoxy; 4NPE, 4-nitrophenol; Ala, alanine; Lys, lysine; Try, tryptophan. (Reproduced from Knox and Kennedy (1988))

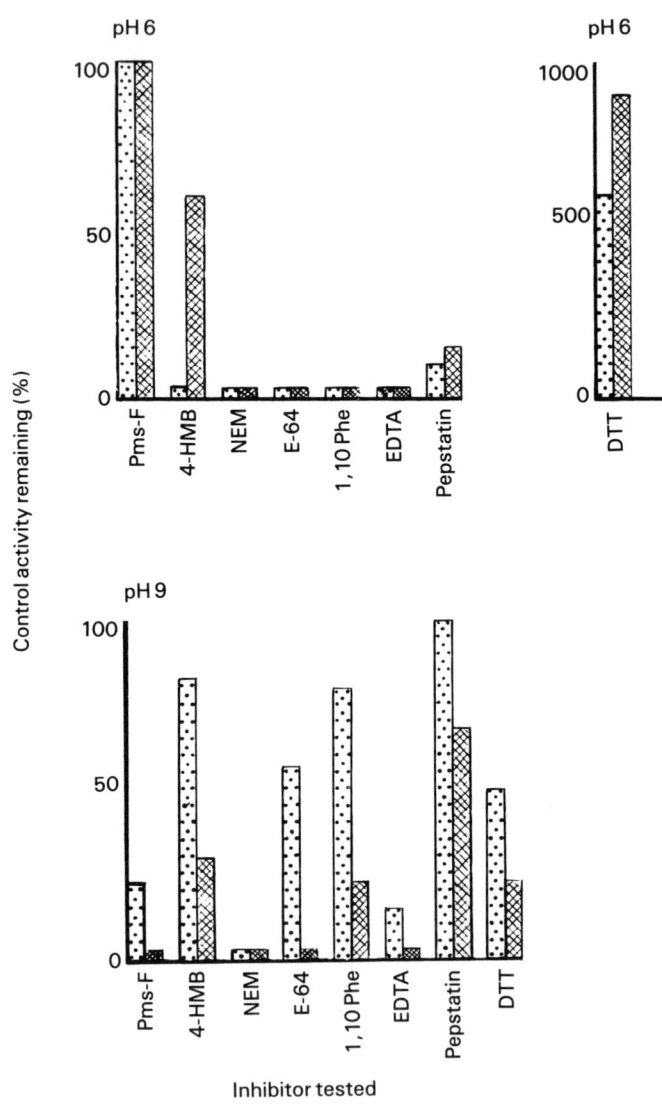

Figure 9.3. The effect of various inhibitors on the hydrolysis of azocoll by L2 ▨ and L3 / L4 IVR ▨ products. Inhibition is expressed as the percentage activity remaining in comparison to a control in the absence of inhibitor. For an explanation of the abbreviations see Table 9.1. (Reproduced from Knox and Kennedy (1988).)

pH 6 and 9, indicated the release by both the L2 and L3/L4 stages of chymotryptic, tryptic, elastolytic and collagenolytic activities, the latter two being consistent with tissue migration (Von Brand, 1973). Stage specificity of secreted proteinases was again demonstrated in terms of relative activity and differential substrate reactivity. Furthermore, proteolysis of azocoll and selected peptide nitrophenol esters by L2 and L3/L4 IVR products is restricted by various inhibitors.

Inhibitor sensitivity

With azocoll as substrate at pH 6, the 'thiol' inactivator, 4HMB, completely inhibited L2 IVR products while only a mild reduction in activity was observed for L3/L4 IVR products (Figure 9.3). However, the specific 'thiol' inhibitor, E64, completely inhibited activity in both IVR products while this activity was markedly enhanced in the presence of the 'thiol' protecting reagent, DTT. By contrast, at pH 9, while azocollytic activity of both IVRs was sensitive to the action of the 'serine' proteinase inhibitor, Pms-F, activity present in L3/L4 IVR products was more sensitive to the action of 4HMB and was inhibited by DTT. Moreover, L2 and L3/L4 IVR showed markedly differing inhibitor sensitivities when determined with the chymotrypsin substrate, carbobenzoxy-L-tryptophan-4-nitrophenol ester.

These experiments indicated that the L2 and L3/L4 IVR products of *A. suum* contained multiple proteolytic activities and that there were distinct differences between the two IVR products. The heterogeneity and developmental regulation of enzyme secretions might be appropriate to a parasite moving through a series of different tissue environments, and overlaps in the metabolic characteristics of the different stages could be anticipated. Similar observations have been reported in relation to proteolytic enzymes secreted by the infective larvae of *Toxocara canis in vitro* (Robertson *et al.*, 1989) and are summarized elsewhere (see Chapter 6).

Reactivity with antibodies from infected serum

The proteinaceous secretions of the L2 and L3/L4 stages of *A. suum* are highly antigenic in the infected host (Kennedy and Qureshi, 1986) and the influence of antibody on their proteinases has relevance to their function *in vivo*. With azocoll as substrate, activity of both L2 and L3/L4 IVR products was totally inhibited by rabbit anti-*A. suum* IgG with L2 IVR products being more sensitive (Figure 9.4). The effect was not due to competition with Ig protein since no measurable inhibition was observed in the presence of normal rabbit IgG. This inhibition could potentially impair migration in sensitized hosts.

Characterization of proteinases from different ruminant gastrointestinal nematodes

In a recent study, Knox and Jones (1990) measured the proteolytic activity present in whole worm homogenates of various ruminant gastrointestinal nematodes and proteinase release when worms were cultured *in vitro*. They also examined

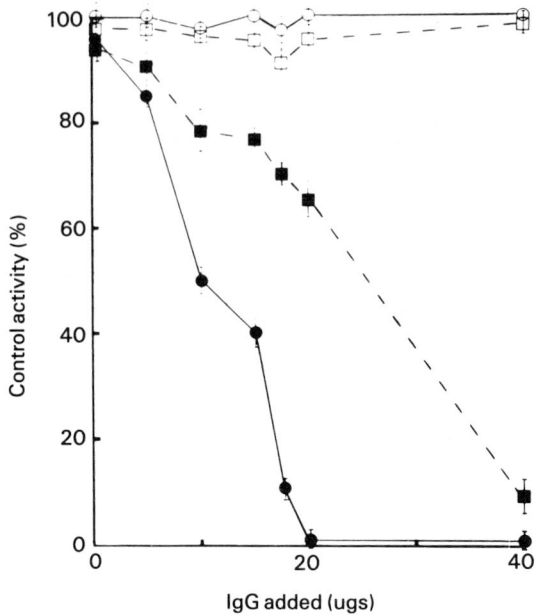

Figure 9.4. The effect of purified IgG, from normal rabbits and rabbits repeatedly infected with *A. suum* on proteolytic activity of L2 and L3/L4 IVR products. Incubations were performed in the presence of normal IgG (○) L2 and (□) L3/L4 IVR products and anti-*A. suum* IgG (●) L2 and (■) for L3/L4 IVR products. Inhibition is expressed as the activity remaining relative to appropriate incubated in the absence of IgG. (Reproduced from Knox and Kennedy (1988).)

proteolytic activity associated with the rat intestinal nematode, *N. brasiliensis*, a commonly used laboratory model for the study of intestinal nematodiasis. The species and stages examined in the culture studies are summarized in Table 9.2.

Whole worm homogenates

Proteinase activity in whole worm homogenates, from the L3 and adult stages of these species, was examined using a general proteinase substrate, azocasein, an elastin-based substrate, elastin-orcein and azocoll. Although all homogenates were active against all three substrates, the degree of proteolysis varied between stages of the same parasite, between parasites and between substrates. These observations could reflect differences in the proteolytic requirements of nematodes and stages occupying different niches along the gastrointestinal tract. For example, the L3 stages of all the parasites examined, with the exception of *N. brasiliensis*, are more intimately associated with the gut mucosa than their adult counterparts (Miller, 1984), and generally contained higher proteinase activities than the adult. The L3 may have a greater requirement for proteinases to facilitate mucosal penetration and to degrade host proteins for nutrient provision. Moreover, L3 stages undergo a period of rapid growth and differentiation prior to the moult to the L4 stage (Whitlock, 1960).

Table 9.2. *In vitro* release of proteolytic activity by gastrointestinal nematodes of ruminants.[a]

Species	Developmental stage	Colour release for substrates		
		Azocasein (380 nm)	Azocoll (520 nm)	Elastin-orcein (550 nm)
N. brasiliensis	L3	0.13	0.11	0.41
	Adult	0.03	0.03	0.28
Nematodirus battus	L3	0.02	0.02	0.20
	Adult	0.00	0.04	0.44
Ostertagia circumcincta	L3	0.53	0.48	0.24
	Adult	22.30	7.20	4.20
T. vitrinus	L3	0.34	0.12	0.24
	Adult	0.55	0.82	4.26
T. colubriformis	L3	0.36	0.11	0.34
	Adult	1.31	1.81	6.00
H. contortus	L3	0.04	0.01	0.13
	Adult	0.36	0.00	6.54

[a] All data are expressed as change in absorbance at the wavelengths indicated / hour culture / hour assay in 1 ml culture ($\times$ 10^2). Parasites were incubated at concentrations of 1000 L3 or 100 adults per millilitre of culture fluid. Incubations were maintained for 8 h at 37 °C.

Proteinases in extracts of adult H. contortus

Preliminary data from current experiments (Knox, 1990) indicate the presence of predominantly acidic 'thiol' proteinase activities in Triton X-100 extracts of the adult parasite. Several zones of activity were visible on gelatin substrate gels ranging in molecular weight from 45 to greater than 200 kDa. The bulk of this activity was sensitive to 'thiol' proteinase inhibitors. The molecular size of these enzymes contrasts with a 35 kDa 'thiol' proteinase with fibrinolytic activity extracted from adult *H. contortus* (Cox *et al.*, 1990). This apparent anomaly might be due to the different methodologies employed in detecting enzyme activity and is currently under investigation.

We now have evidence that some of the enzymes under study in our laboratory are capable of degrading ovine haemoglobin. This observation is in accord with the well-defined blood feeding habits of the adult parasite (Whitlock, 1960) and is consistent with data from other blood feeding nematodes such as *Necator* and *Ancylostoma* spp. described elsewhere in this volume (see Chapter 8). In addition, fibrinolytic activity (Cox *et al.*, 1990) would be consistent with an anti-coagulant role and is likely to contribute to the anaemia observed in acute cases of haemonchosis. It is encouraging to note that Cox *et al.* (1990) describe unpublished observations which suggest that the purified enzyme can confer significant protection to sheep against *H. contortus* infections.

Culture supernatant

Detection of proteinase activity in media (Table 9.2) harvested following *in vitro* culture of L3 and adult stages, provides the first direct biochemical evidence for release of proteinases by these nematodes. The release of multiple proteolytic activities was demonstrated by proteolysis of low molecular weight peptide substrates; the results were similar to those reported for *A. suum* (Knox and Kennedy, 1988) and could again reflect specific stage and environmental requirements. For example, elastase activity released by *N. brasiliensis* may facilitate skin penetration and similar activity released by adult *H. contortus* could act as an anticoagulant to facilitate blood feeding. The release of an anticoagulant by *Ancylostoma* hookworms has previously been described (Hotez and Cerami, 1983).

Molecular biology and proteolytic enzymes

Difficulties of parasite availability and the laborious and time-consuming efforts of devising appropriate purification procedures have, in the past, tended to impair enzyme definition and characterization. These problems can now be minimized by cloning, sequencing and expressing the gene encoding the enzyme(s) of interest. Data obtained by genetic manipulation should provide information on the function of these enzymes and their potential role(s) at the host–parasite interface. Expression of the cloned gene can allow immunogenic potential to be assessed with direct implications for the development of potential vaccines. For illustrative purposes, two approaches to cloning parasite proteinase genes are outlined below.

The primary sequence of a 35 kDa cysteine proteinase expressed by adult *H. contortus* has been described by Cox *et al.* (1990). Near full-length cDNAs for the enzyme were isolated by immunoscreening an adult worm cDNA expression library with a rabbit antiserum prepared against a partially purified enzyme preparation and by rescreening the library with oligonucleotide probes. Following sequence analysis of these cDNA clones and of a genomic DNA clone the authors were able to deduce the entire amino acid sequence of the proteinase, the position of the active site and four potential nitrogen linked glycosylation sites. The enzyme appeared to be glycosylated *in vivo*. Comparison of sequence data with other cysteine proteinases showed that the enzyme was related to the human lysosomal thiol proteinase cathepsin B.

An alternative approach, requiring no enzyme purification, has been applied to the isolation of gene fragments encoding serine and thiol proteinases (Sakanari *et al.*, 1989). Serine proteinase gene fragments were isolated from the parasitic nematode *Anisakis simplex* by using degenerate oligonucleotide primers and the polymerase chain reaction (PCR). Primers were designed based upon the consensus sequence of amino acids flanking the active site serine and histidine residues of eukaryotic serine proteinases. A similar approach has recently been applied to the amplification of cysteine proteinase fragments from the protozoan parasites *Entamoeba histolytica*, *Trypanosoma cruzi* and *Trypanosoma brucei* (Eakin *et al.*, 1990) using PCR primers based on conserved sequences at Cys-25 and Asn-125 which are involved at the active site. In addition, the same authors isolated a cathepsin B-like thiol proteinase gene fragment from *Caenorhabditis elegans* DNA. We are currently applying this approach to the isolation of adult *H. contortus* thiol proteinase genes (D. L. Miller and D. P. Knox, unpublished) and have successfully amplified and sequenced a number of gene fragments. In theory, genes for other enzymes or any protein with conserved structural motifs can be identified and isolated using this technology (Sakanari *et al.*, 1989).

SUPEROXIDE DISMUTASE (SOD)

SOD and the host–parasite relationship

A feature of protective mucosal responses against gastrointestinal nematodes is the development of an inflammatory type reaction normally associated with peripheral blood eosinophilia and extensive granulocyte infiltration of mucosae (Miller, 1984). The 'respiratory burst' (Babior *et al.*, 1973) of activated phagocytes and particularly eosinophils results in sustained release of large quantities of superoxide anion (O_2^-) and other toxic products of oxidative metabolism which may contribute to their ability to destroy extracellular pathogens (Tauber *et al.*, 1976). In rats infected with *N. brasiliensis*, studies of the kinetics of eosinophil infiltration indicated that maximal accumulation occurred at the time of worm expulsion (Wells, 1962; Kelly and Ogilvie, 1972) and that degranulation occurred prior to worm expulsion (Kelly and Ogilvie, 1972). Secreted granule products may be involved in the modulation of local hypersensitivity reactions and the promotion of further tissue damage and helminthotoxicity (Wellar and Goetzl, 1979). Although evidence that eosinophils and other phagocytic cells serve a protective function *in vivo* is limited (Miller, 1984), the possibility that free-radical mediated cytotoxicity plays a significant role in mediating worm damage cannot be ignored. Conversely, secretion of antioxidants such as SOD, an enzyme ubiquitous in aerotolerant organisms (Leid and Suquet, 1986) or their presence at the host–parasite interface may assist the parasite in avoiding the cytotoxic effects of free radicals.

Evidence for a functional role of nematode SOD

Evidence for 'immune evasion' by secretion of SOD with associated inactivation of oxygen derived free radicals has been provided in studies on the nematode *Trichinella spiralis*. Rhoads (1983) identified and purified a SOD from muscle stage larvae and demonstrated its antigenicity by an immunospecific reaction with *T. spiralis* antisera derived from infected rats, mice and pigs. In addition, the enzyme was secreted into culture fluids by muscle stage larvae. A single enzyme was present in muscle stage larvae, but two isoenzymes were found in the adult and Rhoads (1983) suggested that this modification of enzyme structure could occur in response to immune stress. Marked differences in the susceptibility of different life cycle stages of *T. spiralis* to both *in vitro* killing by granulocytes and oxidant mediated damage were associated with differences in SOD content together with another antioxidant enzyme, glutathione peroxidase (Kazura and Meschnick, 1984). Co-incubation of oxidant resistant adult parasites with highly sensitive new-born larvae partially protected larvae against oxidant damage *in vitro* (Kazura and Meshnick,

1984). Smith and Bryant (1986) demonstrated that adult *N. brasiliensis* was markedly more susceptible to free radical damage *in vitro* than *Heligmosomoides polygyrus* (*Nematospiroides dubius*) and this was related to differences in antioxidant enzyme protection since *H. polygyrus* contained two to four times as much SOD, catalase and glutathione reductase as *N. brasiliensis*. The authors concluded that the ability of *H. polygyrus* to persist in the rodent small intestine, whilst *N. brasiliensis* is spontaneously expelled, could be due to a more efficient enzymatic defence system against host-generated free radicals.

SOD in ruminant gastrointestinal nematodes

To date, there has been no description of free radical-scavenging enzymes associated with ruminant gastrointestinal nematodes, but in a recent study (Knox and Jones, 1989) SOD activities in a variety of species at different developmental stages, were compared (Table 9.3). The data show that the invasive larval stages of all the parasites examined contained considerably more SOD than their adult counterparts, and this may be related to differences in oxidative stress experienced by parasite stages living within the epithelium or mucosa compared with lumen-dwelling parasites. Moreover, following *in vitro* maintenance of parasitic larval and adult stages of *T. vitrinus*, *H. contortus* and *O. circumcincta*, significant activities of SOD were detectable in the culture supernatants. Enzyme release appeared to be related both to the numbers of parasites cultured and the duration of the culture. Further studies on the antigenicity of parasite SOD in infected hosts are worthy of consideration. Antibodies to *T. spiralis* SOD were detectable in sera from infected mice, rats and pigs (Rhoads, 1983) and, although it was not possible to establish whether the binding of antibody to SOD inhibited enzyme activity, the author did note that antiserum to spinach SOD was a potent inhibitor of worm enzyme activity

Table 9.3. Superoxide dismutase content of gastrointestinal nematode homogenates.

Nematode	Developmental stage	Location	Site	SOD (IU/mg protein)
T. colubriformis	L3	Small	Intraepithelial	27.3
	Adult	intestine	Intraepithelial (posterior in lumen)	7.8
T. vitrinus	L3	Small	Intraepithelial	16.3
	L4	intestine	Intraepithelial	15.8
	Adult		(posterior in lumen)	6.1
H. contortus	L3	Abomasum	Gastric glands	19.8
	Adult		Mucosal surface	11.8
O. circumcincta	L3	Abomasum	Gastric glands	79.4
	11d†	Abomasum	Gastric glands	52.1
	Adult		Mucosal surface	12.6
N. battus	L3	Small	Submucosal	38.0
	Adult	intestine	Lumen	12.2

† 11d: Harvested 11 days after single infection with 3rd stage larvae.

(Asada *et al.*, 1974). The potential role of SOD in parasite survival, combined with specific antibody responses capable of inhibiting enzyme activity in naturally infected hosts, raises the possibility that parasite SOD might provide a suitable target antigen for vaccine preparation.

SOD isoenzyme variants

In accord with previous reports describing SOD isoenzyme variety, both within stages of the same parasite (Rhoads, 1983) and between closely related parasites (Sanchez-Moreno *et al.*, 1987), we have noted marked differences in SOD activity present in gastrointestinal nematodes following separation by PAGE (Figure 9.5). Whole worm extracts were fractionated in 7.5 per cent gels and SOD activity was detected as clear bands resulting from the inhibition of the photochemical reduction of nitroblue tetrazolium (Beauchamp and Fridovich, 1971).

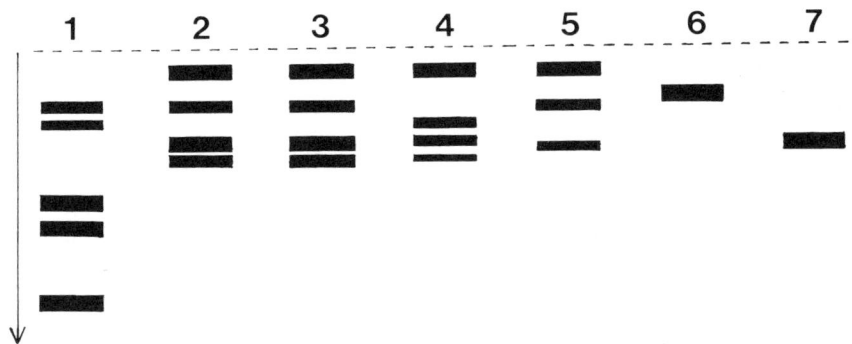

Figure 9.5. Electrophoretic variation in superoxide dismutases of gastrointestinal nematodes. Whole worm extracts were fractionated in 7.5 per cent gels and SOD activity was detected as clear bands resulting from the inhibition of the photochemical reduction of nitroblue tetrazolium (Beauchamp and Fridovich, 1971). Lanes: (1) Adult *N. brasiliensis*; (2, 3) L3 and adult *H. contortus*, respectively; (4, 5) 11-day old and adult *O. circumcincta*, respectively; (6) adult *T. vitrinus*; and (7) SOD released by adult *T. vitrinus* during *in vitro* maintenance. (———) Origin; (←) direction of migration.

The figure illustrates the considerable variety of SOD isoenzymes observed in closely related parasites occupying similar ecological niches and is similar in degree to the range of SOD isoenzymes observed in various *ascaridids* (Sanchez-Moreno *et al.*, 1987). Sanchez-Moreno *et al.* (1988) resolved SOD from *A. suum* into two copper–zinc isoenzymes and one manganese form, on the basis of cyanide and hydrogen peroxide sensitivity. In addition, it is interesting to note the differing electrophoretic mobilities of SOD present in adult *T. vitrinus* homogenates and IVR products. It is tempting to speculate that the less mobile form, present in homogenates, may be concerned with internal free-radical detoxification, while the more rapidly migrating form, present in IVR products, might be released in response to host mediated free radical attack.

Can nematodes alter SOD expression in response to immunological stress?

Smith and Bryant (1986) showed that adult *N. brasiliensis* were more susceptible to *in vitro* free radical damage than *H. polygyrus* (*N. dubius*) and correlated the increased levels of free radical scavenging enzymes in the latter with its enhanced ability to persist in the host. However, this comparison was performed with 'normal' worms. Ogilvie and Hockley (1968) defined 'normal' worms as those harvested from rats 6–9 days after a primary infection before the rats had developed immunity and 'adapted' worms, those harvested from rats 7 days after a second or third infection. Edwards *et al.* (1971) described marked alterations in the isoenzyme profile of the secretory enzyme, AChE, in 'adapted' *N. brasiliensis* and Jones and Ogilvie (1972) related these changes to the longevity of adapted worms in naive rats. In addition, Smith and Bryant (1989a) demonstrated a stringent temporal association between generation of free radicals by leucocytes and the rejection of *N. brasiliensis*. To test the hypothesis that the longevity of adapted worms could also be related to alterations in their total SOD activity or to SOD isoenzymes, we examined SOD variants in 'normal' and 'adapted' *N. brasiliensis*. Superoxide anion can affect DNA synthesis and RNA transcription (Michelson, 1977) and it is possible that this radical could induce altered gene expression and synthesis of abnormal isoenzymes. The results of this pilot experiment are summarized in Figure 9.6.

Specific SOD activity/mg of homogenate protein for adapted worms was three times that observed in normal worms and was very similar to data reported for *H. polygyrus* (Smith and Bryant, 1986). More SOD variants were detected than previously described for parasitic nematodes (Rhoads, 1983; Sanchez-Moreno *et al.*, 1987, 1988) and two additional bands were consistently noted in adapted worms (arrowed bands in Figure 9.6). The results of this study, in addition to previous reports (Rhoads, 1983; Smith and Bryant, 1986) suggest that efficient defence mechanisms against host-generated free radicals play a key role in parasite survival. Recent experiments (Smith and Bryant, 1989a,b) have demonstrated a

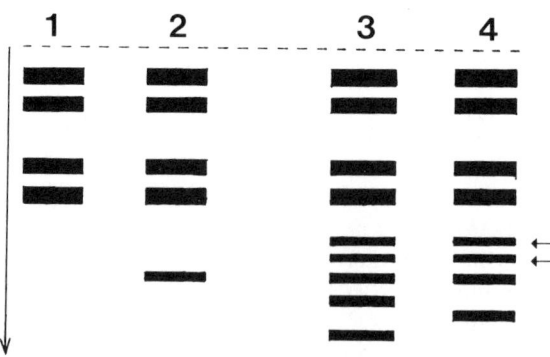

Figure 9.6. Electrophoretic comparison of superoxide dismutases of 'normal' and 'adapted' *N. brasiliensis*. Separations were performed as described in Figure 9.5. Lanes: (1, 2) 'normal' parasites; (3, 4) 'adapted' parasites. Arrows indicate bands uniquely observed in 'adapted' parasites.

stringent temporal association between generation of free radicals by leucocytes and the rejection of *N. brasiliensis*, and this response was susceptible to inhibition by antioxidants.

Effects of host diet at the host–parasite interface — are parasite enzyme activities altered?

Contrasting effects of host dietary molybdenum (Mo) intake on the establishment and survival of ovine gastrointestinal nematodes have been described (Suttle *et al.*, 1988, 1989). Addition of Mo to the diet of lambs reduced the establishment of a trickle larval infection of the intestinal parasite, *T. vitrinus* (Suttle *et al.*, 1988). Poor establishment may have been attributable partly to the direct effects of Mo on the parasite since proteinase activity in, and secreted by, the retrieved parasites was reduced. Similarly, Mo inhibited proteinase secretion when added to cultures of *T. vitrinus* not previously exposed to Mo (Knox and Jones, 1988). There was indirect evidence that Mo enhanced the inflammatory reaction in the intestinal mucosa. A further experiment examined the effects of host dietary Mo intake on the establishment of the ovine abomasal parasite, *H. contortus* (Suttle *et al.*, 1989). In contrast to *T. vitrinus*, proteinase activity in, and secreted by, retrieved parasites was only marginally reduced and azocaseinolytic activity was enhanced by Mo. This pattern was consistent with direct effects of Mo on adult *O. circumcincta* maintained *in vitro* (Knox and Jones, 1988) and indicated the complexity of host–parasite interactions in the gastrointestinal tract. However, although the direct effects of Mo on abomasal parasites appeared to be less severe than those observed for *T. vitrinus*, Mo greatly enhanced the rejection of *H. contortus* from the animals (Suttle *et al.*, 1989) and lambs given Mo had increased intraepithelial mast-cell counts in the abomasal mucosa. One possibility was that Mo had accelerated the cellular response to the parasite. Current work in our laboratory suggests that parasites derived from the Mo fed lambs exhibited increased SOD variety while secreting less SOD *in vitro*. Moreover, mucosal xanthine oxidase activity, an enzyme involved in O_2^- generation, tended to be higher in Mo fed lambs and again increased isoenzyme polymorphism was suggested.

CONCLUDING REMARKS

Enzymes released by parasites are part of a complex, integrated system and have considerable application in monitoring the response of parasites to immunological attack, chemotherapeutic agents (Rapson *et al.*, 1983; Jones and Knox, 1989) and the effect of environmental factors on parasite establishment and maintenance (Suttle *et al.*, 1988, 1989; Knox and Jones, 1988).

To summarize, examination of AChE (Edwards *et al.*, 1971; Jones and Ogilvie, 1972) and SOD (Knox and Jones, 1989) profiles revealed a considerable number of distinct isoenzymes in populations of *N. brasiliensis* capable of withstanding the intestinal immune response. In addition, current work indicates that there is considerable SOD and proteinase heterogeneity both within stages of the same parasite and between closely nematode species. Together with the experiments showing complex interactions between host diet, parasite survival and parasite biochemistry, it would appear that secretory enzymes play a key role in nematode survival. The definition of the immune reactions responsible for triggering these changes, and the enzymes themselves, should facilitate the search for novel immunological and/or chemotherapeutic control strategies. The observations that host antibody can inhibit parasite enzymes offers an important opportunity to examine their function *in vivo*, possibly by selective inhibition induced by specific immunization. Moreover, immunological responses to parasite enzymes may determine strain and individual resistance to infection.

ACKNOWLEDGEMENTS

We are indebted to all our colleagues in the Parasitology department of the Moredun Research Institute for their constant advice and help in the provision of parasite material. Also, we thank Andy McGuigan for providing 'normal' and 'adapted' *N. brasiliensis* and, finally, Neville Suttle for his helpful comments during the preparation of this manuscript.

REFERENCES

Andary, T.J. and Dabich, D., 1974, A sensitive polyacrylamide disc gel method for the detection of proteinases, *Analytical Biochemistry*, 57, 457–66.

Asada, K., Takahashi, M. and Nagate, M., 1974, Assay and inhibitors of spinach superoxide dismutase, *Agricultural and Biological Chemistry*, 38, 471–3.

Auriault, C., Joseph, M., Dessaint, J.P. and Capron, A., 1980, Inactivation of rat macrophages by peptides resulting from cleavage of IgG by schistosomula larval proteases, *Immunological Letters*, 2, 135–9.

Auriault, C., Ovaissi, M.A., Torpier, G., Elsen, H. and Capron, A., 1981a, Proteolytic cleavage of IgG bound to the Fc receptor of *S. mansoni* schistosomula, *Parasite Immunology*, 3, 33–4.

Auriault, C., Pestel, J., Joseph, M., Dessaint, J.P. and Capron, A., 1981b, Interaction between the macrophage and *Schistosoma mansoni* schistosomula: role of IgG peptides and aggregates on the modulation of glucoronidase release and cytotoxity against schistosomula, *Cellular Immunology*, 62, 15–27.

Babior, B.M., Kipnes, R.S. and Curnutte, J.T., 1973, Biological defence mechanisms. The production by leucocytes of superoxide, a potential bacteriocidal agent, *Journal of Clinical Investigation*, 52, 741–4.

Beauchamp, C.O. and Fridovich, I., 1971, Superoxide dismutase: improved assays and assay applicable to polyacrylamide gels, *Analytical Biochemistry*, 44, 276–87.

Blackmore, D.J. and Palmer, A., 1977, Phenylalanine inhibited *p*-nitrophenolphosphate activity as an indication of intestinal cellular disruption in the horse, *Research in Veterinary Science*, 23, 146–52.

Chandler, A.C., 1932, Experiments on resistance of rats to superinfection with the nematode *Nippostrongylus muris*, *American Journal of Hygiene*, 16, 750–82.

Chandler, A.C., 1936, Studies on the nature of immunity to intestinal helminths. III. Renewal of growth of egg production in *Nippostrongylus* after transfer from immune to non immune rats, *American Journal of Hygiene*, **23**, 46–54.

Chapman, C.B. and Mitchell, G.F., 1982, Proteolytic cleavage of immunoglobulin by enzymes released by *Fasciola hepatica*, *Veterinary Parasitology*, **11**, 165–78.

Coop, R.L., Sykes, A.R. and Angus, K.W., 1977, The effect of daily intake of *Ostertagia circumcincta* larvae on body weight, food intake and concentration of serum constituents in sheep, *Research in Veterinary Science*, **23**, 76–82.

Cox, G.N., Pratt, D., Hageman, R. and Boisvenue, R.J., 1990, Molecular cloning and primary sequence of a cysteine protease expressed by *H. contortus* adult worms, *Molecular and Biochemical Parasitology*, **41**, 25–34.

Dresden, M.H. and Asch, H.L., 1972, Proteolytic enzymes in extracts of *Schistosoma mansoni* cercariae, *Biochimica et Biophysica Acta*, **289**, 378–84.

Eakin, A.E., Bouvier, J., Sakanari, J.A., Craik, C.S. and McKerrow, J.H., 1990, Amplification and sequencing of genomic DNA fragments encoding cysteine proteases from protozoan parasites, *Molecular and Biochemical Parasitology*, **39**, 1–8.

Eckersall, P.D., 1988, Protein analysis and veterinary diagnosis, in Blackmore, D.J. (Ed.) *Animal Clinical Biochemistry — The Future*, Cambridge: Cambridge University Press.

Edwards, A.J., Burt, J.S. and Ogilvie, B.M., 1971, The effect of immunity upon some enzymes of the parasitic nematode, *Nippostrongylus brasiliensis*, *Parasitology*, **62**, 339–47.

Fishman, W.H., Green, S. and Inglis, N.I., 1963, L-Phenylalanine: an organ specific stereospecific inhibitor of human intestinal alkaline phosphatase, *Nature (London)*, **198**, 685–6.

Harper, E. and Bloch, K.J., 1974, Collagenolytic activity of infective larvae of *Nippostrongylus brasiliensis*, Proceedings of 3rd International Congress and Parasitology, Munich.

Hartley, B.S., 1960, Proteolytic enzymes, *Annual Reviews of Biochemistry*, **29**, 45–72.

Hill, R.E. and Hastie, N.D., 1987, Accelerated evolution in the reactive centre region of serine proteinase inhibitors, *Nature*, **236**, 96–9.

Holmes, P.H., 1985, Pathogenesis of trichostrongylosis, *Veterinary Parasitology*, **25**, 89–101.

Hotez, P.J. and Cerami, A., 1983, Secretion of a proteolytic anticoagulant by *Ancylostoma* hookworms, *Journal of Experimental Medicine*, **157**, 1594–603.

Hotez, P.J., Le Trang, N., McKerrow, J.H. and Cerami, A., 1985, Isolation and characterisation of a proteolytic enzyme from the adult hookworm *Ancylostoma caninum*, *Journal of Biological Chemistry*, **260**, 7343–8.

Jones, D.G., 1982, Changes in intestinal enzyme activity of lambs during chronic infection with *Trichostrongylus vitrinus*, *Research in Veterinary Science*, **32**, 316–23.

Jones, D.G., 1983, Intestinal enzyme activity in lambs chronically infected with *Trichostrongylus colubriformis*: effect of anthelmintic treatment, *Veterinary Parasitology*, **12**, 79–89.

Jones, D.G. and Knox, D.P., 1984, Distribution of plasma alkaline phosphatase and its isoenzymes in normal and chronically parasitised lambs, *Journal of Comparative Pathology*, **94**, 463–5.

Jones, D.G. and Knox, D.P., 1988, Acetylcholinesterase: a diagnostic marker for ovine *Trichostrongyle* infections?, *Transactions of the Royal Society of Tropical Medicine and Hygiene*, **82**, 937A.

Jones, D.G. and Knox, D.P., 1990, Evidence for the presence of nematode derived acetylcholinesterase in sheep infected with *Trichostrongylus colubriformis*, *Research in Veterinary Science*, **48**, 136–7.

Jones, V.E. and Ogilvie, B.M., 1972, Protective immunity to *Nippostrongylus brasiliensis*, III. Modulation of worm acetylcholinesterase by antibodies, *Immunology*, **22**, 119–29.

Juhasz, S. and Kassai, T., 1981, A protease inhibitor of *Nippostrongylus brasiliensis*, *Molecular and Biochemical Parasitology*, **3**, 83–90.

Karnowsky, M.J. and Roots, L., 1964, A direct colouring thiocholine method for cholinesterases, *Journal of Histochemistry and Cytochemistry*, **12**, 219–20.

Kazura, J.W. and Meshnick, S.R., 1984, Scavenger enzymes and resistance to oxygen-mediated damage in *Trichinella spiralis*, *Molecular and Biochemical Parasitology*, **10**, 1–10.

Kelly, J.D. and Ogilvie, B.M., 1972, Intestinal mast cells and eosinophil numbers during worm expulsion in nulliparous and lactating rats infected with *Nippostrongylus brasiliensis*, *International Archives of Allergy and Applied Immunology*, **43**, 497–509.

Kennedy, M.W. and Qureshi, F., 1986, Stage specific antigens secreted by parasitic larval stages of the nematode *Ascaris*, *Immunology*, **58**, 515–22.

Knox, D.P., 1990, Preliminary characterisation of proteolytic enzymes in extracts of adult *Haemonchus contortus*, communication to The British Society for Parasitology, spring meeting, Aberdeen.

Knox, D.P. and Jones, D.G., 1988, Modulatory effects of molybdenum on proteolytic enzyme activity of two ovine gastro-intestinal nematodes, *Trichostrongylus vitrinus* and *Ostertagia circumcincta*, during *in vitro* culture. (Abstract), *Transactions of the Royal Society for Tropical Medicine and Hygiene*, **82**, 937A.

Knox, D.P. and Jones, D.G., 1989, Superoxide dismutase polymorphism in closely related gastrointestinal nematodes, abstract of communication to joint spring meeting of the British, Netherlands and Belgian Societies for Parasitology with the Belgian Society of Protozoology, Spring 1989.

Knox, D.P. and Kennedy M.W., 1988, Proteinases released by the parasitic larval stages of *Ascaris suum* and their inhibition by antibody, *Molecular and Biochemical Parasitology*, **28**, 207–16.

Knox, D.P. and Jones, D.G., 1990, Studies on the presence and release of proteolytic enzymes (proteinases) in ruminant gastrointestinal nematodes, *International Journal for Parasitology*, **20**, 243–50.

Lee, D.L., 1970, The fine structure of the excretory system in adult *Nippostrongylus brasiliensis* (Nematoda) and a suggested function for the 'excretory' glands, *Tissue and Cell*, **2**, 225–31.

Lee, D.L., 1972, Penetration of mammalian skin by the infective larvae of *Nippostrongylus brasiliensis*, *Parasitology*, **65**, 499–505.

Leid, R.W., 1987, Parasite defence mechanisms for evasion of host attack, a review, *Veterinary Parasitology*, **25**, 147–62.

Leid, R.W. and Suquet, B.M., 1986, A superoxide dismutase of metacestodes of *Taenia taeniaformis*, *Molecular and Biochemical Parasitology*, **18**, 301–11.

Matthews, B.E., 1977, The passage of larval helminths through tissue barriers, in *Parasite Invasion. Symposium of the British Society for Parasitology*, **15**, 103–19.

Matthews, B.E., 1982, Skin penetration by *Necator americanus* larvae, *Zeitschrift für Parasitenkunde*, **68**, 81–6.

Matthews, B.E., 1984, The source, release and specificity of proteolytic enzyme activity produced by *Anisakis simplex* larvae (Nematoda: Ascarida) *in vitro*, *Journal of Helminthology*, **58**, 178–85.

McLaren, D.J., Burt, J.S. and Ogilvie, B.M., 1974, The anterior glands of *Necator americanus* (Nematoda: Strongyloidea) — II. Cytochemical and functional studies, *International Journal for Parasitology*, **4**, 39–46.

Michelson, A.M., 1977, Toxicity of superoxide radical anions, in Michelson, A.M., McCord, J.M. and Fridovich, I. (Eds) *Superoxide and Superoxide Dismutases*, London: Academic Press.

Miller, H.R.P., 1984, The protective mucosal response against gastro-intestinal nematodes in ruminants and laboratory animals, *Veterinary Immunology and Immunopathology*, **6**, 167–259.

North, M.J., Mottram, J.C. and Coombs, G.H., 1990, Cysteine proteinases of parasitic protozoa, *Parasitology Today*, **6**, 270–5.

Ogilvie, B. M. and Hockley, D. J., 1968, Effects of immunity on *Nippostrongylus brasiliensis* adult worms: reversible and irreversible changes in infectivity, reproduction and morphology, *Journal of Parasitology*, **54**, 1073–84.

Ogilvie, B. M., Rothwell, T. L. W., Bremner, K. C., Schnitzerling, H. J., Nolan, J. and Keith, R. K., 1973, Acetylcholinesterase secretion by parasitic nematodes. 1. Evidence for secretion of the enzyme by a number of species, *International Journal of Parasitology*, **3**, 589–97.

Parkhouse, R. M. E., 1987, Filiariasis, *CIBA Foundation Symposium No. 127*, London: John Wiley and Sons.

Petralanda, I., Yarzabal, L. and Piessens, W. F., 1986, Studies on a filarial antigen with collagenase activity, *Molecular and Biochemical Parasitology*, **19**, 51–9.

Pritchard, D. I., 1986, Antigens of gastrointestinal nematodes, *Transactions of the Royal Society of Tropical Medicine and Hygiene*, **80**, 728–34.

Rapson, E. B., Jenkins, D. C. and Topley, P., 1983, *Trichostrongylus colubriformis*: in vitro culture of parasite stages and their use for the evaluation of anthelmintics, *Research in Veterinary Science*, **39**, 90–4.

Rathaur, S., Robertson, B. D., Selkirk, M. E. and Maizels, R. M., 1987, Secretory acetylcholinesterases from *Brugia malayi* adult and microfilarial parasites, *Molecular and Biochemical Parasitology*, **26**, 257–65.

Rhoads, M. L., 1983, *Trichinella spiralis*: identification and purification of superoxide dismutase, *Experimental Parasitology*, **56**, 41–54.

Rhoads, M. L., 1984, Secretory cholinesterases of nematodes: possible functions in the host–parasite relationship, *Tropical Veterinarian*, **2**, 3–10.

Robertson, B. D., Bianco, A. E., McKerrow, J. H. and Maizels, R. M., 1989, *Toxocara canis*: proteolytic enzymes secreted by the infective larvae in vitro, *Experimental Parasitology*, **69**, 30–6.

Rothwell, T. L. W., Ogilvie, B. M. and Love, R. J., 1973, Acetylcholinesterase secretion by parasitic nematodes. II. *Trichostrongylus spp.*, *International Journal for Parasitology*, **3**, 599–608.

Rothwell, T. L. W. and Merritt, G. C., 1974, Acetylcholinesterase secretion by parasite nematodes. IV. Antibodies against the enzyme in *Trichostrongylus colubriformis* infected sheep, *International Journal for Parasitology*, **4**, 63–71.

Rothwell, T. L. W., Anderson, N., Bremner, K. C., Dash, K. M., Le Jambre, L. F., Merritt, G. C. and Ng., B. K. Y., 1976, Observations of the occurrence and specificity of antibodies produced by infected hosts against the acetylcholinesterase present in some common gastrointestinal nematode parasites, *Veterinary Parasitology*, **1**, 221–30.

Sakanari, J. A., Staunton, C. E., Eakin, A. E., Craik, C. S. and McKerrow, J. H., 1989, Serine proteases from nematode and protozoan parasites: isolation of sequence homologs using generic molecular probes, *Proceedings of the National Academy for Sciences (USA)*, **86**, 4863–7.

Sanchez-Moreno, M., Leon, P., Garcia-Ruiz, M. A. and Monteoliva, M., 1987, Superoxide dismutase activity in nematodes, *Journal of Helminthology*, **61**, 229–32.

Sanchez-Moreno, M., Monteoliva, M., Fatou, A. and Garcia-Ruiz, M. A., 1988, Superoxide dismutase from *Ascaris suum*, *Parasitology*, **97**, 345–53.

Smith, N. C. and Bryant, C., 1986, The role of host generated free radicals in helminth infections: *Nippostrongylus brasiliensis* and *Nematospiroides dubius* compared, *International Journal for Parasitology*, **16**, 617–22.

Smith, N. C. and Bryant, C., 1989a, Free radical generation during primary infections with *Nippostrongylus brasiliensis*, *Parasite Immunology*, **11**, 147–60.

Smith, N. C. and Bryant, C., 1989b, The effect of antioxidants on the rejection of *Nippostrongylus brasiliensis*, *Parasite Immunology*, **11**, 161–7.

Soulsby, E. J. L., 1985, Advances in immunoparasitology, *Veterinary Parasitology*, **18**, 303–19.

Suttle, N. F., Knox, D. P. and Jackson, F., 1988, The effect of dietary molybdenum intake on ovine host and intestinal parasite, in Keen, C., Lonnerdahl, L. and Hurley, B. (Eds) *Proceedings of the Sixth International Symposium on Trace Element Metabolism in Man and Animals*, pp. 101–2, New York: Plenum Press.

Suttle, N. F., Knox, D. P., Angus, K. W., Jackson, F. and Coop, R. L., 1989, Dietary molybdenum may enhance the inflammatory reaction to and hence rejection of gut nematodes in lambs, *Proceedings of the Nutrition Society (UK)*, **48**, 71A.

Tauber, A. I., Goetzl, E. J. and Babior, B. M., 1976, The production of superoxide (O_2^-) by human eosinophils, *Blood*, **48**, 968–71.

Thomas, R. J. and Waller, P. J., 1975, Significance of serum pepsinogen and abomasal pH levels in a field infection of *Ostertagia circumcincta* in lambs, *Veterinary Record*, **97**, 468–70.

Thorson, R. E., 1953, Immunity in the rat to *Nippostrongylus muris*, *American Journal of Hygiene*, **58**, 1–15.

Thorson, R. E., 1956, Proteolytic activity in extracts of the oesophagus of *Ancylostoma caninum* and the effect of immune serum on this activity, *Journal of Parasitology*, **42**, 21–5.

Turner, J.C. and Shanks, V., 1982, An enzyme-linked immunoassay for pepsinogen in sheep plasma, *Veterinary Parasitology*, **10**, 79–86.

Von Brand, T., 1973, Proteases, in *Biochemistry of Parasites*, New York: Academic Press. pp. 268–75.

Waller, P.J., 1990, Resistance in nematode parasites of livestock to the benzimidazole anthelmintics, *Parasitology Today*, **6**, 127–9.

Wang, C. C., 1984, Parasite enzymes as potential targets for antiparasitic chemotherapy, *Journal of Medicinal Chemistry*, **27**, 1–9.

Weller, P. F. and Goetzl, E. J., 1979, The regulatory and effector roles of eosinophils, *Advances in Immunology*, **27**, 339–71.

Wells, P. D., 1962, Mast cell, eosinophil and histamine levels in *Nippostrongylus brasiliensis* infected rats, *Experimental Parasitology*, **12**, 82–101.

Whitlock, J. H., 1960, in *Diagnosis of Veterinary Parasitisms*, pp. 141 and 167, London: Henry Kimpton.

10. Molecular approaches to the diagnosis of *Onchocerca volvulus* in man and the insect vector

W. Harnett

INTRODUCTION

The human filarial parasite, *Onchocerca volvulus*, is found predominantly in Africa with a few additional foci in the Middle East, Central America and Northern South America. The number of individuals infected with the worm is in the region of 18 million but the total number at risk may be as many as 100 million (WHO, 1987). About 1 per cent of infected individuals are blind, making onchocerciasis one of the major causes of blindness in the tropical world.

Disease associated with onchocerciasis is almost exclusively due to the larvae of the parasite (microfilariae). The larvae are released by fertilized adult female worms located in the dermis and subcutaneous tissues. Microfilariae migrate within the skin and may also invade the eyes and their presence in both anatomical locations can ultimately lead to tissue destruction (for a detailed clinical account see Anderson *et al.* (1974)). The actual mechanisms involved are still largely unknown, but it is possible that immune destruction or an immune/allergic/inflammatory response to dead or dying microfilariae may play a part. Skin lesions are exacerbated by scratching in response to the presence of microfilariae and/or their destruction. Microfilariae in the skin are taken up by blackflies (*Simulium* spp.) which feeds on humans. The microfilariae undergo a number of developmental changes to become infective larvae within the insect and enter the skin of the human host during subsequent feeding. Over a period of months they develop into adult worms and microfilariae are then released.

DETECTING *ONCHOCERCA VOLVULUS*

The need exists to be able to detect *O. volvulus* both in the insect vector and in man. The problems peculiar to detection in the insect vector are discussed in a separate section. The present section comprises an introduction to the need for molecular approaches to the diagnosis of infection in man.

In many cases, infection with *O. volvulus* is all too obvious. Heavily infected individuals in endemic regions often have such severe alteration and degeneration of the skin that onchocerciasis can be diagnosed on sight with a high degree of certainty. Similarly, the finding of blind or near-blind patients with severe eye pathology in an endemic area is frequently indicative of the disease. Severe destruction of the skin and eyes is, however, usually the product of longstanding infection, prior to which there may have been a long period in which infected individuals appeared disease free.

Clearly there exists a need to be able to diagnose onchocerciasis as soon as possible after initial infection and current methods depend on finding the parasite. This involves either locating microfilariae in skin biopsies ('skin snip') or in the eyes by ophthalmic examination. Adult worms can also be found because they frequently become encapsulated in nodules which are visible at the body surface and can be surgically removed for examination. In practice, such parasitological diagnosis of onchocerciasis is unsatisfactory because the skin-snip approach is of low sensitivity, examination of the eye requires skilled practitioners, and many adult worms are inaccessible. There is also the additional consideration that parasitological diagnosis offers no opportunity for detection of prepatent infection. This is a crucial point, as diagnosis of such infections would enable drug treatment to be initiated such that infected individuals did not (a) become infective for vectors, thereby contributing to the transmission of the disease, and (b) suffer the clinical and pathological consequences of infection.

Attempts to improve the detection of parasites have invariably relied on immunological methods, the majority of which are serological in nature (Table 10.1) and can generally be considered as more simple and convenient than skin tests. This, taken in conjunction with the potential hazard associated with the skin test, that of infection, dictates that a serological assay will remain the method of choice.

The following two sections of this chapter describe the role of molecular approaches in the pursuit of serological assays of high sensitivity and specificity, the first focusing on antibody detection, the second on detection of circulating parasite products.

Table 10.1.   Approaches to the diagnosis of onchocerciasis described to date.[a]

Parasitological
1.   Detection of microfilariae in the skin by biopsy (skin-snip)
2.   Detection of microfilariae by ocular examination
3.   Detection of adult worms in excised nodules

Detection of antibody in serum[b]
1.   Complement fixation
2.   Gel diffusion
3.   Indirect haemagglutination
4.   Immunoelectrophoresis
5.   Indirect immunofluorescence
6.   Enzyme-linked immunosorbent assay
7.   Radioimmunoprecipitation
8.   Radioimmunoassays

Detection of antigen in serum[c]
1.   Gel diffusion
2.   Radio immunoprecipitation
3.   Radio immunoassays

Skin tests
1.   Immediate hypersensitivity
2.   Delayed hypersensitivity

[a]   See the reviews given by Haque and Capron (1986) and Mackenzie *et al.* (1986).
[b]   Generally, heterologous in addition to homologous antigen was used.
[c]   Attempts to detect immune complexes were also undertaken.

DIAGNOSIS OF ONCHOCERCIASIS BY DETECTION OF ANTIBODY

The problem of specificity

The use of serological assays for the detection of antibodies to *O. volvulus* is not a recent idea. As early as 1916, Rodhain and Van den Branden were employing the complement fixation test for such a purpose and were followed by many other workers (reviewed by Mackenzie *et al.*, 1986). Other techniques which have been widely tested include gel diffusion, indirect haemagglutination, immunoelectro-phoresis and indirect immunofluorescence (for reviews see Haque and Capron (1986) and Mackenzie *et al.* (1986)). The results obtained with these approaches have been very variable and such methods are not as sensitive or as simple to perform as the more recently available radioimmunoassays (RIAs) and enzyme-linked immunosorbent assays (ELISAs). ELISAs are particularly suitable as diagnostic tests for onchocerciasis as their simplicity enables them to be performed under field conditions. The high sensitivity which is obtainable with ELISAs and RIAs could lead one to predict that sensitivity ought not be a problem with respect to diagnosis of onchocerciasis. Recent results indicate that this can be the case (Cambell *et al.*, 1983; Lujan *et al.*, 1983; Karam and Weiss, 1985; Cabrera *et al.*, 1989). Unfortunately, this high sensitivity can be associated with low specificity. A striking example of this can be found in some recent work done by Cabrera and colleagues (Cabrera *et al.*, 1989), who described an ELISA which demonstrated 100

Table 10.2. Sensitivity and specificity of ELISAs designed to detect antibody to *Onchocerca volvulus*.

Antigens	Antibody	Mexico (*n* = 25) vs. intestinal (*n* = 25)		Venezuela (*n* = 135) vs. *M. ozzardi* (*n* = 32)		Africa (*n* = 15) vs. *W. bancrofti* (*n* = 15)	
		Sensitivity (%)	Specificity (%)	Sensitivity (%)	Specificity (%)	Sensitivity (%)	Specificity (%)
PBS extract	Ig total	100	0	98	6	100	0
PBS extract	IgG$_4$	100	100	96	75	100	0
Fraction 11	Ig total	100	100	96	91	93	87
Fraction II	IgG$_4$	100	100	96	97	93	87

Plates coated with *O. volvulus* PBS extract or a purified surface antigen extract, fraction II (Cabrera and Parkhouse, 1987), were allowed to react with positive sera from confirmed cases of onchocerciasis and negative control sera from other nematode infections, as indicated. The total Ig and IgG$_4$ antibody levels bound were measured. Sensitivity values were calculated from the control positive sera and specificity values were calculated from the control negative sera (intestinal nematodes for Mexico, *M. ozzardi* for Venezuela and *W. bancrofti* for Ivory Coast). Reprinted from Cabrera *et al.* (1989) with kind permission from the authors.

per cent sensitivity but 0 per cent specificity (see Table 10.2). The reason for the zero specificity value was that the assay employed a whole worm fraction as the source of antigen for detection of antibody. Such preparations are highly cross-reactive (Almond and Parkhouse, 1985) and hence will contain epitopes recognizable by antibodies in the serum of patients harbouring other infections. There are, however, some rare exceptions to this rule (Tada *et al.*, 1987).

DEFINING SPECIFIC ANTIGENS

Clearly, the answer to the cross-reactivity problem is to characterize parasite antigens that bear species-specific epitopes. The idea of using more specific forms of antigen for diagnosis of onchocerciasis has been employed previously. It had, for instance, been considered for many years that parasite excretory–secretory material (ES) might be more specific than whole parasite extracts (Fife, 1971) and ES were therefore employed in skin tests (Schiller *et al.*, 1980; Ngu *et al.*, 1981). With respect to serological assays, one of the first attempts was probably that of Klenk and colleagues who, ironically, employed heterologous antigen material (Klenk *et al.*, 1983, 1984). Antigenic extracts of other nematodes have frequently been used for the diagnosis of onchocerciasis in the past (for reviews, see Haque and Capron (1986) and Mackenzie *et al.* (1986)) as a consequence of problems in obtaining *O. volvulus* (*O. volvulus* will not undergo complete development in convenient laboratory rodents or *in vitro*). These were again usually whole parasite extracts and, as would be expected, such an approach will generate an assay of low specificity. The use of a purified antigen, although still heterologous, gave better results. The antigen concerned was prepared from adult worms of the rodent filarial parasite, *Litomosoides carinii*, by two-step, preparative, flat bed electrofocusing in granulated gel. Analysis by immunodiffusion, latex agglutination and ELISA showed the antigen to be much more specific than the whole extract from which it was derived. This specificity was by no means perfect, however, although the purified antigen appeared to show a particular cross-reactivity with *O. volvulus* (100 per cent sensitivity), some serum samples from loiasis and bancroftian filariasis patients also gave positive results.

About the same time, a purified fraction of another readily available rodent filarial parasite, *Acanthocheilonema viteae*, was also applied to the diagnosis of onchocerciasis (Ouaissi *et al.*, 1983). This extract was purified from a whole worm (adult) homogenate by gel filtration and was shown by radioallergosorbent assay (RAST, which measures immunoglobulin E (IgE) antibody) to demonstrate a preferential cross-reactivity with *O. volvulus* amongst filarial worms. Again, however, in some cases, sera obtained from patients infected with other filarial species produced comparable positive results.

The first attempt at employing a purified antigen from *O. volvulus* was probably that of Philipp *et al.* (1984). The antigen in question was a surface component of adult worms with a molecular weight of 20 kDa. This antigen was purified by gel filtration, radiolabelled, and its specificity investigated in a radioimmunoprecipitation assay employing a range of sera. The assay was highly successful when employing sera from patients from Mexico, being able to differentiate between these sera and samples from residents of nearby, onchocerciasis-free regions, most of whom were infected with the nematodes *Ascaris lumbricoides*, *Trichuris trichuria* or hookworms (specificity 98.3 per cent). Similarly, no cross-reactivity was observed when employing serum samples from Trinidadians infected with the filariae *Mansonella ozzardi* or *Wuchereria bancrofti*. Unfortunately, however, with respect to *W. bancrofti*, the same was not true of serum samples obtained from Indian patients: 68 per cent were found to give a positive reading in the radio-immunoprecipitation assay.

The *L. carinii*, *Acanthocheilonema viteae* and *O. volvulus* antigens described above share the property of having their specificities tested after their selection. A more rational approach would be to undertake specificity analysis as a first step. This was undertaken in a comprehensive manner by Cabrera and Parkhouse (1986). Adult *O. volvulus* were divided into a number of antigen extracts, i.e. surface, phosphate buffered saline (PBS) soluble, detergent soluble (and also glycoprotein fractions of each), and examined by means of the 'Western blot' technique using pooled sera from individuals infected with a variety of filarial and non-filarial nematodes. These studies revealed two areas of specificity: surface molecules and low molecular weight somatic antigens (Figure 10.1). As a result of these findings, Cabrera and Parkhouse (1987) isolated a low molecular weight surface membrane fraction from *O. volvulus* (fraction II) and tested the specificity of this fraction in an ELISA using pooled sera from patients harbouring various nematodes. The results of this analysis indicated that fraction II was indeed restricted to *O. volvulus* amongst human nematode parasites (Figure 10.2).

Fraction II has more recently (Cabrera *et al.*, 1989) been examined for specificity using a series of individual sera from patients harbouring either of the intestinal nematodes *M. ozzardi* or *W. bancrofti*. As can be seen in Table 10.2, the use of fraction II provides a striking increase in specificity over that observed with the commonly employed PBS extract.

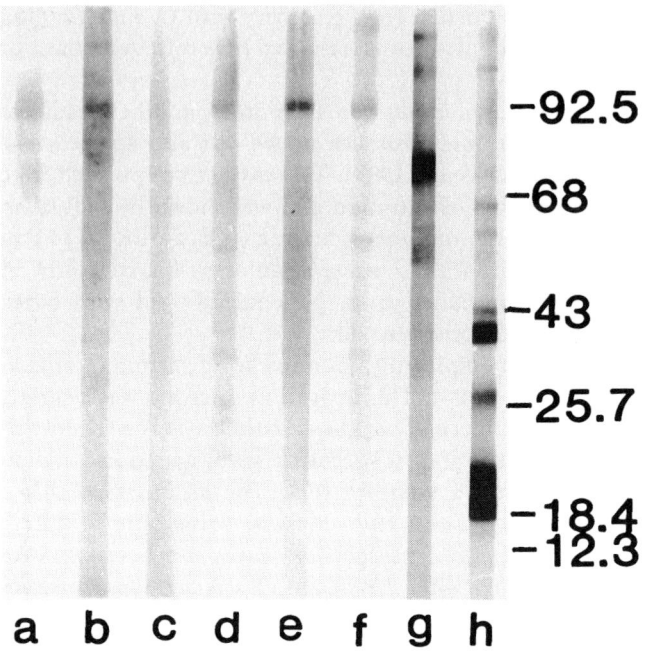

Figure 10.1. A somatic antigen extract of adult *O. volvulus* was prepared, separated by sodium dodecyl sulphate polyacrylamide gel electrophoresis (SDS-PAGE) and electrophoretically transferred to nitrocellulose paper. Antigens were revealed by exposure to a range of pooled human sera followed by ^{125}I labelled, affinity purified goat anti-human immunoglobulin. (a) Mexican, non-endemic. (b—h) Infection with: (b) *Trichinella spiralis*; (c) *Trichuris trichuris*; (d) *Mansonella ozzardi*; (e) *Brugia malayi*; (f) *W. bancrofti*; (g) *Loa loa*; (h) *Onchocerca volvulus* (Mexican isolate). Molecular weight markers (× 10⁻³) are inserted on the right-hand side of the figure. (Reprinted from Cabrera and Parkhouse (1986) with the kind permission of the authors and Elsevier Science Publishers, Amsterdam.)

The finding that *O. volvulus*-specific antigens tend to be of low molecular weight is also consistent with the results of Lobos and Weiss (1986). These workers prepared a soluble whole worm extract, iodinated the antigens and subjected the labelled molecules to immunoprecipitation and two-dimensional gel electrophoresis. By comparing the results obtained with onchocerciasis and lymphatic filariasis serum pools, it was possible to show that specific antigens were restricted to the 20 to 43 kDa range.

Recently, another apparently specific antigen has been described (Lucius *et al.*, 1988a). The rationale for defining it as specific is two-fold: first, a monoclonal antibody directed against it fails to recognize any antigen in extracts of other filarial parasites; second, the purified antigen is not recognized by a range of individual sera from patients infected with *W. bancrofti* or *B. malayi*. Both experiments employed the 'Western blot' technique: when the latter experiment was repeated using ELISA, differences between the homologous and heterologous sera were statistically significant, although not clear enough to differentiate reliably between the infections. The authors suggested that this reflects the presence of cross-reacting

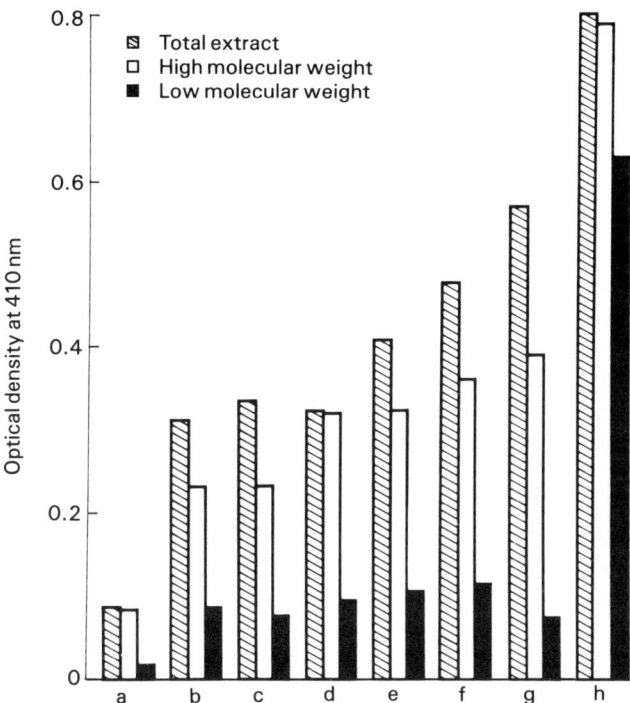

Figure 10.2. Microtitre plates were coated by incubation overnight with 100 μl per well of a total adult *O. volvulus* surface extract or a high or low molecular weight fraction of this (fractions I and II, respectively). The antigen extracts were probed with the following human serum pools: (a) normal; (b) Mexican non-endemic (anti-intestinal nematodes); (c) anti-*T. spiralis*, *T. trichuris*, *Ascaris lumbricoides* and *Ancylostoma duodenale*; (d) Brugian filariasis; (e) bancroftian filariasis; (f) *Mansonella ozzardi* infected; (g) *Loa loa* infected; (h) onchocerciasis. Plates were then developed with goat anti-human Ig alkaline phosphatase followed by appropriate substrate. (Reprinted from Cabrera and Parkhouse (1987) with the kind permission of the authors and Blackwell Scientific Publications Limited, Oxford.)

contaminants in the antigen preparation. The molecular weight of the antigen varies, depending on the stage of the parasite which is examined, and, for adult worms, it is in the low molecular weight range (20, 21, 33 and 39 kDa). The findings of three separate laboratories thus support the idea that low molecular weight *O. volvulus* antigens are particularly specific. Whether each laboratory is describing the same antigen species remains, however, to be established. It seems likely that this is not the case with the two laboratories which have purified the specific antigens, since they differ both in molecular weight as determined by Western blotting (14 and 18 kDa for fraction II (Cabrera and Parkhouse, 1987)) and anatomical origin (the molecule detected by Lucius *et al.* (1988b) is located in the parasite reproductive system and muscles).

Information on the nature of specific *O. volvulus* antigens has enabled Cabrera *et al.* (1989) to design a different form of ELISA which has the advantage of not requiring purified parasite antigen. These workers have raised a monoclonal antibody which binds to two parasite antigens of molecular weight 15 and 25 kDa (by Western blotting) and have developed an assay in which the binding of this antibody is subjected to competition from antibody in serum samples. This assay was compared with the assay employing the fraction II referred to earlier and was found to give even more impressive results with respect to specificity (and sensitivity).

CLONING OF ANTIGENS FOR USE IN ANTIBODY DIAGNOSIS

The first reported molecular cloning of a diagnostic antigen appeared in the literature in 1988 (Lucius *et al.*, 1988b). The antigen concerned was that defined by Lucius and co-workers (Lucius *et al.*, 1988a) referred to above. Cloning of the antigen was achieved by screening an adult *O. volvulus* λgt11 cDNA library with a polyvalent mouse serum prepared by immunization with monoclonal antibody-purified antigen. Sequence analysis has indicated that the cloned fragment is not significantly similar to other documented nucleotide and amino acid sequences. Northern blot analysis indicates that the antigen is expressed in both African and Central American isolates of *Onchocerca*. Hybridization experiments suggest that similar, but not identical, genes are present in *B. malayi*, and *Dirofilaria immitis*, but the differences in amino acid sequence must be sufficient to explain the specificity of the molecule. No work involving the use of the cloned antigen in a diagnostic assay has yet been reported.

A second cloned antigen for diagnostic purposes has been described by Lobos *et al.* (1990). This clone was isolated from an adult female *O. volvulus* (from Mali) λgt11 cDNA library using antibodies affinity purified from an onchocerciasis serum pool (from Mali patients). Specificity analysis by Western blotting has indicated that the clone reacts strongly with antibodies in each of the eight individual sera which make up the pool, but not with antibodies in serum pools from patients infected with *W. bancrofti*, *L. loa*, *M. perstans*, *A. lumbricoides* or *T. trichuria*.

The specific clone has an open reading frame encoding 152 amino acids and the deduced sequence gives a molecular weight of 16 850. Antibodies specific for the clone (prepared by affinity purification from the onchocerciasis serum pool) were shown by immunoprecipitation to recognize an 18 kDa *in vitro* translation product

Table 10.3. Recognition of fusion proteins expressed by selected cDNA clones by individual serum samples (Garate *et al.*, 1990).[a]

cDNA clone	O. volvulus (n = 19)			W. bancrofti (n = 18)		
	+	±	−	+	±	−
I / 1	84	11	5	0	39	61
II / 2	68	6	26	0	17	83
III / 3	37	21	42	0	0	100
IV / 12	21	37	42	6	0	94
V / 2.9	84	16	0	0	17	83

[a] +, Clearly positive; ±, weakly positive; −, negative. Analysis undertaken by Western blotting.

and by Western blotting to recognize two native proteins of molecular weight 24 and 26 kDa (three polypeptides of 17, 22 and 40 kDa were also recognized weakly). Again, therefore, the association between specificity and low molecular weight *Onchocerca* antigens prevailed.

More recently, Garate *et al.* (1990) have applied a variety of differential screening procedures (e.g. detection by anti-*O. volvulus*, but not anti-*W. bancrofti* serum pools) to the isolation of cloned antigens for diagnostic purposes. The β-galacto-sidase fusion proteins of five clones isolated from an adult female *O. volvulus* cDNA library (Donelson *et al.*, 1988) were examined (by Western blotting) using individual serum samples from patients harbouring either *O. volvulus* or *W. ban-crofti*. The results of this analysis are shown in Table 10.3. None of the cloned antigens was found to be universally recognized by all the *O. volvulus* sera tested. However, all individual serum samples but one clearly recognized either clone II/2 or clone V/2.9 or both, the exception giving a weak positive reaction with V/2.9. In addition, clones II/2 and V/2.9 also exhibited promising specificity characteristics, because neither was clearly recognized by any bancroftian filariasis serum sample. Although some weak reactions were detected (approximately 17 per cent in each case) none of the sera involved demonstrated this with respect to both antigens. Thus, a combination of clones II/2 and V/2.9 may merit further analysis by incor-poration into a convenient form of serological assay such as the ELISA.

Variations in the antibody response of individuals to recombinant peptides and the use of a combination of peptides to provide a sensitive assay have also recently been investigated by Maizels *et al.* (1990). These workers isolated a large number (22) of apparently *Onchocerca*-specific clones from a λ-gt11 cDNA library (Donelson *et al.*, 1988) by differential screening using *O. volvulus* and *W. bancrofti* serum pools. The clones were examined for reactivity with 31 individual *O. volvulus* sera using a micro-spot lysis technique and considerable variation was observed. This extended from one cloned polypeptide which was recognized by all 31 sera, to one which was recognized by only three sera. Two of the most frequent recom-binants were subcloned into a plasmid vector (pNGS8+) which is designed for higher production of inserted protein (also with the minimal presence of bacterial protein, thus reducing background). When the cloned antigens were then employed in combination in an ELISA, they were clearly able to show a strong difference in signal between pooled human serum from *O. volvulus* and *W. bancrofti* infection. It is of interest to observe how this system performs with respect to individual samples.

IMPROVED SPECIFICITY OF IMMUNOGLOBULIN ISOTYPE-RESTRICTED ASSAYS

Several workers have demonstrated that the specificity of assays which detect anti-*Onchocerca* antibodies may be improved by focusing on a particular immuno-globulin class or subclass. The first indication of differences in the specificity of different isotypes to parasite antigens was provided by Weiss *et al.* (1982) who

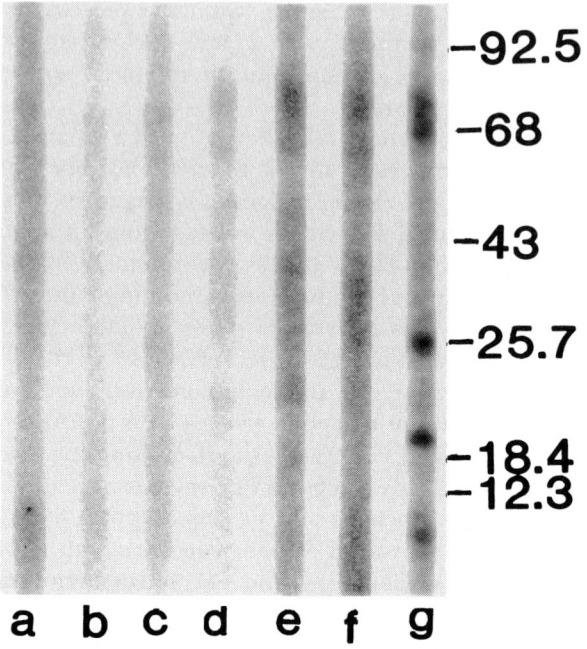

Figure 10.3. Antigens in total extract of adult *O. gibsoni* recognized by IgE antibodies present in a pool of human serum: (a) normal European; (b) Mexican non-endemic; (c) *T. spiralis, T. trichuris, A. lumbricoides, A. duodenale*; (d) *B. malayi*; (e) *W. bancrofti*; (f) *L. loa*; (g) *O. volvulus*. The total parasite extract was sequentially probed with the sera indicated and then ^{125}I-mouse monoclonal anti-human IgE. Molecular weight markers (× 10^{-3}) are shown on the right-hand side of the figure. (Reprinted from Cabrera *et al.* (1986) with the kind permission of the authors and Georg Thième Verlag, Stuttgart.)

demonstrated that the immunoglobulin E (IgE) response to *O. volvulus* infection was more species-specific than the IgG one. The greater specificity of the IgE response to *Onchocerca* has also been noted by Cabrera *et al.* (1986). The approach adopted by this group involved determining the recognition patterns of distinct human antibody classes to individual antigens of *O. gibsoni*. (The antigens of *O. volvulus* and *O. gibsoni* are virtually indistinguishable (Cabrera and Parkhouse, 1986).) The IgE response was found to be both the more restricted and the more specific one (Figure 10.3).

A perfect example of the effect which restricting the isotype of antibody being detected can have on specificity is provided by Cabrera *et al.* (1989) (see Table 10.2). Employing an ELISA to measure antibodies to *O. volvulus* in sera from known Mexican patients, they were able to shift the specificity from 0 to 100 per cent when focusing solely on IgG_4 antibody rather than total antibody to a whole parasite extract. Sera from Mexican individuals infected with intestinal helminths were employed as the negative controls. It should be noted, however, that the use of control serum samples obtained from patients harbouring filarial infections produced inferior results. Thus, when employing sera from Venezuelans infected with *O. volvulus* or *M. ozzardi* the specificity obtained reached only 75 per cent:

when comparing African onchocerciasis and filariasis patients it remained at zero. Nevertheless, with respect to the Mexican situation, the result obtained with the isotype-specific assay is an important one, not least because it negates the need to provide a specific antigen to ensure the necessary level of specificity required.

Returning to IgE, this isotype may provide increased sensitivity in addition to offering the possibility of increased specificity in serological assays. Karam and Weiss (1985) have demonstrated that detection of onchocerciasis in young children is much more readily achieved by measurement of IgE antibodies against *O. volvulus* by RAST than by measurement of IgG by ELISA.

SEROLOGICAL DIFFERENTIATION OF *O. VOLVULUS*

O. volvulus is widely distributed in tropical Africa and a great deal of circumstantial evidence exists (reviewed by Awadzi and Duke, 1984) to suggest that different strains have evolved. The most clearly defined examples of this idea are termed, in relation to their habitat, 'forest' or 'savannah'. The distinction is potentially important because the savannah form of the worm is thought to induce a greater degree of pathology and disease (reviewed by Awadzi and Duke, 1984). This, taken in conjunction with the recent suggestion that blackflies harbouring savannah infective larvae are moving into areas formerly classified as forest (Garms, 1987), demonstrates the requirement for a reliable method of identification of what are morphologically indistinguishable parasites. The recent description of a cloned DNA fragment unique to the forest form (Erttmann *et al.*, 1987) would appear to meet this need. Nevertheless, the idea of serological differentiation based on detection of a forest- or savannah-specific antibody response might be considered as an additional tool. There is some evidence in the literature to suggest that the two strains do possess some antigenic differences (Bryceson *et al.*, 1976; Lobos and Weiss, 1985), but results supporting the contrary also exist (Lucius *et al.*, 1987). In addition, it is possible to find differences in individual worms removed from one patient (Bryceson *et al.*, 1976) and between isolates from different forest regions (Lucius *et al.*, 1987). Clearly, therefore, the situation is complex and a confirmed example of a forest- or savannah-specific antigen remains to be found. Serological differentiation based on a unique antibody response is thus not feasible at present. Nevertheless, one parameter which might aid differentiation and which may deserve further study concerns the IgG response: Lucius *et al.* (1987) have found that, in general, it tends to be stronger and more differentiated in savannah patients.

SEROLOGICAL DIFFERENTIATION OF DISEASE STATUS

The nature of the pathology associated with infection with *O. volvulus* can vary. A common distinction is that of patients suffering from the generalized — as opposed to the localized — form of the disease. The latter form, as its name implies, tends to

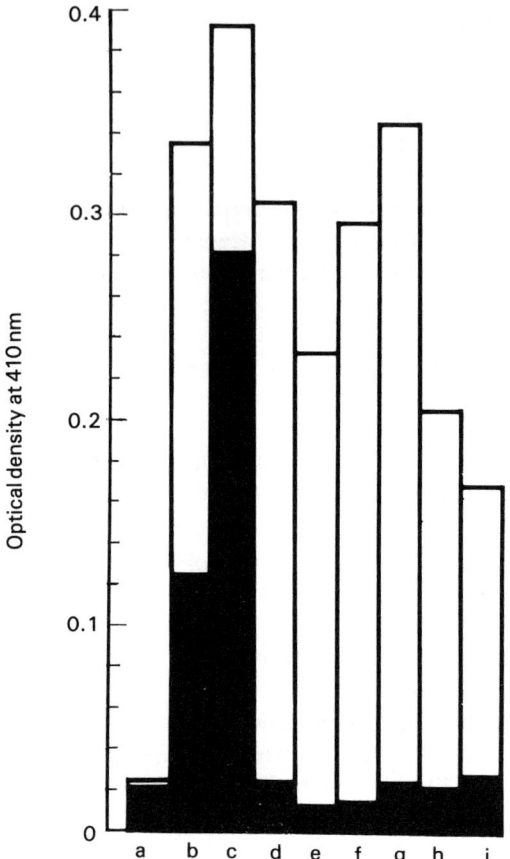

Figure 10.4. Microtitre plates were sensitized with a PBS extract of female *O. volvulus* and then probed with pools of: (a) normal human serum; (b) sowda from the Yemen; (c) sowda from Liberia; (d) generalized onchocerciasias from Venezuela; (e) generalized onchocerciasis from Mexico; (f) generalized onchocerciasis from Liberia; (g) *B. malayi*, *W. bancrofti* and *L. loa*; (h) *M. ozzardi*; (i) intestinal nematodes. The bound human antibodies were revealed by applying either a mouse monoclonal antibody to total IgG (□) or to IgG$_3$ (■). The plate was finally developed with goat anti-mouse Ig-phosphatase and *p*-nitrophenylphosphate. (Reprinted from Cabrera *et al.* (1988) with the kind permission of the authors.)

be associated with distinct, localized areas of skin in which there is a particularly severe pathology and a very low microfilarial density (Buttner *et al.*, 1982; Piessens and Mackenzie, 1982). This may relate to an effective antibody-dependent, anti-microfilaria immune mechanism, since the localized form of the disease is associated with a greater detectable antibody response (Buttner *et al.*, 1982; Lucius *et al.*, 1986). Analysis of the antibody response of sera from the two groups is an attractive idea in relation to finding a qualitative difference in antigen recognition. This might not only provide a clue to a mechanism of microfilaria killing, but might also offer a possibility of diagnosis or prognosis. Such a tool would be of value as infection may progress from localized to generalized with increasing age (Buttner *et al.*, 1982).

A consistent qualitative difference in antibody response was sought by Lucius *et al.* (1986), but was not found. These workers, however, restricted their analysis to antigens recognized by a whole antibody class (IgG or IgM). In contrast, by investigating the antigen recognition profiles of distinct Ig subclasses, Cabrera *et al.* (1988) were able to demonstrate a unique recognition of a low molecular weight antigen by IgG_3 antibodies in sera from patients with localized onchocerciasis. This finding allowed the development of an IgG_3-specific ELISA assay which, when using a crude PBS extract as antigen, provided a signal specific for the 'sowda' syndrome of onchocerciasis (Figure 10.4). It is clearly possible to imagine such an assay being employed to monitor the progression from localized to generalized disease.

DETECTION OF ANTIGEN

A major criticism of serological tests for antibody detection is that they cannot yield absolute proof of a current infection, since a positive result may simply indicate exposure to the parasite or evidence of a past infection. It was shown many years ago by Franks (1946), that circulating parasite antigen was likely to be present in the bloodstream of patients with lymphatic filariasis, and it is generally accepted that such molecules are indicative of an active infection. Clearly, if such molecules were also present in *O. volvulus*-infected individuals, then the opportunity existed to promote their detection and hence offer an alternative approach to diagnosis by measurement of antibody.

ATTEMPTS TO MEASURE CIRCULATING ANTIGENS

Although the presence of circulating antigens in the bloodstream of onchocerciasis patients was earlier suggested by the finding of immune complexes (Paganelli *et al.*, 1980; Steward *et al.*, 1982), the first clear demonstration was published by Ouaissi *et al.* (1981). These workers employed a rabbit antiserum against a soluble extract of adult worms to demonstrate the presence of circulating antigen in African onchocerciasis patients. The sensitivity obtained depended on the nature of the assay employed, the maximum being in the region of 75 per cent with techniques such as the radioimmunoprecipitation-PEG assay (RIPEGA). No relationship between the presence of circulating antigen and skin microfilaria density was observed. The specificity of the assay was rather poor, the rabbit antiserum also detecting antigen in serum samples from patients harbouring other filarial worms.

An improvement in specificity was generated by the introduction of a monoclonal antibody into the RIPEGA (Des Moutis *et al.*, 1983), although a few individuals infected with other filarial parasites were still found to give positive results. The level of sensitivity was found to be similar to that obtained using a polyclonal serum. Again, no association between circulating antigen presence and skin micro-filaria density was observed.

More recently, antigen detection has been attempted using serum samples from individuals present in the onchocerciasis endemic region in Chiapas, Mexico (Schlie-Guzman and Rivas-Alcala, 1989). The technique employed was the indirect ELISA and the results obtained indicated a high degree of sensitivity (92.3 per cent). In addition, antigen could be detected in urine samples (85.9 per cent sensitivity), thereby raising the possibility of future analysis avoiding the incon-venience of blood sampling. The correlation between test values for serum and urine samples was, moreover, found to be good. Furthermore, unlike the two studies described above, good correlation was also observed between the amount of antigen in serum or urine and the estimated skin microfilaria density.

These impressive results would appear to indicate that the indirect ELISA employed by Schlie-Guzman and Rivas-Alcala (1989) is a strong candidate for a routine screening test for onchocerciasis. It should be noted, however, that the specificity of the antiserum used was not defined and, since it was raised against a crude soluble antigen extract, one must conclude that it is likely to have multiple specificities, some of which will be directed against epitopes common to many nematodes (Cabrera *et al.*, 1989). An example of such an epitope is phosphoryl-choline (see Maizels and Selkirk, 1989) and it has been shown, using monoclonal antibodies, that phosphorylcholine can be detected in the serum of Mexican oncho-cerciasis patients (W. Harnett, unpublished observations). The value of non-specific assays in areas where more than one human filarial parasite co-exists is likely to be limited. Before the assay of Schlie-Guzman and Rivas-Alcala (1989) can be considered for use in Africa and South America, therefore, its specificity must be defined.

IMPROVEMENT OF ASSAY SENSITIVITY

With the exception of the indirect ELISA employed by Schlie-Guzman and Rivas-Alcala (1989), the levels of sensitivity observed in the above assays result in approxi-mately one-quarter of known onchocerciasis cases being scored as negative. One possible explanation for this unsatisfactory situation concerns the interfering role which the host antibody can play. It has been demonstrated by competitive RIAs that host antibodies inhibit binding of ^{125}I-labelled monoclonal antibodies to parasite antigens present in serum samples obtained from a chimpanzee experi-mentally infected with *O. volvulus* (Weiss *et al.*, 1986). Interference can also be demonstrated when measuring antigen in serum samples obtained from lymphatic filariasis patients (Hamilton *et al.*, 1984) and in jirds infected with *A. viteae* (Harnett *et al.*, 1990b) (Figure 10.5). In addition, the rate at which filarial antigens are removed from the circulation is in some cases increased by the presence of complementary antibody (Harnett *et al.*, 1989b). Clearly, therefore, the presence of a host antibody response has the potential to lower the sensitivity of antigen detection assays.

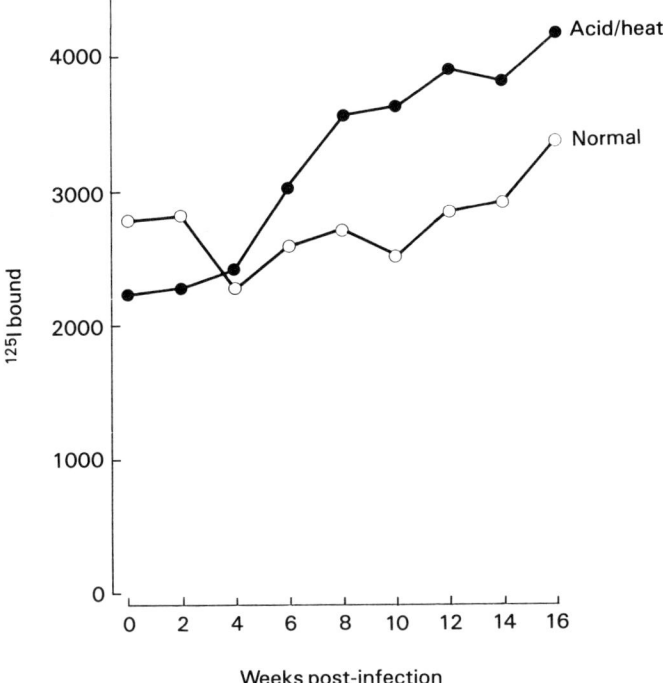

Figure 10.5. A jird was infected with *A. viteae* and serial bleeds taken. Parasite antigen present in each bleed was measured by one-site sandwich RIA in which one of the antibodies was iodinated. Samples were measured for antigen content both in their unaltered form (normal) and after treatment with acid and heat to dissociate immune complexes and destroy antibody binding capacity, respectively.

This suggests that individuals with low levels of parasites might escape detection and this has in fact been demonstrated using the *A. viteae*/jird model system (Harnett *et al.*, 1990b). Here, phosphorylcholine containing antigens can be detected (and correlated with worm burden) in the presence of an antibody response, but only if a certain level of infection is present. Animals having a worm burden below this level score as negative, unless their serum samples are treated to dissociate antigen from antibody. It should be noted that dissociating antigen from antibody is an approach to eliminating the problem of antibody interference, but, since such procedures usually involve denaturation of antibody by boiling, the target antigen must be heat stable.

Another factor which could theoretically play a role in generating low assay sensitivity is the employment of an antibody whose target molecule is restricted to a certain stage of parasite development, or whose rate of release during the life cycle is variable. For instance, patients harbouring only adult worms and microfilariae would not be detected using an antibody whose target is only released by L3s. *Litomosoides carinii*, a filarial parasite of the cotton rat, exemplifies this phenomenon (both qualitative and quantitative changes) at least as determined by characterization of antigens released *in vitro* (ES) (Harnett *et al.*, 1986). These findings imply that the ideal target molecule for antigen detection would be a non-immunogenic product, continuously released during development and adulthood, and which persists in the circulation. Again, therefore, a rational strategy would appear to be careful analysis of worm products prior to consideration for use in assays. At this point, however, the problem of obtaining adequate supplies of parasite material must once again be faced. Thus, although some attempt to characterize the ES of adults (Parkhouse *et al.*, 1987) and microfilariae (Ngu *et al.*, 1981) has been possible, it is fair to say that the nature of the ES of *O. volvulus* is still largely unknown. With respect to L4 and young adult stages, this situation is likely to persist until the problem of inducing development *in vitro* has been solved. Nevertheless, it should be stated that the ES of adult worms contain components not recognized by antibodies in infection serum (Parkhouse *et al.*, 1987). It cannot be ruled out at present, however, that these are of host, rather than parasite, origin, as both the ES of *O. volvulus* microfilariae (Ngu *et al.*, 1981) and adult worms of other filarial species (Kaushal *et al.*, 1982; Maizels *et al.*, 1985; Harnett *et al.*, 1986) have been shown to contain host products.

IMPROVEMENT OF ASSAY SPECIFICITY

As with antibody diagnosis, the key to specificity lies in the analysis of ES for reactivity with antibodies directed against a range of nematodes. That this analysis has not been carried out in any depth probably reflects the dearth of available ES. Nevertheless, the presence of at least one specific molecule is suggested by immunoelectrophoretic analysis (Ouaissi *et al.*, 1981). In addition, the finding that an *O. volvulus*-specific, monoclonal antibody is targeted against the parasite intestine, an area of ES production (Nogami *et al.*, 1988), has led to the suggestion that it could be employed for detection of circulating antigen. Other monoclonal antibodies which may be of value are those prepared by Harnett *et al.* (1990a) against non-PC epitopes of *Onchocerca* ES. It is not yet fully clear as to whether the epitopes in question are *Onchocerca*-specific, but preliminary work is consistent with this view in some cases.

DETECTION OF INFECTIVE LARVAE IN BLACKFLIES

Transmission of onchocerciasis takes place in areas in which *Onchocerca* species which are parasitic in animals are prevalent. Like *O. volvulus*, the animal parasites may be transmitted by blackflies and, although there is a tendency for distinct species of blackfly to be either man-biting or zoophilic, this is by no means absolute (Nelson, 1970; Baker *et al.*, 1985; Porter and Collins, 1988). A situation thus exists in which blackfly species which predominantly feed on man may acquire microfilariae of animal *Onchocerca* species. Similarly, zoophilic flies may take up *O. volvulus* microfilariae. Although, in many cases, development to the L3 stage does not take place (Porter and Collins, 1988), this is not always the case. Thus anthropophilic blackflies may harbour 'animal' infective larvae and vice versa (Duke, 1967; Nelson, 1970; Omar *et al.*, 1979; Porter and Collins, 1988).

An urgent need exists to be able to assess the extent of infection with *O. volvulus* in blackfly populations such that epidemiological surveys can be accurately undertaken and control programmes satisfactorily monitored. Two areas in which this is particularly relevant at present are:

1. in the monitoring of the onchocerciasis control programme in West Africa. This is a large, multinational effort, the strategy of which is based on blocking transmission by the elimination of blackfly larvae using insecticides, but which is compromised by reinvasion by flies from outside the control area; and

2. in assessing the impact of chemotherapy, using the new drug, Ivermectin, on disease transmission.

Unfortunately, however, the possible presence of infective larvae of *Onchocerca* species which are parasites of animals makes these goals difficult to attain, as the 'animal' infective larvae may be morphologically indistinguishable from *O. volvulus* (Nelson and Pester, 1962; Duke, 1967; Omar *et al.*, 1979). Clearly, therefore, an alternative method of differentiation is required.

DNA PROBES FOR DIFFERENTIATION OF INFECTIVE LARVAE

It is immediately possible to conceive of two molecular approaches which would provide the specificity necessary to differentiate infective larvae of *O. volvulus* from those of closely related species. These are the production of a monoclonal antibody to a species-specific epitope, and the generation of a DNA probe. Both of these approaches have been successfully applied to differentiating L3s of *B. malayi* from other filarial species, including the closely related *B. pahangi* (Carlow *et al.*, 1987; Williams *et al.*, 1988). With respect to *Onchocerca*, experimental efforts have been restricted to the production of a DNA probe. This may again reflect problems in the acquisition of *Onchocerca* infective larvae. The production of DNA probes can be achieved using material from any stage of the life cycle, but preparing a monoclonal antibody may necessitate the use of infective larvae themselves (the monoclonal antibody specific for *B. malayi* is also stage specific (Carlow *et al.*, 1987)).

The first two *O. volvulus* cloned DNA fragments which were used to probe DNA samples from other *Onchocerca* species were described by Perler and Karam (1986). These clones were able to differentiate *O. volvulus* from other *Onchocerca* species by restriction fragment length polymorphisms but not by dot blot hybridization.

A further series of clones was prepared by Shah *et al.* (1987). This series was found to be interrelated, although individual members varied in their specificities. None was able to differentiate DNA of *O. volvulus* from that of the closely related *O. ochengi* (Bain, 1981) by dot blot hybridization, although successful results were obtained with Southern blot anlaysis. Being able to distinguish between *O. volvulus* and *O. ochengi* is considered to be of particular importance in view of the close morphological similarity of infective larvae of the two parasites (Omar *et al.*, 1979).

An oligonucleotide DNA probe (Figure 10.6) which appears to have this latter property was prepared by Harnett *et al.* (1989a). Initially, a clone which failed to hybridize with total DNA from *O. gibsoni* and *O. gutturosa* was isolated from a genomic DNA library in λgt10. The insert of the clone was subcloned into the filamentous phage vector, M13mp18, and sequenced. Two oligonucleotides, each corresponding to a unique region of 60 nucleotides (out of a total of 154) were synthesized and examined for hybridization with three different geographical isolates of *O. volvulus* (including forest and savannah strains) and six other *Onchocerca* species, including *O. ochengi*. One of the oligonucleotides (C1A1–2) was found to hybridize to the three *O. volvulus* isolates with an intensity in the region of 300-fold greater than to any other *Onchocerca* species. This oligonucleotide would thus appear to represent the first *O. volvulus*-specific DNA probe. It is currently being tested directly on infective larvae spotted onto nitrocellulose filters and then broken open by chemical means.

A second, apparently *O. volvulus*-specific DNA probe has also been described (Meredith *et al.*, 1989). This clone consists of 12 copies of a 150 base pair sequence, the consensus sequence of which is nearly identical to the clone isolated by Harnett *et al.* (1989a). It is not surprising, therefore, that it possesses species-specific properties.

DNA probes specific for the forest (Erttmann *et al.*, 1987) and savannah

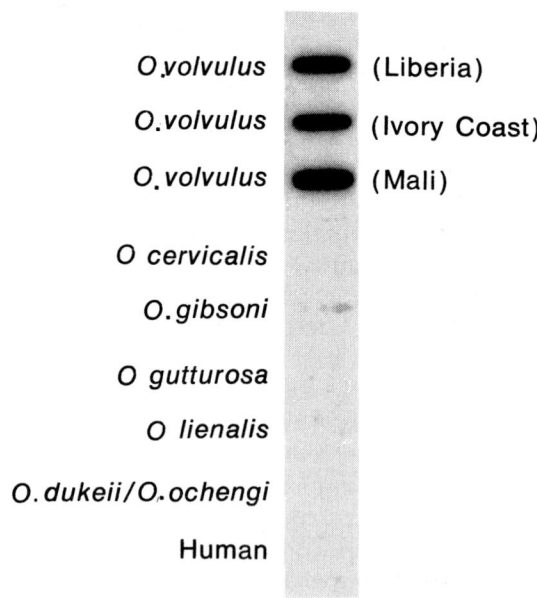

Figure 10.6. Genomic DNA samples were bound to nitrocellulose and exposed to an oligonucleotide derived from a cloned *O. volvulus* DNA fragment. The oligonucleotide hybridizes solely with *O. volvulus* DNA. (From Harnett *et al.*, 1989a).

(Erttmann *et al.*, 1990) forms of *O. volvulus* have also been described. The insert of the former consists of 153 nucleotides and is related in sequence to the *O. volvulus*-specific clones described above. There would thus appear to be a family of closely related repeat sequences in the *O. volvulus* genome and small differences in the composition of these repeats can lead to differences in hybridization properties. This was also evident when attempting to isolate a savannah-specific probe. Thus the specificity of the latter probe is dependent upon a 67 base pair fragment derived from a clone containing two copies of the repeat, originally isolated from an *O. volvulus* savannah form DNA library.

CONCLUSIONS AND FUTURE PROSPECTS

Antibody detection

The introduction of molecular approaches in the form of antigen characterization and monoclonal antibody production has led to the development of ELISAs which can specifically detect infection with *O. volvulus* in virtually all individuals examined under laboratory conditions. Some success has also been achieved by focusing on particular antibody isotypes. The ELISAs are simple to perform and do not require any specialized equipment: they are therefore likely to be suitable for use under field conditions.

The only disadvantage of the ELISAs is their requirement for parasite material as a source of antigen. It should theoretically be possible to circumvent this problem by using monoclonal antibody inhibition assays or by molecular cloning, to which end a number of antigens which are possibly unique to *O. volvulus* have now been cloned. Further specificity testing on these antigens is awaited.

The possibility of detecting antibody responses indicative of certain 'disease forms' of onchocerciasis or restricted to particular strains of the parasite has received some attention. Success has been obtained in relation to the former with the finding of a unique IgG_3 antibody response to a distinct parasite antigen associated with patients having the severe localized form of the disease. Concerning the latter, as yet parasite strains cannot be differentiated by the antibody response they elicit and indeed the concept of a strain-specific antigen is not clearly established.

As a final thought, a differential form of antibody diagnosis which is amenable to a molecular approach and which may receive increasing attention in the near future, is stage-specific diagnosis. This would be particularly useful with respect to detecting reinfection, i.e. prepatent infection in control areas.

Antigen detection

At present, no assay exists which has been shown to possess the necessary specificity. Locating an antigen which has the properties to provide such specificity is hampered by the lack of parasite material which is available to researchers. There are, however, one or two indicators which suggest that ES products with specific epitopes do exist and that monoclonal antibodies directed against them have been prepared. It is to be hoped that these latter reagents will stand up to further extensive analysis of specificity. More extensive ES characterization, perhaps enabled by advances in the maintenance of the parasite under laboratory conditions, is also desirable, particularly in view of the need to investigate the stage-specificity of *Onchocerca* ES.

Infective larva detection

Probes now exist that are specific for *O. volvulus* forest and savannah strains. C1A1–2, an oligonucleotide probe which is species specific, is currently being tested directly on infective larvae spotted onto nitrocellulose and broken open by chemical means. This, in combination with a non-radioactive indicator system, should permit the use of the probe under field conditions.

ACKNOWLEDGEMENTS

I would like to thank Franz Conraths and Michael Worms for critical reading of the manuscript.

REFERENCES

Almond, N. M. and Parkhouse, R. M. E., 1985, Nematode antigens, in Parkhouse, R. M. E. (Ed.) *Current Topics in Microbiology*, Vol. 120, pp. 173–203, Heidelberg: Springer-Verlag.

Anderson, J., Fuglsang, H., Hamilton, P. J. S. and Marshall, T. F. de C., 1974, Studies on onchocerciasis in the United Cameroon Republic. 1. Comparison of populations with and without *Onchocerca volvulus*, *Transactions of the Royal Society of Tropical Medicine and Hygiene*, **68**, 190–208.

Awadzi, K. and Duke, B., 1984, Onchocerciasis, in Gilles, H. M. (Ed.) *Recent Advances in Tropical Medicine*, No. 1, pp. 153–70, Edinburgh: Churchill Livingstone.

Bain, O., 1981, Le genre *Onchocerca*: hypothesis sur son evolution et cle dichotomique des espèces., *Annales de Parasitologie*, **56**, 503–26.

Baker, R. H. A., Mustafa, M. B. and Abdelnur, O. M., 1985, The current status of Simulium species in the Sudan with reference to onchocerciasis, *Sudan Medical Journal*, **21** (Suppl.), 19–27.

Bryceson, A. D. M., van Keen, K. S., Adulojo, A. J. and Duke, B. O. L., 1976, Antigenic diversity among *Onchocerca volvulus* in Nigeria, and immunological differences between onchocerciasis in the savanna and forest of Cameroon, *Clinical and Experimental Immunology*, **24**, 168–76.

Buttner, D. W., V. Laer, G., Mannweiler, E. and Buttner, M., 1982, Clinical, parasitological and serological studies on onchocerciasis in the Yemen Arab Republic, *Tropenmedizin und Parasitologie*, **33**, 201–12.

Cabrera, Z. and Parkhouse, R. M. E., 1986, Identification of antigens of *Onchocerca volvulus* and *Onchocerca gibsoni* for diagnostic use, *Molecular and Biochemical Parasitology*, **20**, 225–31.

Cabrera, Z. and Parkhouse, R. M. E., 1987, Isolation of an antigenic fraction for diagnosis of onchocerciasis, *Parasite Immunology*, **9**, 39–48.

Cabrera, Z., Cooper, M. D. and Parkhouse, R. M. E., 1986, Differential recognition patterns of human immunoglobulin classes to antigens of *Onchocerca gibsoni*, *Tropical Medicine and Parasitology*, **37**, 113–16.

Cabrera, Z., Buttner, D. W. and Parkhouse, R. M. E., 1988, Unique recognition of a low molecular weight *Onchocerca volvulus* antigen by IgG$_3$ antibodies in sowda-type onchocerciasis, *Clinical and Experimental Immunology*, **74**, 223–9.

Cabrera, Z., Parkhouse, R. M. E., Forsyth, K., Gomez Priego, A., Pabon, R. and Yarzabal, L., 1989, Specific detection of human antibodies to *Onchocerca volvulus*, *Tropical Medicine and Parasitology*, **40**, 454–9.

Cambell, C. C., Figueroa, H., Collins, M. R. C., Lujan, R. and Collins, W. E., 1983, Diagnosis of *Onchocerca volvulus* infection in Guatemalan children, *American Journal of Tropical Medicine and Hygiene*, **32**, 760–3.

Carlow, C. K. S., Franke, E. D., Lowrie, R. C., Partono, F. and Philipp, M., 1987, Monoclonal antibody to a unique surface epitope of the human filaria *Brugia malayi* identifies infective larvae in mosquito vectors, *Proceedings of the National Academy of Sciences (USA)*, **84**, 6914–18.

Des Moutis, I., Ouaissi, A., Grzych, J. M., Yarzabal, L., Haque, A. and Capron, A., 1983, *Onchocerca volvulus*: detection of circulating antigen by monoclonal antibodies in human onchocerciasis, *American Journal of Tropical Medicine and Hygiene*, **32**, 533–42.

Donelson, J. E., Duke, B. D. L., Moser, D., Zeng, W., Erondu, N. E., Lucius, R., Renz, A., Karam, M. and Zea Flores, G., 1988, Construction of *Onchocerca volvulus* cDNA libraries and partial characterization of the cDNA for a major antigen, *Molecular and Biochemical Parasitology*, **311**, 241–50.

Duke, B. O. L., 1967, Infective filarial larvae other than *Onchocerca volvulus* in *Simulium damnosum*, *Annals of Tropical Medicine and Parasitology*, **61**, 200–5.

Erttmann, K. D., Unnasch, T. R., Greene, B. M., Albiez, E. J., Boateng, J., Denke, A. M., Ferraroni, J. J., Karam, M., Schulz-Key, H. and Williams, P. N., 1987, A DNA sequence specific for forest form *Onchocerca volvulus*, *Nature*, **327**, 415–17.

Erttmann, K. D., Meredith, S. E. O., Greene, B. M. and Unnasch, T. R., 1990, Isolation and characterisation of form-specific DNA sequences of *O. volvulus*, *Acta Leidensia*, **59**, 253–60.

Fife, E. H., Jr, 1971, Advances in methodology for immunodiagnosis of parasitic diseases, *Experimental Parasitology*, **30**, 132–63.

Franks, M. B., 1946, Specific soluble antigen in the blood of filarial patients, *Journal of Parasitology*, **32**, 400–6.

Garate, T., Conraths, F. J., Harnett, W., Buttner, D. W. and Parkhouse, R. M. E., 1990, Cloning of specific diagnostic antigens for *Onchocerca volvulus*, *Tropical Medicine and Parasitology*, **41**, 245–50.

Garms, R., 1987, Occurrence of the savanna species of the *Simulium damnosum* complex in Liberia, *Transactions of the Royal Society of Tropical Medicine and Hygiene*, **81**, 518.

Hamilton, R. G., Hussain, R. and Ottesen, E. A., 1984, Immunoradiometric assay for detection of filarial antigens in human serum, *Journal of Immunology*, **133**, 2237–42.

Haque, A. and Capron, A., 1986, Filariasis: antigens and host–parasite interactions, in Pearson, T. W. (Ed.) *Parasite Antigens*, pp. 317–402, New York: Marcel Dekker.

Harnett, W., Meghji, M., Worms, M. J. and Parkhouse, R. M. E., 1986, Quantitative and qualitative changes in production of excretions/secretions by *Litomosoides carinii* during development in the jird (*Meriones unguiculatus*), *Parasitology*, **93**, 317–31.

Harnett, W., Chambers, A. E., Renz, A. and Parkhouse, R. M. E., 1989a, An oligonucleotide probe specific for *Onchocerca volvulus*, *Molecular and Biochemical Parasitology*, **35**, 119–26.

Harnett, W., Worms, M. J., Kapil, A., Grainger, M. and Parkhouse, R. M. E., 1989b, Origin, kinetics of circulation and fate *in vivo* of the major excretory–secretory product of *Acanthocheilonema viteae*, *Parasitology*, **99**, 229–39.

Harnett, W., Patterson, M., Preece, G. and Parkhouse, R. M. E., 1990a, Excretions–secretions of adult male *Onchocerca gibsoni*, submitted for publication.

Harnett, W., Worms, M. J., Grainger, M., Pyke, S. D. M. and Parkhouse, R. M. E., 1990b, Association between circulating antigen and parasite load in a model filarial system, *Acanthocheilonema viteae* in jirds, *Parasitology*, in press.

Karam, M. and Weiss, N., 1985, Seroepidemiological investigations of onchocerciasis in a hyperendemic area of West Africa, *American Journal of Tropical Medicine and Hygiene*, **34**, 907–17.

Kaushal, N. A., Hussain, R., Nash, T. E. and Ottesen, E. A., 1982, Identification and characterization of excretory–secretory products of *Brugia malayi*, adult filarial parasites, *Journal of Immunology*, **129**, 338–43.

Klenk, A., Geyer, E., Zahner, H. and Trojan, H. J., 1983, Isolation of antigen from *Litomosoides carinii* macrofilariae detecting serum antibodies due to *Onchocerca volvulus*, *Zeitschrift für Parasitenkunde*, **69**, 377–86.

Klenk, A., Geyer, A. and Zahner, H., 1984, Serodiagnosis of human onchocerciasis: evaluation of sensitivity and specificity of a purified *Litomosoides carinii* adult worm antigen, *Tropenmedizin und Parasitologie*, **35**, 81–4.

Lobos, E. and Weiss, N., 1985, Immunochemical comparison between worm extracts of *Onchocerca volvulus* from savanna and rain forest, *Parasite Immunology*, **7**, 333–47.

Lobos, E. and Weiss, N., 1986, Identification of non-cross-reacting antigens of *Onchocerca volvulus* with lymphatic filariasis serum pools, *Parasitology*, **93**, 389–99.

Lobos, E., Altmann, M., Mengod, G., Weiss, F. N., Rudin, W. and Karam, M., 1990, Identification of an *Onchocerca volvulus* cDNA encoding a low-molecular-weight antigen uniquely recognized by onchocerciasis patient sera, *Molecular and Biochemical Parasitology*, **39**, 135–46.

Lucius, R., Buttner, D. W., Kirsten, C. and Diesfield, H. J., 1986, A study on antigen recognition by onchocerciasis patients with different clinical forms of disease, *Parasitology*, **92**, 569–80.

Lucius, R., Prod'Hon, J., Kern, A., Hebrard, G. and Diesfield, H. J., 1987, Antibody responses in forest and savanna onchocerciasis in Ivory Coast, *Tropical Medicine and Parasitology*, **38**, 194–200.

Lucius, R., Schulz-Key, H., Buttner, D. W., Kern, A., Kaltmann, B., Prod'Hon, J., Seeber, F., Walter, R. D., Saxena, K. C., and Diesfield, H.-J., 1988a, Characterisation of an immunodominant *Onchocerca volvulus* antigen with patient sera and a monoclonal antibody, *Journal of Experimental Medicine*, **167**, 1505–10.

Lucius, R., Erandu, N., Kern, A. and Donelson, J. E., 1988b, Molecular cloning of an immunodominant antigen of *Onchocerca volvulus*, *Journal of Experimental Medicine*, **168**, 1199–204.

Lujan, R. L., Collins, W. E., Stanfill, P. S., Cambell, C. C., Collins, R. C., Brogdon, W. and Huong, A. Y., 1983, Enzyme-linked immunoabsorbent assay (ELISA) for serodiagnosis of Guatemalan onchocerciasis: comparison with the indirect fluorescent antibody (IFA) test, *American Journal of Tropical Medicine and Hygiene*, **32**, 747–52.

Mackenzie, C. D., Burgess, P. J. and Sisley, B. M., 1986, Onchocerciasis, in Walls, K. M. and Schantz, P. M. (Eds) *Immunodiagnosis of Parasitic Diseases*, Vol. 1, pp. 255–89, London: Academic Press.

Maizels, R. M., Denham, D. A. and Sutanto, I., 1985, Secreted and circulating antigens of the filarial parasite *Brugia pahangi*: analysis of *in vitro*-released components and detection of parasite products *in vivo*, *Molecular and Biochemical Parasitology*, **17**, 277–88.

Maizels, R. M. and Selkirk, M. E., 1989, Biology and immunochemistry of nematode antigens, in Englund, P. T. and Sher, F. A. (Eds) *Biology of Parasitism*, pp. 285–308, New York: Alan R. Liss Inc.

Maizels, R. M., Bradley, J. E., Helm, R. and Karam, M., 1990, Immunodiagnosis of onchocerciasis: circulating antigens and antibodies to recombinant peptides, *Acta Leidensia*, **59**, 261–70.

Meredith, S. E. O., Unnasch, T. R., Karam, M., Piessens, W. and Wirth, D. F., 1989, Cloning and characterisation of an *Onchocerca volvulus*-specific DNA sequence, *Molecular and Biochemical Parasitology*, **36**, 1–10.

Nelson, G. S., 1970, Onchocerciasis, in Dawes, B. (Ed.) *Advances in Parasitology*, Vol. 8, pp. 173–224, London: Academic Press.

Nelson, G. S. and Pester, F. R. N., 1962, The identification of infective larvae in *Simuliidae*, *Bulletin of the World Health Organisation*, **27**, 473–81.

Ngu, J. L., Ndumbe, P. M., Titanji, V. and Leke, R., 1981, A diagnostic skin test for *Onchocerca volvulus* infection, *Tropenmedizin und Parasitologie*, **32**, 165–70.

Nogami, S., Hayashi, Y., Korenaga, M., Tada, I. and Tanaka, H., 1988, Monoclonal antibodies specific for *Onchocerca volvulus* as determined by immunofluorescence, *International Journal for Parasitology*, **4**, 503–7.

Omar, M. S., Denke, A. M. and Raybould, J. N., 1979, The development of *Onchocerca ochengi* (Nematoda: Filarioidea) to the infective stage in *Simulium damnosum* s.l. with a note on the histochemical staining of the parasite, *Tropenmedizin und Parasitologie*, **30**, 157–62.

Ouaissi, A., Kouemeni, L.-E., Haque, A., Ridel, P.-R., Saint Andre, P. and Capron, A., 1981, Detection of circulating antigens in onchocerciasis, *American Journal of Tropical Medicine and Hygiene*, **30**, 1211–18.

Ouaissi, A., des Moutis, J., Cornette, J., Pierce, R. and Capron, A., 1983, Detection of IgE antibodies in onchocerciasis using a semi-purified fraction from *Dipetalonema viteae* total antigen, *International Archives of Allergy and Applied Immunology*, **70**, 231–7.

Paganelli, R., Ngu, J. L. and Levinsky, R. J., 1980, Circulating immune complexes in onchocerciasis, *Clinical and Experimental Immunology*, **39**, 570–5.

Parkhouse, R. M. E., Cabrera, Z. and Harnett, W., 1987, Onchocerca antigens in protection, diagnosis and pathology, *CIBA Foundation Symposium No. 127*, pp. 125–45, Chichester: John Wiley.

Perler, F. B. and Karam, M., 1986, Cloning and characterisation of two *Onchocerca volvulus* repeated DNA sequences, *Molecular and Biochemical Parasitology*, **21**, 171–8.

Philipp, M., Gomez-Priego, A., Parkhouse, R. M. E., Davies, M. W., Clark, N. W. T., Ogilvie, B. M. and Beltran-Hernandez, F., 1984, Identification of an antigen of *Onchocerca volvulus* of possible diagnostic use, *Parasitology*, **89**, 295–309.

Piessens, W. F. and Mackenzie, C. D., 1982, Immunology of lymphatic filariasis and onchocerciasis, in Cohen, S. and Warren, K. S. (Eds) *Immunology of Parasitic Infections*, 2nd edn, pp. 622–53, Oxford: Blackwell Scientific Publications.

Porter, C. H. and Collins, R. C., 1988, Transmission of *Onchocerca volvulus* by secondary vectors in Guatemala, *American Journal of Tropical Medicine and Hygiene*, **39**, 559–66.

Rodhain, J. and Van den Branden, F., 1916, Recherches diverses sur la Filaria *Onchocerca volvulus*, (Some investigations on the filarial nematode *Onchocerca volvulus*), *Bulletin de la Société de Pathologie Exotique*, **9**, 186–98.

Schiller, E. L., D'Antonio, R. and Marroquin, H. F., 1980, Intradermal reactivity of excretory and secretory products of onchocercal microfilariae, *American Journal of Tropical Medicine and Hygiene*, **29**, 1215–19.

Schlie-Guzman, M. A. and Rivas-Alcala, A. R., 1989, Antigen detection in onchocerciasis: correlation with worm burden, *Tropical Medicine and Parasitology*, **40**, 47–50.

Shah, J. S., Karam, M., Piessens, W. F. and Wirth, D. F., 1987, Characterisation of an *Onchocerca*-specific DNA clone from *Onchocerca volvulus*, *American Journal of Tropical Medicine and Hygiene*, **37**, 376–84.

Steward, M. W., Sisley, B., Mackenzie, C. D. and El-Sheik, H., 1982, Circulating antigen–antibody complexes in onchocerciasis, *Clinical and Experimental Immunology*, **48**, 17–24.

Tada, I., Korenaga, M., Shiwaku, K., Ogunba, E. O., Ufomado, G. O. and Nwoke, B. E. B., 1987, Specific serodiagnosis with adult *Onchocerca volvulus* antigen in an enzyme-linked, immunosorbent assay, *American Journal of Tropical Medicine and Hygiene*, **36**, 383–6.

Weiss, N., Hussain, R. and Ottesen, E. A., 1982, IgE antibodies are more species-specific than IgG antibodies in human onchocerciasis and lymphatic filariasis, *Immunology*, **45**, 129–37.

Weiss, N., van Den Ende, M. C., Albiez, E. J., Barbiero, V. K., Forsyth, K. and Prince, A. M., 1986, Detection of serum antibodies and circulating antigens in a chimpanzee experimentally infected with *Onchocerca volvulus*, *Transactions of the Royal Society of Tropical Medicine and Hygiene*, **80**, 587–91.

WHO, 1987, *TDR Newsletter No. 24*, Geneva: World Health Organization.

Williams, S. A., De Simone, S. M. and McReynolds, L. A., 1988, Species-specific oligonucleotide probes for the identification of human filarial parasites, *Molecular and Biochemical Parasitology*, **28**, 163–70.

11. The antibody repertoire in nematode infections

M. W. Kennedy

INTRODUCTION

It is probably safe to say that most workers dealing with immune responses to transmissible disease agents would expect that an infected individual would respond to most if not all the foreign proteins and glycoproteins with which it is presented by the infection. Many immunologists might agree with this view despite their knowledge of immune response (Ir) genes which define whether or not mammals of a particular genotype will respond to a given purified antigen (Klein, 1986). Moreover, nematodes will present their hosts with considerable quantities of foreign biological material, so antigenic stimulation is unlikely to be subthreshhold. The finding that there are substantial limitations to antigen recognition in infection with these organisms was, therefore, contrary to expectations.

Differential immune responsiveness to the antigens of pathogens in populations at risk of infection has several implications. Firstly, particular antigen recognition patterns might reflect or determine the pathology, resistance or susceptibility to infection. Secondly, serodiagnostic assays will need to take account of the variability in antigen recognition between subjects. This will be especially so where there is heterogeneity in responsiveness to abundant components of antigen preparations used for serodiagnosis. Finally, new generation vaccines comprising single or a few cloned polypeptide species might not generate protective responses in certain individuals. An example of this has already arisen in the response to hepatitis B vaccine, which contains purified surface antigen of the virus. In this case, people of certain major histocompatibility complex (MHC) types tended to respond poorly (Varla-Leftherioti *et al.*, 1990).

The purpose of this chapter is to illustrate the genetic control of the antibody repertoire in nematode infections, the part played by the MHC, and to discuss some of the implications of genetic restriction of the immune repertoire for the outcome of infection.

THE MAJOR HISTOCOMPATIBILITY COMPLEX AND IMMUNE RESPONSE GENES

The original characterization of Ir genes came through experimentation with two distinct classes of antigen (Klein, 1986). The first of these was co-polymers of two to four types of amino acid, the molecular masses of these artificial antigens being between 1000 and 1 000 000. These will have repeating structures and, consequently, a limited array of potential antigenic determinants (epitopes). The second class of antigens was proteins such as insulin, myoglobin, lysozyme and cytochrome c, all of which were from mammalian or avian sources. Such proteins have homologues in mammalian tissue and the immune response to them might consequently be tempered by tolerance to self components. It might not be surprising, therefore, that genetic control of immune recognition for both these classes of antigen, would be apparent.

With pathogens, however, it might be expected that their constituents would be sufficiently foreign for the immune system to recognize several epitopes per molecular species and that responses to a given molecule as a whole would obscure genetic control of responses to individual determinants on it. It has recently become clear, however, that humans, experimental and domestic animals do not produce an immune response to all the potential antigens of nematode parasites with which they are infected and that there is considerable heterogeneity in responsiveness between infected individuals.

The adaptive immune system has the capacity to respond to an almost inconceivable array of antigens, including those which had never been encountered until the advent of synthetic chemistry. Despite this, there are strict limitations which mean that every individual probably has its own unique immunological repertoire. There are two principal constraints. The first is the requirement for tolerance to an individual's own tissues, to harmless or beneficial organisms such as those occurring on the skin or in the alimentary tract, and to potentially antigenic materials inhaled or present in food. The second constraint is that an immune response to most antigens depends on the successful presentation of processed antigen to lymphocytes, and their subsequent activation. The cell-surface proteins encoded by the major histocompatibility gene complex (originally discovered as a barrier to tissue transplantation) are at the heart of this process.

The MHC comprises a set of closely linked genes in all species so far investigated. Its structure and function are best understood in humans and mice, in which the complexes are known as the 'HLA' and 'H-2', respectively. The precise arrangement of genes varies from species to species, but the complex can always be split into class I and class II regions in terms of the structure and function of the polypeptides that they encode (Figure 11.1). The products of the class I region are involved in direct cell-mediated killing of, for example, virus-infected cells. The class II region is primarily associated with antigen presentation and cooperation between T and B lymphocytes in the induction of an antibody response. It is thought that a response depends on whether an individual's MHC molecules have an affinity for given processed fragments of an antigen, and the manner in which

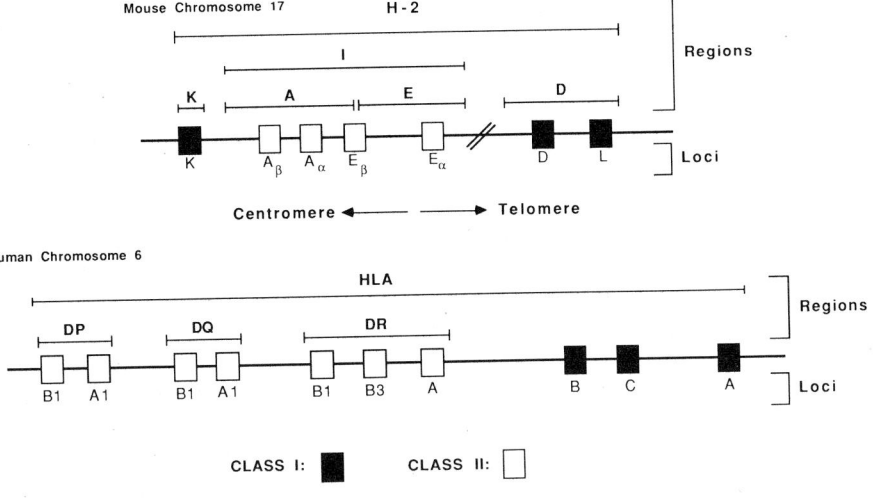

Figure 11.1. The MHC genes of mice and humans. Simplified, comparative linkage maps of the MHC of mice (H-2) and humans (HLA). Both contain pseudo-genes or genes whose functional significance is unknown, and these are not included in the diagram. Genes for some of the components of the complement system, some enzymes and cytokines are also closely linked to, or embedded within, the MHC region, and these are also excluded from the diagram. Class I MHC molecules appear on the surface of most nucleated cells, and are found non-covalently associated with a small protein, β_2-microglobulin which is encoded by an unlinked locus. Processed fragments of antigen, e.g. from viruses in an infected cell, are thought to locate in a groove of the class I molecules. It is this complex which is thought to constitute the recognition unit for a specific T cell receptor. Class II molecules are found as non-covalently linked heterodimers on the cell membrane. Class II molecules are thought to present processed antigen to T cells in a groove distal to the membrane, similar to that of class I molecules. Class I molecules appear, therefore, to present antigens which are endogenous to the present cell and class II molecules present exogenous antigen. Class II molecules have a more limited tissue distribution than for class I, but appear constitutively on several tissues and many cells of the immune system, and can be induced in others under conditions of immunological or infection stress. In mice, A_α and A_β chains preferentially associate, as do E_α and E_β, although A_α/E_β and A_β/E_α associations possibly occur. In humans, there are more types of class II molecules, and more genes for β than α chains, although preferential pairing is thought to occur with the appropriate α chain. The β chains tend to be more polymorphic than α chains in both mice and humans (Adapted from Kennedy (1989).)

the fragments are presented to T-cells (Allen *et al.*, 1987; Allen, 1987; Townsend and McMichael, 1987; Davis and Bjorkman, 1988). This, then, would be the basis for class II associated immune response Ir genes controlling the antibody response (Klein, 1986).

The polymorphism of these genes means that mammals have diverse immunological repertoires and that some potential antigens will not elicit responses in certain individuals within a species. Added to this is the requirement for tolerance to self components including the polymorphic proteins of the MHC themselves (alone and, presumably, also in association with fragments of other self components), non-MHC polymorphic loci and even the products of somatic mutation and recombination such as the T cell receptor and immunoglobulin. It is likely, for example, that situations will arise in which the combination of antigen + MHC molecule will resemble a self component and that this immune specificity will be deleted from the repertoire — the 'cross-tolerance' hypothesis (Schwartz, 1978; Matzinger, 1981).

The existence of all these constraints has led some authors to express amazement that mammals have a functional immune system at all (Vidovic and Matzinger, 1988). What it does mean, though, is that the immune repertoire of an individual is, to a large degree, predictable by their class II MHC alleles but that other factors influence the final outcome and will always lead to uncertainty in an outbred population.

HUMANS INFECTED WITH NEMATODES VARY SUBSTANTIALLY IN ANTIBODY REPERTOIRE

There are now several examples in which people infected with nematodes are known to vary considerably in antigen recognition. These include infections with *Necator americanus, Trichinella spiralis, Brugia malayi, Trichuris trichiura, Loa loa* and *Ascaris lumbricoides* (Almond *et al.*, 1986; Pritchard *et al.*, 1986; Maizels *et al.*, 1987; Pinder *et al.*, 1988; Roach *et al.*, 1988; Haswell-Elkins *et al.*, 1989; Kennedy *et al.*, 1990b). An example of this for *Ascaris* infection is illustrated in Figure 11.2.

In the case of *Brugia* infections, there appears to be some correlation, albeit loose, between antibody recognition patterns and infection status. It remains to be seen, though, whether these factors are linked or controlled independently. It has been argued that the analysis must be taken to the level of the individual antigenic determinants on parasite antigens in order to establish which factors reflect or

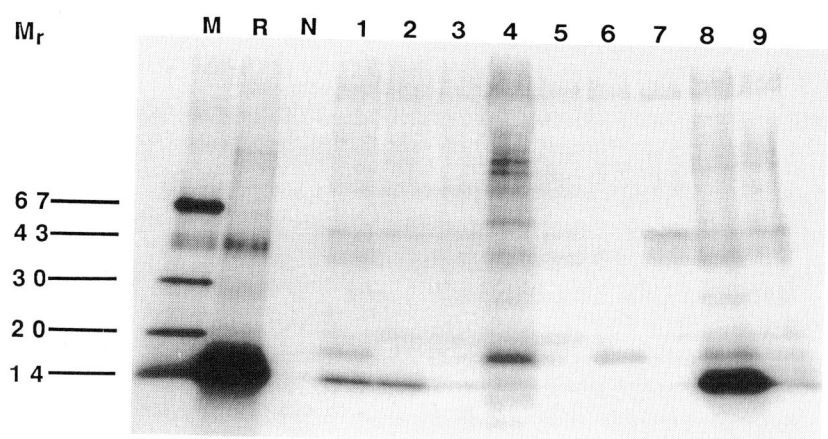

Figure 11.2. Variability in antigen recognition among people infected with *Ascaris*. Body (pseudo-coelomic) fluid from adult *A. lumbricoides* parasites (ABF) was labelled with ^{125}I and immunoprecipitated with serum from an uninfected European (track N) and from children living in an endemic area of Nigeria (numbered tracks). The immunoprecipitates were analysed by gradient SDS-PAGE, along with a sample of the iodinated antigen (track R). Track M was loaded with iodinated standard marker proteins (Pharmacia 17-0446-01), and their relative molecular masses (M_r, in kilodaltons) as given by the manufacturers are indicated. (Data from Kennedy *et al.* (1990b).)

determine the outcome of infection (Maizels *et al.*, 1987). Experience with T and B cell responses to viral antigens (Mills, 1986; Mills *et al.*, 1986a,b, 1988; Morrison *et al.*, 1987; Thomas *et al.*, 1987; Wiley and Skehel, 1987; Barnett *et al.*, 1989; Burt *et al.*, 1989; Graham *et al.*, 1989) could be taken to reinforce this idea but increases our need for an understanding of genetic control of immune specificity.

In none of the above examples have any correlations yet been made between genetic factors and antigen recognition in humans or domestic animals. Several seroepidemiological studies are currently underway which will shed light on this, but we are currently left with having to rely on laboratory models in order to identify the genetic loci controlling the antibody repertoire and the mechanisms by which this might affect the outcome of infection.

MODEL SYSTEMS

Work with model systems requires genetically defined animals and has been carried out in inbred mice, rats and guinea-pigs. Mice are clearly the species of choice because of the wealth of detail known of their genetics and the availability of a large set of inbred strains (Klein and Klein, 1987). The disadvantage of laboratory rodents is that they are not definitive hosts of parasites which are of medical or veterinary importance and full-course infections cannot usually be achieved without manipulations such as immunosuppression. An alternative is to adapt parasites to laboratory animals but this is likely to produce strains of parasite which are atypical of the species concerned. It could be argued, therefore, that the results of immunogenetic studies using non-definitive host–parasite combinations are artifactual. While it is difficult to eliminate this possibility in all situations, the use of systems in which these limitations do not apply have shown that genetic control of the immune response is a general phenomenon.

MAJOR HISTOCOMPATIBILITY COMPLEX CONTROL OF REPERTOIRE

The first indication that the antibody response to the different components of nematodes is under genetic control came from work on *Ascaris suum* infection of mice. The serum antibody responses of a range of inbred strains were analysed by immunoprecipitation of parasite antigens and sodium dodecylsulphate–polyacrylamide gel electrophoresis (SDS–PAGE) analysis of immunoprecipitates (Kennedy *et al.*, 1987a). The original data are presented in Figure 11.3 and the points to note are as follows. First, no strain responded to all the potential antigens of the parasite. Second, some components elicited a response in all strains. Third, only strains of identical H-2 haplotypes responded with similar antigen recognition profiles. Fourth, antigen recognition was sequential and the kinetics of this was only similar between MHC-identical strains.

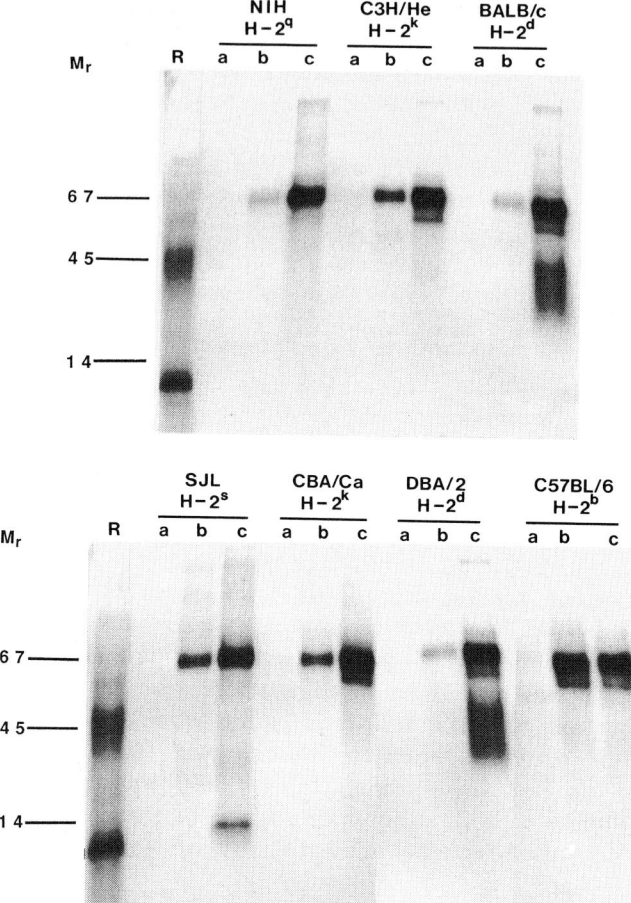

Figure 11.3. MHC-restricted recognition of *Ascaris suum* antigens in mice. Sera from a range of *Ascaris*-infected mouse strains was reacted with radiolabelled excretory–secretory (ES) material from infective larvae of the parasite, and immunoprecipitates were analysed by SDS-PAGE. The strains of mice used and their MHC (H-2) haplotypes are given above the triplets of tracks. Serum was samples before infection (a), after one infection (b), and after three infections (c). The reference tracks (R) give the profile of the iodinated antigen used in the assay. The relative molecular masses (M_r, in kilodaltons) of the major ES components are indicated. (Data from Kennedy *et al.* (1987a).)

To confirm that the repertoire restriction could properly be ascribed to the MHC, it was then necessary to use strains of mice which were congenic for the H-2 region. That is, inbred strains which have been bred and selected to differ only at the MHC and closely linked loci but bearing a similar or identical genetic background (see Table 11.1). When such mice were examined, strains were again found to differ in response patterns according to their H-2 alleles (Tomlinson *et al.*, 1989). The converse of this experiment was to analyse the response of strains which had a common H-2 haplotype but different genetic backgrounds; if the MHC were in control of the repertoire, then identical antigen recognition patterns would result. This proved to be the case.

Table 11.1. MHC haplotypes of commonly used strains of mice.

Strain	H-2 haplotype[a]
C57BL10 (= B10)	b
BALB/b	b
DBA$_2$	d
BALB/c	d
B10.D2	d
CBA	k
C3H/He	k
BALB/k	k
B10.BR	k
NIH	q
B10.G	q
SJL	s
B10.S	s

A haplotype is a set of genetic determinants located on a single chromosome. The above strains are inbred, and are, therefore, homozygous for all MHC genes. The single letter designations mean that the combination of alleles expressed by mice of a given haplotype are unique to that haplotype. Congenic strains are those which have been bred to have similar or identical genomes apart from a closely linked set of genes such as the MHC. The most commonly used congenic strains have the C57BLh10 (= B10) or BALB genetic backgrounds. Recombinant haplotypes are those in which a genetic cross-over has occurred within the MHC. An example would be a mouse which had the H-2^b allele of the H-2K protein and the H-2q allele of the H-2D protein are encoded on a single homologous chromosome. The rules for the naming of recombinant and other haploytpes are given in Klein (1986).

It is likely that these MHC effects explain the differential responsiveness by humans infected with *A. lumbricoides* illustrated in Figure 11.2. Another important parasite of humans for which the effect is particularly clear is *B. malayi*, responses to major surface-associated (glyco)proteins of which is MHC-restricted in the mouse (Kwan-Lim and Maizels, 1990). Among parasites of domestic animals, evidence for MHC control has also been found for infection of guinea-pigs with *Dictyocaulus viviparus* (C. Britton, G. Canto, G. M. Urquhart and M. W. Kennedy, unpublished results).

None of the above parasites are native to the experimental rodents used and cannot complete their life cycles within them. However, MHC effects have also been found in definitive host–parasite combinations such as *Trichuris muris* in mice (Else and Wakelin, 1989) (although non-MHC genes were also thought to be influential), *Nippostrongylus brasiliensis* in the rat (Kennedy *et al.*, 1990a) and *Trichinella spiralis* infection in mice (Figure 11.4 and Kennedy *et al.*, 1991b). It is particularly propitious that the effect operates with the latter parasite because more is known of the genetic effects on immunity to this nematode than for any other.

CLASS II CONTROL

While the control of the antibody response would conventionally be ascribed to the class II region of the MHC, it is nevertheless important to confirm that this is so for

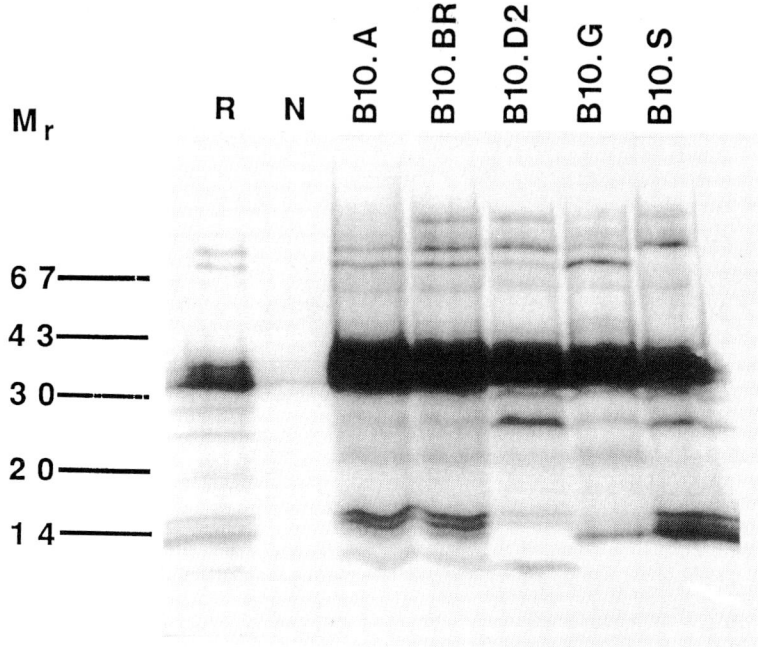

Figure 11.4. MHC control of the antibody repertoire to *T. spiralis* infection in mice. Animals of the indicated strains were infected three times with the parasite and their sera used to immunoprecipitate biosynthetically labelled ([^{35}S]methionine) secretions of adult worms. Immunoprecipitates were analysed on gradient SDS-PAGE gels. All the strains used have a common genetic background (B10) and supposedly only differ at the MHC and closely linked loci. It is almost certain, therefore, that the differences are due to the H-2 complex. The B10.A and B10.BR strains share class II loci but differ in some class I alleles. They had, however, identical antigen recognition profiles. This and other evidence indicates that it is the class II loci which are central to the control of the repertoire. See Tables 11.1 and 11.2 for the H-2 haplotypes of the strains used. (See also Kennedy *et al.*, 1991b)

responses to nematode antigens. Moreover, there is some evidence that loci within this region have differing roles in regulation of the immune response. The I-E of the mouse is particularly notable in this respect in that it has been associated with suppression of I-A-restricted T cell responses (Baxevanis *et al.*, 1981; Oliveira *et al.*, 1985) and with susceptibility to *T. spiralis* (Wassom *et al.*, 1987). There is, however, no strong evidence to date that the human homologue of I-E, HLA-DR, has a similar activity.

The conventional way of establishing which locus is in control of responses to a given antigen is to use inbred strains which have recombinant MHC haplotypes. These are strains which have been developed by crossing inbred mice of differing H-2 haplotypes and selecting for animals which represent chromosome cross-overs within the MHC. These are then inbred, backcrossing if necessary to produce mice with as near as possible common genetic backgrounds (Klein and Klein, 1987). By examining the responses of strains in which recombinations had occurred at different sites, it is then possible to focus on a particular locus by a process of elimination. This is made easier in the mouse because certain haplotypes are defective in the cell surface expression of the class II I-E molecule (Murphy *et al.*, 1980; Jones *et al.*, 1981). It can hence be shown that the response to a major allergen (ABA-1) of *Ascaris* is I-A restricted (Kennedy *et al.*, 1991) as is the response of several elements of the repertoire to *T. spiralis*. I-A restriction with the latter parasite is particularly clear and can be illustrated as follows.

Animals possessing the H-2^k haplotype have a characteristic antigen-recognition profile in that they produce antibody to a triplet of polypeptides in the low molecular weight region, whereas mice of all other haplotypes tested respond to only a subset of these (Figure 11.4). The influential locus/loci can then be identified using recombinant mouse strains such as those listed in Table 11.2. When those bearing only the k allele at I-A in common were compared, their repertoires were found to be identical (Table 11.2; and Kennedy *et al.*, 1991b).

This kind of analysis cannot, however, provide information on the loci controlling responses to those parasite components which elicit responses in all strains, but it would now be reasonable to assume that the class II region is involved here also.

ALLERGENS

One of the characteristics of immune responses to nematodes is the generation of hypersensitivity responses. Immunoglobulin E (IgE) mediated Type I hypersensitivities are particularly notable as in, for example, Loeffler's syndrome in ascariasis (Ogilvie and de Savigny, 1982). The components of a parasite which elicit an IgE response would, therefore, be allergens by definition. If the MHC controls the

Table 11.2. Role of the class II (I-A) region in determining the antibody repertoire in *T. spiralis* infection.[a]

Strain		K	A_β	A_α	E_β	E_α	S	D
					Haplotype			
A.TL	t1	s	k	k	k	k	k	d
B10.A	a	k	k	k	k	k	d	d
B10.AQR	y1	q	k	k	k	k	d	d
B10.A(4R)	h4	k	k	k	(k/b	b)	b	b
B10.A(5R)	i5	b	b	b	b/k	k	d	b

[a] H-2^k mice have a characteristic repertoire to infection with *T. spiralis* (see Figure 11.4) which can be used to identify the locus/loci which control it. This involves the use of MHC-recombinant inbred strains of mice. These are strains which have been developed by crossing inbred strains of differing H-2 haplotypes and selecting for animals which represent chromosome cross-overs within the MHC. In the selection dealt with in the table, all the strains produced responses identical to H-2^k animals and quite unlike that for those of haplotypes b, d, q and s with the exception of B10.A(5R) which displayed an H-2^b-like repertoire. The brackets in the allele list for the B10.A(4R) strain indicates defective in expression of the I-E molecule on the cell surface. A role for this class II molecule can, therefore, be eliminated in this case. By a process of elimination it can be argued that the repertoire is, therefore, controlled by the I-A heterodimer. Allele listings taken from Klein *et al.* (1983) and Klein (1986).

specificity of the response to parasite allergens, then it could also determine the immunological specificity of some allergic reactions, although non-MHC genes would control the magnitude of response.

More allergens have been reported from *Ascaris* than from any other species of nematode (Jarrett and Miller, 1982) and the one about which most is known is the most abundant protein present in the pseudo-coelomic fluid of adult worms. The allergen has been obtained in purified form, has been variously sized at between 10 and 14 kDa, partially N-terminal sequenced (Figure 11.5) and named ABA-1 (Christie *et al.*, 1990; McGibbon *et al.*, 1990). The genetic control of IgE responses to it has been examined by testing sera from inbred mouse strains infected with the parasite in a passive cutaneous anaphylaxis (PCA) assay (Table 11.3). Only those mice of one particular haplotype (H-2^S) responded. Among these, SJL mice produced strong IgG antibody to the molecule under the conditions of the experiment but failed to produce IgE antibody (Figure 11.3; and Tomlinson *et al.*, 1989). This strain is known to have unusual IgE responses (Watanabe *et al.*, 1976; Azuma *et al.*, 1987) and we have recently found evidence that SJL mice will produce intense IgE responses which are of brief duration and easily missed if blood sampling is not carried out at several time points after *Ascaris* infection.

The IgE response of humans infected with *Ascaris* has also been examined by allergoblots (E. M. Fraser, J. F. Christie and M. W. Kennedy, unpublished results) and the immune recognition of the protein appears to be as uncommon among humans as it is among strains of mice. While our results would predict the involvement of the HLA complex in humans, no data on this is yet available.

The apparent rarity of immune recognition of ABA-1 might be explained if the molecule comprised repeated epitopes which are the subject of a response in only a limited number of hosts. The partial sequence of the protein, however, shows no evidence of such a structure (Figure 11.5). Another possibility would be that the protein has a homologue in mammalian tissue which induces tolerance to it in most people and strains of rodents. The tolerogen concerned would need to be either MHC-linked or that the specificity of tolerance be MHC-restricted. It would be

	1	5	10	15	20	25	30	35	40	45	50	55
Ascaris suum	HHFTLESSLDTHLKWLSQEQKDELLKMKKDGKAKKELEAKILHYYDELEGDAKKE											

Ascaris lumbricoides   HHFTLESSLDTHLKWLSQEQKDELLKMKKDGKAKKELEAKI

Figure 11.5. *N*-terminal amino acid sequences of the ABA-1 allergens from *Ascaris suum* and *Ascaris lumbricoides*. The allergen was purified from the pseudo-coelomic fluid of adult parasites by immunoaffinity chromatography or preparative SDS-PAGE. The sequence is expressed in the single-letter code for amino acids. The protein has been variously sized at 10 to 14 kDa, possibly occurs naturally as a non-covalently linked dimer, and is thought to be similar or identical to the previously described allergen A of the parasite (McGibbon *et al.*, 1990; Christie *et al.*, 1990). No significant similarity with any other protein sequence has been found to date. No repetitive sections of the protein have been found, indicating that the observed MHC-restricted antibody response is not merely due to recognition of repeated epitopes. The known sequence of the ABA-1 molecule of *A. lumbricoides* of humans is also shown, and this is identical to that of the ABA-1 of *A. suum* up to 41 residues. (Data from Christie *et al.* (1990).) © Blackwell Scientific Publications.

Table 11.3. H-2 restricted IgE antibody responses to the 14 kDa allergen of *Ascaris*.[a]

| Strain of mouse | H-2haplotype | PCA titre | |
		Purified 14 kDa	ABF
B10	b	0	256
B10.D2	d	0	256
B10.BR	k	0	>512
B10.G	q	0	256
B10.S	s	128	512
SJL	s	0	16

[a] IgE was assayed in a passive cutaneous anaphylaxis assay (PCA) in which serum from mice of the strains listed was tested in rats. The assay was carried out against two sources of allergen: the purified 14 kDa allergen and the body fluid of adult worms (ABF) from which the 14 kDa allergen was obtained. The assay with ABF was carried out in parallel as a positive control to ensure that the mice used as serum donors were able to produce an IgE response to other allergens of the parasite. The SJL mice are included to show control of the amplitude of the IgE response by non-H-2 genes. Hence, H-2 controls the specificity of the response, but other genes regulate its level. We have since found that SJL mice will produce substantial IgE responses, but that the kinetics of these is quite different from that of other strains. Data from Tomlinson *et al.* (1989).

difficult to eliminate these possibilities but we have failed to find any significant similarity between ABA-1 and any protein of known sequence. We are left with the probability, therefore, that non-responsiveness is due to the failure of most allelomorphs of class II proteins to properly present ABA-1 to T cells. It remains a remarkable fact, then, that a molecule which is apparently so foreign to mammalian tissues, and presented in such relative bulk, can fail to induce a response in so large a proportion of infected individuals.

INFLUENCE OF NON-MHC LOCI

As mentioned above, tolerance to self components should further limit the immune repertoire in addition to the limits imposed by the MHC. Such influences possibly explain how strains can differ in their antigen-recognition patterns despite being MHC identical. Examples of this have been recorded in infections with *A. suum*, *T. muris* and *B. malayi*, and apply to both excretory–secretory (ES) and surface antigens (Roach *et al.*, 1988; Haswell-Elkins *et al.*, 1989; Kennedy *et al.*, 1990b; Kwan-Lim and Maizels, 1990). The effect appears to extend to a minority of nematode antigens, but potentially important ones are clearly involved. Whatever the causes(s) might be, it clearly introduces an unpredictable element into the immune response of outbred species in that knowledge of an individual's MHC cannot predict their immune repertoire with certainty.

There are several possible ways in which to examine whether or not tolerance is the explanation for additional deletions in the repertoire. For example, radiation chimaeras can be constructed using non-responder lymphocytes transferred to otherwise responder hosts. An example of this would be where bone marrow cells from normal animals of a strain which responds to a given antigen are transferred to, and develop within, otherwise non-responder, but MHC identical, mice. Recipient animals will usually have been irradiated to ablate their own immune systems, the transferred lymphocytes will then establish to replace the recipient's immune system and must develop tolerance to its tissues. If the resultant chimaera is unresponsive to the antigen in question, then this is probably due to immuno-logical tolerance to the recipient's tissues. This experiment can, of course, be carried out *vice versa*.

Establishing which non-MHC genes are involved in tolerance is a considerably more difficult matter, but the use of recombinant inbred (RI) strain sets can at least provide information on whether or not the genes concerned are linked with the MHC and which chromosome they might be on (Klein and Klein, 1987).

UNPREDICTABLE RESPONSES IN HETEROZYGOTES

All the above analyses involve inbred strains of mice which are homozygous throughout their MHC. This condition is atypical of most mammals, including humans and domestic animals. The question arises, therefore, of whether heterozy-gotes have repertoires which are merely combinations of parental antigen recognition profiles. It appears that in many instances they are not.

Figure 11.6 shows the results of an examination of sera from hybrid rats infected with *N. brasiliensis*. This shows that in the (AGUS × PVG)F_1 hybrid there can be both dominant and recessive inheritance of recognition of particular ES components, exemplified by the 29 and 18kDa molecules, respectively (arrowed).

A dominant response to a particular parasite molecule in an F_1 argues for an inability of the class II gene products to present processed antigen in the non-responder parent. This could be due to low affinity of crucial peptides derived from processed antigen for the class II heterodimer (Allen, 1987; Allen *et al*., 1987; Nagy *et al*., 1989). In hybrids the genes for competent class II molecules (and/or antigen processing mechanisms) from the responder parent will be co-expressed with the non-responder type molecule from the other parent. Upon exposure to parasite antigen, therefore, the hybrid will respond.

Recessive inheritance would indicate that in the non-responder parent the complex of MHC plus antigen cross-reacts with a self component and that this specificity is consequently deleted from the repertoire (see preceding section). In an F_1 hybrid, genes for that self component will be carried over and lead to tolerance in the progeny, as manifested by non-responsiveness to the parasite component concerned.

An even more unexpected effect has been found in F_1 hybrid mice infected with *Ascaris* in which both parents of a particular cross were responders to a 43 kDa ES molecule of the parasite, but the hybrid was not (Tomlinson *et al*., 1989). This requires a more elaborate hypothesis which at its simplest would state that there are

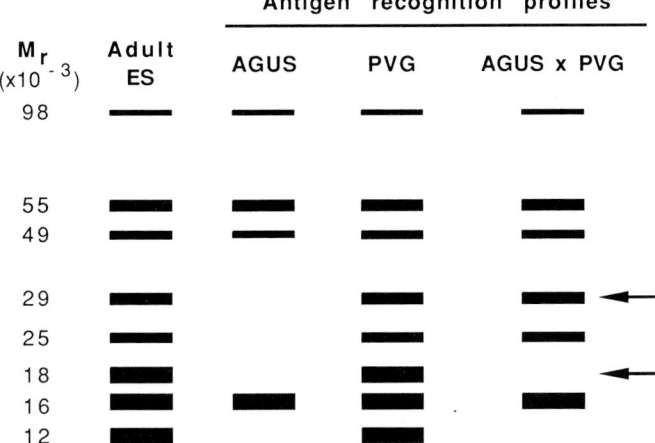

Figure 11.6. Antibody repertoires of F_1 hybrid animals are not merely summations of parental response phenotypes. Rats of the indicated strains were infected three times with *N. brasiliensis* and their sera analysed by immunoprecipitation of ^{125}I labelled ES material of adult parasites and analysis by gradient SDS-PAGE. The figure is a schematic representation of an immunoprecipitation experiment given in Kennedy *et al.* (1991). The 'Adult ES' track represents all the iodinatable ES components resolved by gradient SDS-PAGE and the remaining tracks are the antigen recognition profiles of the parental strains and hybrid used. The components marked with arrows indicate molecules which are discussed in the text as examples of unexpected responses in hybrid animals.

two epitopes on the antigen molecule concerned, one parent being tolerant to one epitope but reactive to the other, and the other parent having the complementary responsiveness. Tolerance to both epitopes will then prevail in the hybrid, and overall non-responsiveness to the molecule will result. While this is conceivable under immunological theory, it remains for this to be demonstrated with the purified molecule. Most parasite antigens are available in such small quantities that this is currently not possible and will have to await molecular cloning of the polypeptides concerned — this, however, assumes that carbohydrate epitopes are not involved.

GENETIC CONTROL OF RESPONSES TO ARTIFICIAL IMMUNIZATION

It is clear, then, that the MHC is primarily responsible for restricting the repertoire, even if other factors serve to reduce it further. It is important to know, however, whether this is absolute or whether it can be circumvented by artificial immunization. This would seem particularly important to the application of vaccines comprising recombinant or purified parasite components where the degree of responsiveness could make the difference between the success or failure of immunization.

Whether this is likely to prove a problem can be tested by comparing the antigen-recognition profiles of animals presented with parasite material through infection or adjuvant-assisted immunization. With *Ascaris* it has been found that whilst adjuvant-assisted immunization of rats or mice elicits antibody responses which are much stronger than by infection, the antigen-recognition profiles do not differ markedly between the two situations (J. F. Christie, E. M. Fraser and M. W. Kennedy, unpublished results). There was also evidence, however, that this MHC restriction can be overruled if large quantities of parasite antigen are administered.

Recombinant polypeptides for use in potential vaccines must, therefore, be selected to elicit immune responses in most if not all of the population to be protected, or that they be manipulated to increase their immunogenicity without inducing deleterious immune responses such as autoimmunity or hypersensitivity.

With regard to the induction of IgE responses, work done on the ABA-1 allergen (E. M. Fraser, J. F. Christies and M. W. Kennedy, unpublished results) shows that artificial immunization elicits strong IgG antibody responses without detectable IgE antibody. We found that exceptions to this existed and the final outcome will almost certainly be influenced by the adjuvant used. This will necessitate that any candidate polypeptide would, nevertheless, need to be tested carefully in the species of mammal for which the immunization is intended in order to avoid deleterious hypersensitivity responses when a vaccine is eventually used for mass immunization.

DOES THE MHC AND IMMUNE REPERTOIRE RELATE TO RESISTANCE OR SUSCEPTIBILITY TO INFECTION?

Whether or not the MHC relates to resistance to infection with parasitic nematodes or to the ensuing pathology is controversial, but there is clear evidence that it, or closely linked loci, is influential in experimental animals (Wakelin, 1988). It must be emphasized, however, that non-MHC loci can be equally if not more potent. The genetics of resistance/susceptibility is best understood in the *T. spiralis*/mouse model and this provides an opportunity to test the predictive value of the immune repertoire. One approach would be to compare the repertoires of strains which have the same MHC but which differ markedly in susceptibility. When such a comparison was made there were only minor differences in antigen-recognition profiles (Kennedy *et al.*, 1991b). Resistant strains did, however, tend to respond to one or two more components of the parasite's ES material than did susceptible strains. This was probably an effect rather than a cause of slow reactivity because the particular antigens to which this applied were subject to a response in poor responder strains of other H-2 haplotypes. This does not, however, exclude repertoire effects which might operate through differences in the isotype bias of antibody responses in the different strains. Such differences have already been observed in MHC-identical strains which differ in susceptibility and these biases in turn correlate with lymphokine responses of T cells from immunized animals (Pond *et al.*, 1989).

A central role for T cells in immune control of nematodes is now beyond doubt (Miller, 1984; Riedlinger *et al.*, 1986; Grencis *et al.*, 1985) and it is unfortunate that so little is known of the T cell repertoire in nematodiases. This can potentially be examined by 'T cell Westerns' in which SDS-PAGE separated parasite material is transferred to a solid support (usually nitrocellulose) and serial slices of this used to stimulate T cells from infected animals *in vitro*. This can potentially provide a direct comparison of the antigen-recognition profiles of T and B cells, although it must be borne in mind that transfer to nitrocellulose can in some cases nullify antigenic determinants, although this is more likely to affect determinants for antibody.

If the immune repertoire is not responsible, there remains, therefore, no obvious mechanism for MHC control of resistance to infection. There are, however, several other genes important to the immune response which are closely linked with the MHC, e.g. those encoding lymphotoxin and tumour necrosis factor/cachectin, and it is conceivable that many MHC associations are apparent rather than real.

CONCLUDING REMARKS

The implications of genetic control of the immune repetoire in nematode infections for serodiagnosis, vaccination, immunopathology and resistance to infection have already been mentioned. There is, however, an additional point which could be made regarding the evolution of parasite antigens. Conventional wisdom would dictate that, if the specificity of the immune response mounted against it has some influence on the ability of a parasite to survive in its host, then there might be selection for parasites which produce exoantigens which are not antigenic. However, since the host population is polymorphic in terms of immune responsiveness through its MHC, then it might be impossible for a universally non-antigenic parasite molecule to evolve. There might instead be selection for balanced polymorphism in the antigens which are crucial to parasite survival. This could mean that an individual parasite would succeed or fail in a given host but that a polymorphic parasite population would therefore be more successful than one in which there is limited genetic variation in its antigens. There is already evidence for heterogeneity in surface antigen expression among *Ascaris* larvae (Fraser and Kennedy, 1991) and in the size polymorphism of a secreted antigen of infective larvae of *Onchocerca lienalis* (Bianco *et al.*, 1990). Such polymorphisms will have fundamental implications not only for our understanding of the dynamics of parasite populations but also for the application of recombinant vaccines.

ACKNOWLEDGEMENTS

Our work is supported by the Wellcome Trust, the Medical Research Council and the World Health Organization.

REFERENCES

Allen, P. M., 1987, Antigen processing at the molecular level, *Immunology Today*, **8**, 270–3.

Allen, P. M., Matsueda, G. R., Evans, R. J., Dunbar, J. B., Marshall, G. R. and Unanue, E. R., 1987, Identification of the T-cell and Ia contact residues of a T-cell antigenic epitope, *Nature*, **327**, 713–15.

Almond, N. M., Parkhouse, R. M. E., Chapa-Ruiz, M. R. and Garcia-Ortigoza, E., 1986, The response of humans to surface and secreted antigens of *Trichinella spiralis, Tropical Medicine and Parasitology*, **37**, 381–4.

Azuma, M., Hirano, T., Miyajima, H., Watanabe, N., Yagita, H., Enomoto, S., Furusawa, S., Ovary, Z., Kinashi, T., Honjo, T. and Okumura, K., 1987, Regulation of murine IgE production in SJA/9 and nude mice: potentiation of IgE production by recombinant interleukin 4, *Journal of Immunology*, **139**, 2538–44.

Barnett, B. C., Graham, C. M., Burt, D. S., Skehel, J. J. and Thomas, D. B., 1989, The immune response of BALB/c mice to influenza hemagglutinin: commonality of the B cell and T cell repertoires and their relevance to antigenic drift, *European Journal of Immunology*, **19**, 515–21.

Baxevanis, C. N., Nagy, Z. A. and Klein, J., 1981, A novel type of T-T cell interaction removes the requirement for I-B region in the H-2 complex, *Proceedings of the National Academy of Sciences (USA)*, **78**, 3809–13.

Bianco, A. E., Robertson, B. D., Kuo, Y.-M., Townson, S. and Ham, P. J., 1990, Developmentally regulated expression and secretion of a polymorphic antigen by *Onchocerca* infective-stage larvae, *Molecular and Biochemical Parasitology*, **39**, 203–12.

Burt, D. S., Mills, K. H. G., Skehel, J. J. and Thomas, D. B., 1989, Diversity of the class II (I-A^k/I-E^k)-restricted T cell repertoire for influenza hemagglutinin and antigenic drift. Six nonoverlapping epitopes on the HA1 subunit are defined by synthetic peptides, *Journal of Experimental Medicine*, **170**, 383–97.

Christie, J. F., Dunbar, B., Davidson, I. and Kennedy, M. W., 1990, N-terminal amino acid sequence identity between a major allergen of *Ascaris lumbricoides* and *Ascaris suum*, and MHC-restricted IgE responses to it, *Immunology*, **69**, 596–602.

Davis, M. M. and Bjorkman, P. J., 1988, T-cell antigen receptor genes and T-cell recognition, *Nature*, **334**, 395–402.

Else, K. and Wakelin, D., 1989, Genetic variation in the humoral immune responses of mice to the nematode *Trichuris muris*, *Parasite Immunology*, **11**, 77–90.

Fraser, E. M. and Kennedy, M. W., 1991, Heterogeneity in the expression of surface-exposed epitopes among larvae of *Ascaris lumbricoides*, *Parasite Immunology*, in press.

Graham, C. M., Barnett, B. C., Hartlmayr, I., Burt, D. S., Faulkes, R., Skehel, J. J. and Thomas, D. B., 1989, The structural requirements for class II (I-A^d)-restricted T cell recognition of influenza hemagglutinin: B cell epitopes define T cell epitopes, *European Journal of Immunology*, **19**, 523–8.

Grencis, R. K., Riedlinger, J. and Wakelin, D., 1985, L3T4-positive T lymphoblasts are responsible for transfer of immunity to *Trichinella spiralis* in mice, *Immunology*, **56**, 213–18.

Haswell-Elkins, M. R., Kennedy, M. W., Maizels, R. M., Elkins, D. B. and Anderson, R. M., 1989, The antibody recognition profiles of humans naturally infected with *Ascaris lumbricoides*, *Parasite Immunology*, **11**, 615–27.

Jarrett, E. E. E. and Miller, H. R. P., 1982, Production and activities of IgE in helminth infection, *Progress in Allergy*, **31**, 178–233.

Jones, P. P., Murphy, D. B. and McDevitt, H. O., 1981, Variable synthesis and expression of E$_a$ and A$_e$ (E$_\beta$) Ia polypeptide chains in mice of different H-2 haplotypes, *Immunogenetics*, **12**, 321–37.

Kennedy, M. W., 1989, Genetic control of the immune repertoire in nematode infections, *Parasitology Today*, **5**, 316–24.

Kennedy, M. W., Gordon, A. M. S., Tomlinson, L. A. and Qureshi, F., 1987a, Genetic (major histocompatibility complex?) control of the antibody repertoire to the secreted antigens of *Ascaris*, *Parasite Immunology*, **9**, 269–73.

Kennedy, M. W., Qureshi, F., Haswell-Elkins, M. and Elkins, D. B., 1987b, Homology and heterology between the secreted antigens of the parasitic larval stages of *Ascaris lumbricoides* and *Ascaris suum*, *Clinical Experimental Immunology*, **67**, 20–30.

Kennedy, M. W., McIntosh, A. E., Blair, A. J. and McLaughlin, D., 1990a, MHC (RT1)-restriction of the antibody repertoire to infection with the nematode *Nippostrongylus brasiliensis* in the rat, *Immunology*, **71**, 317–22.

Kennedy, M. W., Tomlinson, L. A., Fraser, E. M. and Christie, J. F., 1990b, The specificity of the antibody response to internal antigens of *Ascaris*: heterogeneity in infected humans, and MHC (H-2) control of the repertoire in mice, *Clinical and Experimental Immunology*, **80**, 219–24.

Kennedy, M. W., Fraser, E. M. and Christie, J. F., 1991a, MHC class II (I-A) region control of the IgE antibody repertoire to the ABA-1 allergen of the nematode *Ascaris*, *Immunology*, in press.

Kennedy, M. W., Wasson, D. L., McIntosh, A. E. and Thomas, J. C., 1991b, H-2 (I-A) control of the antibody repertoire in *Trichinella spiralis* infection and its relevance to resistance and susceptibility, *Immunology*, in press.

Klein, J., 1986, *Natural History of the Major Histocompatibility Complex*, New York: Wiley.

Klein, J. and Klein, D., 1987, Mouse inbred and congenic strains, *Methods in Enzymology*, **150**, 163–96.

Klein, J., Figueroa, F. and David, C. S., 1983, *H-2* haplotypes, genes and antigens: second listing, *Immunogenetics*, **17**, 553–96.

Kwan-Lim, G. and Maizels, R. M., 1990, MHC and non-MHC restricted recognition of filarial surface antigens in mice transplanted with adult *Brugia malayi* parasites, *Journal of Immunology*, **145**, 1912–20.

Maizels, R. M., Selkirk, M. E., Sutanto, I. and Partono, F., 1987, Antibody responses to human lymphatic filarial parasites, in *Filariasis — Ciba Foundation Symposium 127*, pp. 189–199, Chichester: Wiley.

Matzinger, P., 1981, A one-receptor view of T-cell behaviour, *Nature*, **292**, 497–501.

McGibbon, A. M., Christie, J. F., Kennedy, M. W. and Lee, T. D. G., 1990, Identification of the major *Ascaris* allergen and its purification to homogeneity by HPLC, *Molecular and Biochemical Parasitology*, **39**, 163–72.

Miller, H. R. P., 1984, The protective mucosal response against gastrointestinal nematodes in ruminants and laboratory animals, *Veterinary Immunology and Immunopathology*, **6**, 167–259.

Mills, K. H. G., 1986, Processing of viral antigens and presentation to class II-restricted T cells, *Immunology Today*, **7**, 260–3.

Mills, K. H. G., Skehel, J. J. and Thomas, D. B., 1986a, Conformation-dependent recognition of influenza virus hemagglutinin by murine T helper clones, *European Journal of Immunology*, **16**, 276–80.

Mills, K. H. G., Skehel, J. J. and Thomas, D. B., 1986b, Extensive diversity in the recognition of influenza virus hemagglutinin by murine T helper clones, *Journal of Experimental Medicine*, **163**, 1477–90.

Mills, K. H. G., Burt, D. S., Skehel, J. J. and Thomas, D. B., 1988, Fine specificity of murine class II-restricted T cell clones for synthetic peptides of influenza virus hemagglutinin. Heterogeneity of antigen interaction with the T cell and the Ia molecule, *Journal of Immunology*, **140**, 4083–90.

Morrison, R. P., Earl, P. L., Nishio, J., Lodmell, D. L., Moss, B. and Chesebro, B., 1987, Different H-2 subregions influence immunization against retrovirus and immunosuppression, *Nature*, **329**, 729–31.

Murphy, D. B., Jones, P. P., Loken, M. R. and McDevitt, H. O., 1980, Interaction between I region loci influences the expression of a cell surface Ia antigen, *Proceedings of the National Academy of Sciences (USA)*, **77**, 5404–8.

Nagy, Z. A., Lehmann, P. V., Falcioni, F., Muller, S. and Adorini, L., 1989, Why peptides? Their possible role in the evolution of MHC-restricted T-cell recognition, *Immunology Today*, **10**, 132–8.

Ogilvie, B. M. and de Savigny, D., 1982, Immune response to nematodes, in Cohen, S. and Warren, K. S. (Eds) *Immunology of Parasitic Infections*, pp. 715–57, Oxford: Blackwell Scientific Publications.

Oliveira, D. B. G., Blackwell, N., Virchis, A. E. and Axelrod, R. A., 1985, T helper and T suppressor cells are restricted by the A and E molecules, respectively, in the F antigen system, *Immunogenetics*, **22**, 169–75.

Pinder, M., Dupont, A. and Egwang, T. G., 1988, Identification of a surface antigen on *Loa loa* microfilariae the recognition of which correlates with the amicrofilaremic state in man, *Journal of Immunology*, **141**, 2480–6.

Pond, L., Wassom, D. L. and Hayes, C. E., 1989, Evidence for differential induction of helper T cell subsets during *Trichinella spiralis* infection, *Journal of Immunology*, **143**, 4232–7.

Pritchard, D. I., Behnke, J. M., Carr, A. and Wells, C., 1986, The recognition of antigens on the surface of adult and L4 *Necator americanus* by human and hamster post-infection sera, *Parasite Immunology*, **8**, 359–67.

Riedlinger, J., Grencis, R. K. and Wakelin, D., 1986, Antigen-specific T-cell lines transfer protective immunity against *Trichinella spiralis in vivo*, *Immunology*, **58**, 57–61.

Roach, T. I. A. Wakelin, D., Else, K. J. and Bundy, D. A. P., 1988, Antigenic cross-reactivity between the human whipworm, *Trichuris trichiura*, and the mouse trichuroids *Trichuris muris* and *Trichinella spiralis*, *Parasite Immunology*, **10**, 279–91.

Schwartz, R. H., 1978, A clonal deletion model for Ir gene control of the immune response, *Scandinavian Journal of Immunology*, **7**, 3–10.

Thomas, D. B., Skehel, J. J., Mills, K. H. G. and Graham, C. M., 1987, A single amino acid substitution in influenza hemagglutinin abrogates recognition by monoclonal antibody and a spectrum of subtype-specific L3T4 + T cell clones, *European Journal of Immunology*, **17**, 133–6.

Tomlinson, L. A., Christie, J. F., Fraser, E. M., McLaughlin, D., McIntosh, A. E. and Kennedy, M. W., 1989, MHC restriction of the antibody repertoire to secretory antigens, and a major allergen, of the nematode parasite *Ascaris*, *Journal of Immunology*, **143**, 2349–56.

Townsend, A. and McMichael, A., 1987, Those images that yet fresh images beget, *Nature*, **329**, 482–3.

Varla-Leftherioti, M., Papanicolaou, M., Spyropoulou, M., Vallindra, H., Tsiroyianni, P., Tassopoulos, N., Kapasouri, H. and Stavropoulos-Giokas, C., 1990, HLA-associated non-responsiveness to hepatitis B vaccine, *Tissue Antigens*, **35**, 60–3.

Vidovic, D. and Matzinger, P., 1988, Unresponsiveness to a foreign antigen can be caused by self-tolerance, *Nature*, **336**, 222–5.

Wakelin, D., 1988, Helminth infections, in Wakelin, D. and Blackwell, J. M. (Eds) *Genetics of Resistance to Bacterial and Parasitic Infection*, pp. 153–224, London: Taylor and Francis.

Wassom, D. L., Krco, C. J. and David, C. S., 1987, I-E expression and susceptibility to parasite infection, *Immunology Today*, **8**, 39–43.

Watanabe, N., Kojima, S. and Ovary, Z., 1976, Suppression of IgE antibody production in SJL mice. I. Nonspecific suppressor T cells, *Journal of Experimental Medicine*, **143**, 833–45.

Wiley, D. C. and Skehel, J. J., 1987, The structure and function of the hemagglutinin membrane glycoprotein of influenza virus, *Annual Reviews in Biochemistry*, **56**, 365–94.

Index

Note: Page references in *italic* refer to tables and/or figures